After Darwin
Animals, Emotions, and the Mind

Clio Medica: Perspectives in Medical Humanities

93

After Darwin
Animals, Emotions, and the Mind

Edited by
Angelique Richardson

Rodopi

Amsterdam - New York, NY 2013

Cover illustration: Robert Braithwaite Martineau, *Kit's Writing Lesson* (1852), reproduced courtesy of the Tate Gallery.

The paper on which this book is printed meets the requirements of "ISO 9706:1994, Information and documentation - Paper for documents - Requirements for permanence".

ISBN: 978-90-420-3747-2
E-Book ISBN: 978-94-012-0998-4

Printed in The Netherlands

At last I fell asleep on the grass, & awoke with a chorus of birds singing around me, & squirrels running up the trees, & some Woodpeckers laughing, & it was as pleasant & rural a scene as ever I saw, & I did not care one penny how any of the beasts & birds had been formed.

Charles Darwin to Emma Darwin, April 1858

Contents

Acknowledgements

I would like to thank the Wellcome Trust for sponsoring a symposium at Exeter on Darwin, Medicine and the Humanities in 2009 (WT089531MA) which provided the impetus for several of the essays in this volume. I am also grateful to the British Academy for a research grant in 2010 (SG090832) which allowed further collaboration with scientists and humanities scholars, and to the University of Exeter. I am especially grateful to the Wellcome Trust for a Research Leave Award (WT096507AIA) which allowed me to bring this project to completion. I would also like to thank the editors of the Darwin Correspondence Project and the librarians of Cambridge University Library. I am grateful to a number of scholars, either for participating in the symposium, or for helping to shape this collection in other ways; these include Marc Bekoff, Carolyn Burdett, Thomas Dixon, John Dupré, Regenia Gagnier, Lee Grieveson, Mark Jackson, Tim Kendall, Phillip Mallett, Nick McDowell, Andrew McRae, John Plunkett, Dorothy Porter, Richard Seaford, Jane Spencer and Paul White. I would also like to thank my colleagues in Exeter's Wellcome Trust-funded Centre for Medical History; the Centre for Genomics in Society (Egenis); and the Centre for Victorian Studies; Brian Dolan, general editor of this series, and co-director of the University of California Medical Humanities Consortium, for valued support, and for design and production; and, for excellent editorial assistance, Demelza Hookway. My special thanks go to the anonymous referees and to the contributors who made collaboration across and within disciplines such an illuminating and rewarding experience.

This publication in the Clio Medica series would not be possible without the generous support of the University of California Office of the President which funded the Multi-Campus Research Program grant for the University of California Medical Humanities Consortium, and the Dean of the School of Medicine at the University of California, San Francisco. I also recognise the valuable support provided by the Department of Anthropology, History and Social Medicine at UCSF and the Center for Humanities and Health Sciences for making the book series possible.

Acknowledgements

Notes on Contributors

David Amigoni is Professor of Victorian Literature at Keele University. He is the editor (with Jeff Wallace) of *Charles Darwin's* Origin of Species: *New Interdisciplinary Essays* (1995); and author of *Colonies, Cults and Evolution: Literature, Science and Culture and Nineteenth-Century Writing* (2007). He is presently engaged in interdisciplinary, collaborative research on gerontology and culture. Longer term, he is working on a study of ideas of inheritance in literary and scientific intellectual families in the nineteenth and twentieth centuries.

Dame Gillian Beer is Professor Emeritus at the University of Cambridge. She is a Fellow of the British Academy and of the Royal Society of Literature and is a Foreign Honorary Member of the American Academy of Arts and Sciences and an Honorary International Member of the American Philosophical Society. From 2009-2011 she was the Andrew W. Mellon Senior Scholar at the Yale Center for British Art. She is President of the British Literature and Science Society. Among her books are *Darwin's Plots* (1983; 2000; third edition, 2009), *George Eliot* (1986), *Arguing with the Past* (1989), *Open Fields: Science in Cultural Encounter* (1996), *Virginia Woolf: The Common Ground* (1996). She is completing a study of Lewis Carroll's Alice books in the context of nineteenth-century intellectual controversies and her collected and annotated edition of the poetry of Lewis Carroll was published by Penguin Classics in 2012. She is General Editor of Cambridge Studies in Nineteenth-Century Literature and Culture.

Marc Bekoff is Professor Emeritus of Ecology and Evolutionary Biology at the University of Colorado and a former Guggenheim Fellow. In 2000 he was awarded the Exemplar Award from the Animal Behavior Society for major long-term contributions to the field of animal behavior and in 2009 he was presented with the Saint Francis of Assisi Award by the Auckland (New Zealand) SPCA. Marc has published more than 300 scientific and popular essays and twenty-two books including *Minding Animals*, *The Ten Trusts* (with Jane Goodall),

The Emotional Lives of Animals, Animals Matter, Animals at Play: Rules of the Game, Wild Justice: The Moral Lives of Animals. The Animal Manifesto: Six Reasons for Expanding Our Compassion Footprint, and two editions of the *Encyclopedia of Animal Rights* and *Animal Welfare*. In 2005 Marc was presented with The Bank One Faculty Community Service Award for the work he has done with children, senior citizens, and prisoners. His websites are marcbekoff.com and, with Jane Goodall, www.ethologicalethics.org.

Rhodri Hayward is a Senior Lecturer in the History of Medicine at Queen Mary, University of London. He has published on the history of dreams, Pentecostalism, demonology, cybernetics, and the relations between psychiatry and primary care. His current research examines the rise and political implications of psychiatric epidemiology in modern Britain. His book *Resisting History: Popular Religion and the Invention of the Unconscious* was published by Manchester University Press in 2007. His new books, *Psychiatry in Modern Britain* and *Between History and Chemistry: Remaking the Self in Modern Primary Care* will both be published by Bloomsbury in 2013.

Stephen Jacyna is Reader in the History of Medicine and Director of the Centre for the History of Medicine at University College London. He has written on various aspects of medical science and clinical medicine in the nineteenth and twentieth centuries with special reference to the history of neurology. His publications include: *Medicine and Modernism: A Biography of Sir Henry Head* (2008) and *The Neurological Patient in History* (2012), edited with Stephen Casper. His current research centres upon the negotiation of the animal/human boundary in the nineteenth-century sciences of mind and brain.

Michael Lewis is University Distinguished Professor of Pediatrics and Psychiatry and Director of the Institute for the study of Child Development at Robert Wood Johnson Medical School, Rutgers University. He is also Professor of Psychology, Education, Nursing and Biomedical Engineering at Rutgers where he is also on the Executive Committee of the Cognitive Science Center. In 2009 he won the Urie Bronfenbrenner Award for Lifetime Contribution to Developmental Psychology in the Service of Science and Society from the American Psychological Association, and more recently the New York Zero-to-Three Hedi Levenback Pioneer Award. He has written over 350 articles in professional journals and edited and written 45 books including *Social Cognition and the Acquisition of Self; Shame: The Exposed Self*; and *Altering Fate: Why The Past Does Not Predict the Future.* He has been the senior editor of the *Handbook of Emotions,* which is now in its third edition. His latest book, *The Development of Emotional Life*, will be published in 2013.

Monika Pietrzak-Franger is a Lecturer in the English Department at Siegen University, Germany. Her books include, as author, *The Male Body and Masculinity: Representations of Men in British Visual Culture of the 1990s* (2007) and, as co-editor, *Adaptations. Performing across Media and Genres* (2009). She is currently working on her post-doctoral project which focuses on the visualisation and gendering of diseases in late Victorian culture, for which she has received funding from the Volkswagen Foundation. In 2011-2012 she was a Visiting Fellow in the Department of Anthropology at Washington University in St. Louis.

Angelique Richardson is a Senior Lecturer in English at the University of Exeter, where she is also a member of the Advisory Board of the Centre for Medical History, and an Associate Research Fellow of the ESRC Centre for Genomics in Society (Egenis). Her research and teaching interests embrace Victorian Studies, medical humanities and animal studies, and she has published widely on aspects of Victorian science and culture. Her books include *Love and Eugenics in the Late Nineteenth Century: Rational Reproduction and the New Woman* (Oxford University Press, 2003); as editor *Women Who Did: Stories by Men and Women, 1890–1914* (Penguin Classics, 2005) and *Essentialism in Science and Culture, Critical Quarterly* Special Issue (2011); and, as co-editor, *Victorian Literature: A Sourcebook* (Palgrave, 2012). She is now completing *Thomas Hardy and Biology.*

Harriet Ritvo is the Arthur J. Conner Professor of History at the Massachusetts Institute of Technology and a fellow of the American Academy of Arts and Sciences. She has published on British history, environmental history, the history of human-animal relations, and the history of natural history. She is the author of *The Dawn of Green:*

Manchester, Thirlmere, and Modern Environmentalism (Chicago University Press, 2009), *The Platypus and the Mermaid, and Other Figments of the Classifying Imagination* (Harvard University Press, 1997), *The Animal Estate: The English and Other Creatures in the Victorian Age* (Harvard University Press, 1987), and *Noble Cows and Hybrid Zebras: Essays on Animals and History* (Virginia, 2010); she is also the co-editor of *Macropolitics of Nineteenth-Century Literature: Nationalism, Imperialism, Exoticism* (University of Pennsylvania Press, 1991), and the editor of Charles Darwin's *The Variation of Animals and Plants under Domestication* (Johns Hopkins University Press, 1998). In addition, she is the editor of the Animals, History, Culture series (Johns Hopkins University Press).

Jane Spencer is Professor of English at the University of Exeter. She has written extensively on the novel, drama, women's literary history and the literary canon from the seventeenth to the nineteenth centuries: her books include *The Rise of the Woman Novelist* (1986), *Aphra Behn's Afterlife* (2000) and *Literary Relations: Kinship and the Canon, 1750-1830* (2005). In recent years she has been researching into the treatment of nonhuman animals in literary history. Her articles about animal representation in children's fiction and in feminist polemic appear in *Journal of Eighteenth-Century Studies* (2010), *Intellectual History Review* (2012), *and Eighteenth-Century Fiction* (2013). She is currently at work on *Writing About Animals in the Age of Revolution,* a book about the way natural history, the philosophy of sympathy and the discourse of rights transformed the literary treatment of nonhuman animals in the eighteenth century.

Paul White is an editor of the Correspondence of Charles Darwin (Cambridge University Press) and teaches in the Department of History and Philosophy of Science, Cambridge. He is the author of *Thomas Huxley: Making the 'Man of Science'* (Cambridge 2003) and articles on Victorian science and culture. He is currently working on a book on Darwin and the evolution of emotion.

List of Illustrations

Introduction

Darwin and Interdisciplinarity: A Historical Perspective

Angelique Richardson

> 'Mr Darwin has the peculiarity of opening up in each of his books new fields'[1]

'He who understands the baboon', declared Darwin in 1838, '~~will~~ would do more towards metaphysics than Locke'.[2] Even shifted to the more measured conditional this was no small claim. Biology, or, more precisely, the study of animals, could contribute more to our knowledge of the world and the nature of being than philosophy. The mix of register and discipline, from the striking specificity of the baboon to the heights of Lockean philosophy, is as telling as it is audacious. By the end of the short sentence the ignoble, even comedic, figure of the baboon is dignified, reorientated in the cosmos. It was an affront to the Cartesian tradition in philosophy of mind. Two centuries earlier, Descartes had declared animals automata: 'although many animals show more skill than we do in some of their actions … It shows rather that they have no intelligence at all, and that it is nature which acts in them according to the disposition of their organs'.[3] 'In the same way', he had added, 'a clock, consisting only of wheels and springs, can count the hours and measure time more accurately than we can with all our reason!' For Descartes reason didn't come into the equation.

But for Darwin the minds of nonhuman animals had much to impart, as his carefully chosen words suggest, with 'understands' evoking both a mirroring and an exploration of complex mental states.[4] The aphoristic remark recognises not only the importance of biology, grounded in empiricism, but its cultural centrality. In his next sentence he draws parallels, contentious even today, between dogs and humans, seeming to liberate them from hierarchy. Both whine, he observes. And the dog laughs and barks alike for joy. Darwin's psychological researches were assisted both by an acute sense of observation and an extraordinary capacity for fellow-feeling, not only with humans but with all life. In the words of one Down House guest, 'he catches every expression that flits over a face & reads it...Full of warmest feelings & quick sympathy'.[5]

After Darwin: Animals, Emotions, and the Mind explores questions of mind, emotion and the moral sense which Darwin opened up through his research on the physical expression of emotions and the human–animal relation. The complexity of these areas of study makes them prime examples of subjects that need perspectives from a variety of disciplines in order to be understood. The project began as a series of conversations with colleagues in other places – other disciplines, institutions, and countries. It aims to integrate ideas and ways of thinking from several disciplines to provide an advance in knowledge and an engagement with issues in the world which are unlikely to have been possible through a single disciplinary approach.[6] In doing so it also aims to produce plural ways of seeing, developing understanding through a range of new perspectives on questions relating to animals, emotions, and the mind, and seeks to bridge a gap between science and humanities subjects as they relate to Darwin, biology and allied sciences. It also reflects on Darwin's ideas and methods, including his use of anecdote and anthropomorphism. Bringing together scholars from biology, literature, history (including environmental history and the history of science and medicine), psychology, psychiatry and paediatrics, from North America, the UK and mainland Europe, it explores Darwin's relation to Enlightenment and Victorian thought, and to medicine, culture and the arts, considering new understanding of the moral sense and emotional life, and the urgent scientific and ethical implications of his ideas today. In doing so, it expresses and reflects on the value and challenges of interdisciplinarity.

While this volume is centrally concerned with questions of biology and evolution, it assumes the relation between science and culture is reciprocal, characterised by what Gillian Beer has called 'two-way' traffic,[7] and rejects reductive biologistic thinking which seeks to explain human behaviour along biological lines without reference to social, economic and political factors, and which bears little relation to Darwin's own enworlded sense of development.[8] The part played by language in science, including the ways in which linguistic norms and cultural assumptions can shape scientific practices, offers a pertinent reason why collaboration between scientists and scholars in the humanities is particularly fruitful, and has been insightfully explored by scholars such as the historian Londa Schiebinger and the feminist philosopher Evelyn Fox Keller.[9] Likewise, the interconnectedness of science and literature as discursive formations has been eloquently demonstrated by scholars working in literature, most notably Gillian Beer, George Levine and David Amigoni.[10] But, equally, and even as a number of the contributors work in disciplines which privilege language and representation, the collaboration and subject matter remind us of the limits of language, recalling us to an appreciation that it is part of a larger world.[11]

Darwin was interdisciplinary by practice and temperament. He read Milton and Wordsworth and had been a Romantic traveller on the *Beagle*, with an intense and wondrous sense of the sublime. He would lend to science not only meticulous observation but a vital imagination. Quintessentially Victorian, he worked at a time before scientific thought would become linked to occupational roles, situations and practices, divided by the internal barriers of professionalisation. Nonetheless, disciplinary formation had been underway from the Enlightenment.[12] Biology had meant something along the lines of a biographical history until the English physician Thomas Beddoes, known for his radical politics, defined it in 1799 as '*the doctrine of the living system in all its states*', [13] and Lamarck used 'biologie' in 1802 to mean the study of living things.[14] It would be another three decades before the word scientist appeared. Coined as an analogy with the term artist by the polymathic Cambridge mineralogist Reverend William Whewell at the 1833 meeting of the British Association for the Advancement of Science, it appeared in print in his review of Mary Somerville's popularising *On the Connexion of the Physical Sciences*.[15] It seemed at first a restricted term and did not prove popular with Victorian men (as they mainly were) of science, gaining currency with the rise of laboratory science at the end of the century. Whewell, indefatigably interdisciplinary, had intended a rounder meaning to the term, as he had argued for a broad undergraduate education, founded on mathematics – in particular geometry and mechanics – and classical languages.[16] Whewell had coined the term consilience as part of the phrase 'the consilience of inductions'. Crucially, for him it meant 'jumping together'; *coincidence*, not unity, of knowledge, though Edward O. Wilson, the father of sociobiology, would use it in this latter sense at the end of the twentieth century.[17]

A new self-consciousness about difference and relation, in both the subjects and objects of study, was emerging. The American botanist Asa Gray remarked to his fellow-biologist J.D. Hooker in 1863 that biology definitely was not natural history,[18] and a couple of years later Darwin wrote to Hooker defending Britain's contribution to biology and geology: 'in Biology Owen & R. Brown ought to save us, & in Geology we are most rich'.[19] But, as Darwin lent biology a new impetus, publishing views that by his own admission would produce 'a considerable revolution in natural history',[20] he would collaborate even more intently with individuals in the areas of science from which he was increasingly distinguishing his own work. Contemplating the natural world, his babies and children, retrievers and terriers, Darwin corresponded with members of an extensive, global network. He needed the input of natural historians, of pigeon fanciers and entomologists, dog breeders, botanists, geologists and physicians. And, in turn, he shared his ideas with a wider community, not only with men of science, but with country vicars who were working hard finding and describing new species, with explorers and writers of novels. Within a few years of his reading George Eliot's first novel, *Adam Bede* (1859) – 'entire rest & the douche & Adam Bede have together done me a world of good', he had written to his friend Hooker[21] – Darwin, Eliot and her scientist partner George Henry Lewes (who was also a drama critic) were calling at each other's houses. It was one of Darwin's chief pleasures to be read a novel as he broke from his science at Down House.

After Darwin, something changed. The disciplines, never far apart to begin with, were now brought into newly self-conscious dialogue. Grant Allen, author of the notorious *The Woman who Did* (1895), and a biologist and populariser of science who well exemplifies the Victorian ease of interdisciplinarity, declared in the multidisciplinary *The Academy: A Monthly Record of Literature, Learning, Science and Art* (founded in 1869), that *The Origin of Species* was 'nothing less than a revolution; it marks the year 1 of a new era, not for science alone, but for every department of human thought – even of human action'.[22] He continued:

> Yet this much may even now be said with certainty, that the influence of his thought upon the thought of the age has far outweighed any influence ever before exerted by a single man during his own lifetime. He has revolutionised, not biology alone, but all science; not science alone, but all philosophy; not philosophy alone, but human life.[23]

The potential of biology, which deeply affects how we see both ourselves and our relations in the world, had been understood, and its profound relevance to all subjects and areas of life acknowledged. 'Mr Darwin's book', George Henry Lewes declared in *John Bull* weeks after its appearance, will 'aid in general culture by opening to many minds new tracts of thought'. [24] For the essayist and physician Havelock Ellis, writing in 1890 in *The New Spirit,* 'the conception of evolution has penetrated every department of organic science, especially where it touches man'. He saw Darwin's character and scientific method as having been vital to this process: 'Darwin personally, to whom belongs the chief place of honour in the triumph of a movement which began with Aristotle, has been a transforming power by virtue of his method and spirit, his immense patience, his keen observation, his modesty and allegiance to truth; no one has done so much to make science – that is to say, all inquiry into the traceable causes or relations of things – so attractive'.[25] Interrelations were stressed, and so too the importance of encountering difference. The philosopher and liberal politician John Stuart Mill had remarked in the middle of the nineteenth century 'it is hardly possible to overrate the value... of placing human beings in contact with persons dissimilar to themselves, and with modes of thought and action unlike those with which they are familiar'.[26] He saw such communication as a major source of progress. There was increasing recognition that complex problems require complex approaches.[27] In the words of Lewes, 'sympathy is one of the great psychological mysteries – and as a psychologist I am bound to explain it, but can't'.[28]

In the *Origin* Darwin had remarked 'In the distant future I see open fields for far more important researches. Psychology will be based on a new foundation, that of the necessary acquirement of each mental power and capacity by gradation'.[29] His seminal study, *The Expression of the Emotions in Man and Animals*, appeared in 1872, establishing the adaptive function of emotions, and their overriding relation to the physical, sensory environment. He increasingly emphasised mutual aid, as he concluded that origins of morality lie in the social instincts of animals, arguing, in *The Descent of Man, and Selection in relation to Sex* (1871), that the moral sense evolved from the 'social instincts' in humans as in some animals, aided by 'active intellectual powers'.[30] His work on the instincts and the moral sense, on human–animal continuity, and on the physical expression of the emotions, made – and continues to make – a vital contribution to psychology. According to the psychologist Paul Ekman, the *Expression* 'should be regarded as the book that initiated the scientific study of human behavior – the first book in psychology'.[31] Darwin looked forward to, and beyond, Freud, exploring mind as a function of body, and investigating non-rational and instinctual impulses. His emphasis on the physical basis of emotion would subsequently be taken up by leading thinkers in psychology and psychoanalysis, notably the American pragmatist psychologist William James, as well as Freud.[32] James argued in 'What is an Emotion?' in *Mind* (1884), that emotions are caused by the physiological changes associated with emotion, rather than the emotions causing those changes, and, in a follow-up piece, 'The Physical Basis of Emotion', he concluded 'the name 'emotion' should be suffered to connote organic excitement as the distinctive feature of the state'.[33]

Darwin's research included developmental psychology, the subject of Michael Lewis's chapter in this volume; comparative psychology, the subject of Marc Bekoff's and Harriet Ritvo's chapters; and abnormal psychology and anthropological psychology. He also provided a historical account of the study of emotion. In the psychology of facial expressions he pioneered the method of showing photographs to people and asking them what emotion they see,[34] and while the question of the universality of expression in humans remains a matter of ongoing debate, it formed an important part of his aim of establishing the unity of races.[35] Darwin's emphasis on the formative role of community in the expression and development of emotion has been borne out in recent research which reveals that facial expressions are constituted in part by social norms and motives; thus, social norms and biological emotion processes combine.[36]

Darwin's research continues to inform new understanding of the relation between mind and emotion. The leading contemporary neuroscientist Antonio Damasio offers a fundamental challenge to Cartesian ideas about the primacy of reason, establishing the inseparability of emotion from reason,[37] and arguing that emotion is integral to the process of reasoning and decision-making.[38] He makes clear the debt these new insights owe to Darwin who, alongside James and Freud, gave emotion a privileged place in scientific discourse, which the English physician John Hughlings Jackson (discussed by Stephen Jacyna in chapter eight) began to develop along neuroanatomical lines. Neuroscience, with new techniques of brain-scanning, also emphasises the close relation between thinking and feeling. New understanding of this relation makes it difficult to dismiss emotions, or sideline such qualities as hope and compassion, and has led to a profound interest in emotions among scientists as well as humanities scholars.[39] As part of these reciprocities the developing field of medical humanities takes medicine, its practice and its patients, as the subject of co-enquiring humanities disciplines, which range from literary studies, history, theology, anthropology and philosophy. One aspect of this diverse field of enquiry is the encouragement of medical students and practitioners to read and write fiction to promote reflection and critical intellectual engagement with their feelings.[40]

As Damasio observes, philosophy has been distrustful of emotion, notwithstanding Hume and the tradition that started with him (discussed by Jane Spencer in chapter one); and, as the Romantics had placed emotion in the body and reason in the brain, Damasio traces the omission of the body from twentieth-century science, and a return of emotion to the brain, but a simultaneous relegation of it to lower neural strata, not worthy of study, or a dismissal of it as subjective, irreducible to logical structures. The mid-twentieth century saw a rise in the psychology of behaviourism, which considered it unscientific to impute mental states to animals because these states could not be observed. As animals were viewed externally, experience, and the possibility of animal subjectivity, disappeared from view. The focus of cognitive science was on computational aspects of mind, such as problem solving, language and memory. Damasio welcomes the evolutionary perspective brought to bear on the study of brain and mind in the final decade of the twentieth century, and the admission of the relation of the brain to the body, of an integrated organism.[41] We might see this, from a Darwinian perspective, as a readmission.

The implications of Darwin's research for our understanding of animals remain equally profound. Arguing, against Cartesian tradition and dominant Victorian thought, for their continuity with humans, he declared in the *Descent*, 'the difference in mind between man and the higher animals, great as it is, is certainly one of degree and not of kind'. He observed 'We have seen that the senses and intuitions, the various emotions and faculties, such as love, memory, attention, curiosity, imitation, reason, &c., of which man boasts, may be found in an incipient, or even sometimes in a well-developed condition, in the lower animals',[42] expanding on abstraction and other mental qualities in animals in the second edition, published three years later. His emphasis on continuity culminated in his study of emotions in humans and animals in the *Expression*. Conceding in the *Descent* that 'the habitual use of articulate language is, however, peculiar to man', language was to extent 'an instinctive tendency'.[43] New comparative research supports the argument for communication continuity with nonhuman animals; human communication, while extraordinarily complex, and so traditionally instanced as evidence of human separateness, bears significant resemblances to simpler systems in other animals.[44]

As we consider historical attitudes to animals, and shifting conceptions of the human–animal relation, we might also ask how Darwinian insights can inform the health and welfare of animals. Here, interdisciplinarity is crucial. As Harriet Ritvo notes, animals have been edging towards the academic mainstream.[45] Correspondences between the sciences and humanities are in evidence in animal studies, but so too are failings in communication. While animal studies in the humanities concerns itself primarily with representation, what of the animals themselves? An overemphasis on language can lead to a blindness to materiality beyond language. Animal studies offers a space for questioning the prioritising of human concerns in academia. It may also enable us to interrogate ways in which academic research can be complicit with the inhumane treatment of animals.[46] Arguing in the *Descent* for the continuity of mental states, Darwin's research provided evidence of animal consciousness, but in 2012 the animal behaviourist Marian Dawkins questioned whether animals have consciousness, going against the scientific research not only of recent decades but centuries.[47] However, later that year, a prominent, international group of scientists ranging from cognitive neuroscientists to neurophysiologists signed a declaration stating that 'non-human animals have the neuroanatomical, neurochemical, and neurophysiological substrates of conscious states along with the capacity to exhibit intentional behaviours. Consequently, the weight of evidence indicates that humans are not unique in possessing the neurological substrates that generate consciousness'.[48] Intellectual and practical progress on ethical and evolutionary questions requires collaboration across disciplines and historically-informed understanding.

Darwin valued animal subjectivity. His contribution to our understanding of animals remains a significant subject of debate, and his research continues to have important ramifications for the well-being of animals. For example, an article in *Behavioral and Brain Sciences* by Derek C. Penn, Keith J. Holyoak and Daniel J. Povinelli, claiming, against Darwin's argument for differences only of degree, discontinuity between human and nonhuman minds, has received wide-spread criticism from scientists working on this relation.[49] The evolutionary biologist Gordon Burghardt points out that the analysis separates the mind from biology, offering a stereotypical human mind and a single type of nonhuman mind and confusing function with underlying mechanism. He observes that research over the last forty years shows Darwin in fact underestimated the mentality of apes, for example in tool making and numeracy, as well as communication.[50] Others point out that nonhuman animals may think about alternative futures outside the realm of perception,[51] and demonstrate striking similarities between behaviour in the early stages of human development and other species.[52] A number of biologists underline the limitations of lab work; the importance of the environmental context is repeatedly stressed – a context Darwin was fully aware of, and worked with himself – and seen as overlooked in arguments against continuity.[53] The paradox is also noted of assuming that reasoning is orthogonal to language and enculturation, while using language to compare what humans and nonhumans know.[54]

Darwin's ideas, new and challenging in the middle of the nineteenth century, continue to stimulate historical and literary debate, no less than urgent practical and ethical questions. The contributors to this volume seek to reflect on these from a range of disciplinary and interdisciplinary perspectives. In the opening essay, Jane Spencer considers ideas of animal abilities and human–animal kinship in intellectual and scientific discourse in the century before Darwin, and explores eighteenth-century attributions of feeling to animals which encompass both the new term emotion and the more established terms passion, sentiment and affection. She argues that a new emphasis on animal feeling became the basis for reformed social attitudes to them, and opens the way for seeing both the historical significance and contemporary value of this emphasis for animal welfare. Gillian Beer reinvestigates the value of reciprocity between science and culture and suggests new ways in which this was possible, and even essential, for Darwin's research. She takes as her focus Darwin's intense aesthetic responsiveness in his youth and through the period of the formation of his evolutionary ideas, asking what happened when he lost his pleasure in the arts. Did it get re-made in his later work, especially the *Descent* and *Expressions*? David Amigoni focuses on biographical and autobiographical accounts of Darwin, his father, Robert Waring Darwin, and his grandfather Erasmus Darwin, the famous Lichfield physician, observing how Darwin emphasised the sympathy that each was able to bring to their medical practice, and saw it both as an emotional state that brought men of science together (as it had bonded three generations of Darwins), and as a naturally selected emotional foundation to morality, linking biological and cultural stories about heredity. Amigoni also explores new interest in Darwin's emotional life, including the place of sympathy in assessing his personal life, and notes that biographical images that we have of Darwin's emotional and intellectual disposition have been products both of history and selection. Monika Pietrzak-Franger examines perceptions of Darwin in twenty-first century culture, contrasting these with dominant twentieth-century representations. Focusing on recent multimedia projects, she argues that academic and popular discourses have become increasingly engaged with his emotional life, and its influence on his work, and, in turn, they offer us scope for engaging with nature, and our own place within it, in new experiential and ethical ways.

Paul White explores ways in which the human–animal connection was configured by Darwin as a relationship of sympathy. He situates this within a larger question about the nature of sympathy, exploring what happens to it as emotion becomes more physiological, and animal, in the nineteenth century. My chapters explore the conception and reception of the *Expression of the Emotions*, considering its fascination and narrative appeal for contemporary readers – and writers – and its underlining of a deep reciprocity between science and culture. I offer a comparative analysis of Eliot and Lewes's engagement with Darwin, and examine ways in which, in emphasising instinct and the unconsciousness of expression, he opened up new possibilities for fiction, challenging the epistemological certainty which underpinned the realist novel. I argue that Eliot gave an increasingly physiological treatment to the emotions, emphasising kinship with animals and integrating her characters into the natural world in new ways. Stephen Jacyna asks whether the roots of a separation of reason and feeling in twentieth-century neuroscience can be traced to developments in neurological thought in the period following the publication of the *Expression of the Emotions*, and argues that there was at this time an appreciation of the need to accommodate the emotions in order to provide a coherent account of the evolution of the human nervous system. Rhodri Hayward examines how Darwin's principle of emotions as 'serviceable associated habits' was transformed in the twentieth century. Taken up by rival schools in psychoanalysis and psychosomatic medicine, he argues that Darwin's theory of *Expression* was used to ground competing models of the expressive nature of illness and emotional signification, and was reinvented in Paul Ekman's publication of the third edition of the work (1996).

New research suggests that even the blush has a social function, communicating apology or appeasement, and signalling sincerity, and Michael Lewis sheds important new light on the self-conscious emotions.[55] Discussing Darwin's work on infant/child emotional development, and ways in which he foresaw many of the issues which came to occupy the study of emotional development, Lewis examines research on the role of cognition or consciousness in the emergence of complex emotions. Harriet Ritvo offers a study of the disjunction between Darwin's analysis of the evolution of wild animals and their domesticated relatives. She argues that the combination of similarity and difference that made the comparison of wild and domesticated animals both effective and problematic for Darwin continues today, and suggests that a valorisation of wildness emerged at a time when its very existence was becoming compromised. And, in the afterword, Marc Bekoff focuses on the evolution of emotions and moral behaviour. Cognitive ethology, evolutionary biology, and social neuroscience (along with common sense) demonstrate that animals are emotional and empathic beings and that they display moral sensibility. Bekoff explores the details of social play behaviour, the many ways in which animals play fairly and honestly, and argues that we owe it to all animals to make every attempt to come to a greater understanding and appreciation of who they are: emotional, empathic, and often moral beings. When we're not sure about what they're feeling, we should let them be. Bekoff's argument that denying animals emotional and moral intelligence is bad biology owes its force and impetus to Darwin. He reminds us that by minding animals we can increase our compassion footprint and make the world a better place for all beings.

That Darwin's research remains the subject of continued debate and interest in areas ranging from psychology and neuroscience to ethics, history, English literature, animal studies and animal welfare is itself remarkable. It is also testimony to the continuing importance of reciprocity between the sciences and the humanities, which allows new questions to be formulated and explored. As the essays in this volume demonstrate from different disciplinary perspectives, Darwin's research, his scrutiny, imagination, and recognition of the limits of knowledge, led to new possibilities, new ways of thinking and feeling which are open today.

I am grateful to the Syndics of Cambridge University Library for permission to quote from the Darwin archive.

Notes

1 Anton Dohrn, 'The Expression of the Emotions in Man and Animals', *Academy*, 4 (1873), 209. See Darwin's scrapbook of reviews, DAR226, Darwin Archive, Cambridge University Library.

2 Notebook M, entry 84e, in Paul H. Barrett, Peter J. Gautrey, Sandra Herbert, David Kohn, Sydney Smith (eds), *Charles Darwin's Notebooks* (1836–1844; London and Cambridge: Natural History Museum and Cambridge University Press, 1987), 561, Darwin's deletion. Unless

otherwise stated, notebook page references are to this volume.

3 René Descartes, *Discourse on the Method of Rightly Conducting One's Reason and Seeking Truth in the Sciences* (1637), in *The Philosophical Writings of Descartes*, trans. by John Cottingham, Robert Stoothoff and Dugald Murdoch, 3 vols (Cambridge: Cambridge University Press, 1985), i. 141.

4 On the complex society of this highly intelligent animal see Dorothy Cheney and Robert Seyfarth, *Baboon Metaphyics: The Evolution of a Social Mind* (Chicago: University of Chicago Press, 2007), and Shirley C. Strum, *Almost Human: A Journey into the World of Baboons* (London: Random House, 1988). See also Robert Sapolsky, *A Primate's Memoir: A Neuroscientist's Unconventional Life Among the Baboons* (London: Simon and Schuster; New York: Touchstone, 2002).

5 Jane Gray to Susan Loring, 28 October–2 November 1868. http://www.darwinproject.ac.uk/jane-gray-on-down#JGrayfulltext (accessed 3 July 2012). The letters of Darwin and many of his correspondents are now available online either in summary form or as full transcriptions through the Darwin Correspondence Database, http://www.darwinproject.ac.uk/. On Darwin's ill-health, which had a bearing on both his sympathy and science, see Ralp Colp *Darwin's Illness* (Gainesville: University Press of Florida, 2008), and Adrian Desmond and James Moore, *Darwin* (Harmondsworth: Penguin, 1992).

6 For an insightful study of interdisciplinarity, see Myra Strober, *Interdisciplinary Conversations: Challenging Habits of Thought* (Stanford: Stanford University Press, 2011). On the related practices of multi-disciplinarity and cross-disciplinarity see Strober, 16, and on the value of interaction more generally between disciplines, see Lisa Lattuca, *Creating Interdisciplinarity: Interdisciplinary Research and Teaching Among College and University Faculty* (Nashville: Vanderbilt University Press, 2001), 18. On the importance of acknowledging disciplinary differences, see Elizabeth Wilson, 'Another Neurological Scene', *History of the Present: A Journal of Critical History*, 1 (2011), 152–69, and E. Virginia Demos, 'Psychoanalysis and the Human Sciences: The Limitations of Cut-and-Paste Theorizing', *American Imago*, 58 (2001), 649–84.

7 Gillian Beer, *Darwin's Plots: Evolutionary Narrative in Darwin, George Eliot and Nineteenth-Century Fiction*, 3rd edn (1983; Cambridge: Cambridge University Press, 2009), 5.

8 Cf. literary Darwinism which argues for deep and unchanging biological imperatives or elemental biological motives, claiming that discourse and representation are subordinated to genes, and narratives underpinned by reproductive drives. Examples include Jonathan Gottschall and David Sloan Wilson (eds), *The Literary Animal: Evolution and the Nature of Narrative* (Evanston: Northwestern University Press, 2005); Joseph Carroll, *Reading Human Nature: Literary Darwinism in Theory and Practice* (Albany, NY: SUNY Press, 2011); and Jonathan Gottschall, *Literature, Science, and a New Humanities* (Basingstoke: Palgrave Macmillan, 2008). For critical assessments of literary Darwinism see John Holmes, 'Victorian Evolutionary Criticism and the Pitfalls of Consilience', in Nick Saul and Simon J. James (eds), *The Evolution of Literature: Legacies of Darwin in European Cultures* (Amsterdam: Rodopi, 2011); Jonathan Kramnick, 'Against Literary Darwinism', *Critical Inquiry*, 37 (2011), 315– 47 and 'Literary Studies and Science: A Response to My Critics', *Critical Inquiry*, 38 (2012), 431-60; David Amigoni, 'A Consilient Canon? Bridges to and from Evolutionary Literary Analysis', *English Studies in Canada*, 32 (2006), 173–85; and Gowan Dawson, 'Literature and Science under the Microscope', *Journal of Victorian Culture*, 11 (2006), 301–15.

9 See, for example, Evelyn Fox Keller, *The Century of the Gene* (Cambridge, MA: Harvard University Press, 2000) and *Refiguring Life: Metaphors of Twentieth-Century Biology* (New York, Columbia University Press, 1995); and Londa Schiebinger, *Nature's Body: Sexual Politics and the Making of Modern Science* (London: Pandora, 1993). See also Theodore L. Brown, *Making Truth: Metaphor in Science* (Urbana and Chicago: University of Illinois Press, 2003).

10 See Beer, 2009; George Levine, *Darwin and the Novelists: Patterns of Science in Victorian Fiction* (Cambridge, MA: Harvard University Press, 1988); and David Amigoni, *Colonies, Cults and Evolution: Literature, Science and Culture in Nineteenth-Century Writing* (Cambridge: Cambridge University Press, 2007).

11 Ten years after the hostile clash between cultural studies and science, which culminated in *Science Wars*, a special issue of *Social Text*, the leading physicist Alan Sokal reiterated his call for the centrality of evidence in all matters of public debate, arguing that right-wing politicians and industry executives have co-opted the language of postmodernist relativity to challenge scientific consensus on questions

of climate change and evolution, undermining all claims to truth. See Sokal, *Beyond the Hoax, Science, Philosophy and Culture* (Oxford: Oxford University Press, 2008). Relations between scientists and humanities scholars have dramatically improved since the 1990s, finding expression in a number of collaborative projects. See, for example, a project at the University of Reading, 'Cultivating Common Ground: Biology and the Humanities', and at the University of Southampton, 'Beyond the Gene', which formed part of the AHRC 'Science in Culture' theme in 2012, and, at the University of Exeter, the British Academy-sponsored 'Science and Culture' project http://blogs.exeter.ac.uk/scienceandculture/. See also David Stack and Nick Battey, 'Reflections on designing a Biology/Humanities interdisciplinary module', *Bioscience Education* 21 (2013), and Jay Clayton, 'Collaboration across the Disciplines: An Experiment in Interdisciplinary Pedagogy', *Literature and Medicine*, 29 (2011), 127–31, and 'Literature and Science Policy: A New Project for the Humanities', *PMLA*, 124 (2009), 947–9.

12 On the rise of biology see Ralf-Dietmar Hegel, 'August Weismann: One of the First Synthetic Theorists of Evolutionary Biology', *Boston Studies in the Philosophy of Science* 134 (1991), 259–67. On evolutionary biology as a source of scientific authority, see Dorothy Ross, 'Changing Contours of the Social Science Disciplines', *Cambridge Modern Science*, 8 vols (Cambridge: Cambridge University Press, 2005), vii. 210. Strober, 2011, details the emergence of academic disciplines from three Aristotelian categories – the theoretical (including theology, mathematics, and physics), the practical (including ethics and politics), and the productive (including engineering, poetics and the fine arts), with philosophy bringing all knowledge together – through to the development of disciplines in the medieval period, and their further differentiation during the enlightenment, 13–14.

13 Thomas Beddoes, *Contributions to Physical and Medical Knowledge, principally from the West of England, collected by Thomas Beddoes M.D.* (Bristol: Biggs & Cottle, 1799), 4, emphasis in original.

14 Lamarck, *Hydrogéologie* (Paris: Agasse, 1802). See also Gottfried Reinhold Treviranus, *Biologie, oder Philosophie der lebenden Natur für Naturforscher und Ärzte*, 6 vols (Göttingen: Röwer, 1802–1822), i.

15 Anon [William Whewell], '*On the Connexion of the Physical Sciences. By Mrs Somerville*', *Quarterly Review*, 51(1834), 54–68: 59. See Sidney Ross, '"Scientist": The Story of a Word', *Annals of* Science,18 (1962), 65–85.

On Whewell's concern to define science in relation to philosophy see Richard Yeo, *Defining Science: William Whewell, Natural Knowledge and Public Debate in Early Victorian Britain* (Cambridge: Cambridge University Press, 1993), 32–8.

16 See Whewell, *Thoughts on the Study of Mathematics as a Part of a Liberal Education* (Cambridge: Deighton, 1835). Paul White notes Huxley's objection to the term, which he considered a vulgar and utilitarian Americanism, in *Science-Gossip*, 1 (1894), 243: White, *Thomas Huxley, Making the 'Man of Science'* (Cambridge: Cambridge University, 2003), 1–5. For a more recent call for an integrated arts–science culture and curriculum, see the physical chemist C.P. Snow's now famous public lecture, 'The Two Cultures and the Scientific Revolution' (1959). In a recent revisiting of Snow's lecture Lisa Jardine urges the need to realise Snow's appeal, if the public, including governments and senior administrators, are to be able judge the scientific arguments and data around GM crops, nuclear energy and climate change. Jardine argues that the 'two cultures' debate was not, ultimately, about a clash between aesthetic value and scientific knowledge, but the result of a confusion among British administrators and educationalists on the value of science. She underlines as the context for Snow's lecture an increasing post-war loss of confidence in the importance of science in education and a return to the canon of the humanities. Lisa Jardine, 'C. P. Snow's Two Cultures Revisited', The 2009 C.P. Snow Lecture, *Christ's College Magazine* (Cambridge: Christ's College, 2010), 48–57. On Snow's model of the complementarity of literature and science see Brian Dolan, 'Speaking about the Other: On the Relevance of the Humanities to Occupational Health', in Paul D. Blanc and Brian Dolan (eds), *At Work in the World* (San Francisco and Berkeley: University of California Medical Humanities Press, 2012), 215–6. Dolan notes that Snow called for 'a literary component throughout the course of medical education', see Snow, Letter to the Editor, *Journal of the American Medical* Association, 225 (1973), 619.

17 'Accordingly the cases in which inductions from classes of facts altogether different have thus *jumped together*, belong only to the best established theories which the history of science contains. And, as I shall have occasion to refer to this particular feature in their evidence, I will take the liberty of describing it by a particular phrase; and will term it the *Consilience of Inductions*': William Whewell, *Philosophy of the Inductive*

Sciences: Founded Upon Their History, 2 vols (Cambridge: Deighton, 1840), ii. 230 (emphasis in original). This instance is given as first use in the *Oxford English Dictionary*. For Wilson's use, see *Consilience: The Unity of Knowledge* (London: Knopf, 1998).

18 Asa Gray to Hooker, 6 July 1863, http://www.darwinproject.ac.uk/entry-4232f (accessed 3 July 2012).

19 Darwin responds to an article entitled `British and Foreign Science' in the *Reader*, 15 July 1865, 61–2 in a letter to Hooker, 29 July 1865, http://www.darwinproject.ac.uk/entry-4874 (accessed 3 July 2012).

20 Darwin, *On the Origin of Species by Means of Natural Selection, or the Preservation of Favoured Races in the Struggle for Life* (London: John Murray, 1859), 484.

21 Darwin to Hooker, http://www.darwinproject.ac.uk/entry-2464 (accessed 3 July 2012).

22 Grant Allen, 'Obituary: Charles Darwin', *Academy*, 21 (1882), 306. For a complication of the one-culture interaction of science and culture see Gowan Dawson, *Darwin, Literature, and Victorian Respectability* (Cambridge: Cambridge University Press, 2007), chapter 5. Dawson argues that aesthetic art and literature sit more uneasily with this model.

23 Allen, 1882, 306.

24 G.H. Lewes, 'Studies in Animal Life', *John Bull*, 24 December 1859; reprinted in *Cornhill Magazine*, 1 (1860), 438–47, http://darwin-online.org.uk/supplementary_works.html (accessed 3 July 2012); and Lewes, 'Mr Darwin's Hypothesis', *Fortnightly Review*, 3 (1868), 353–73, 611–28; *Fortnightly Review*, 4 (1868), 61–80, 492–509.

25 Havelock Ellis, *The New Spirit* (1890; London: Walter Scott, 1892), 5.

26 John Stuart Mill, *Principles of Political Economy, with Some of Their Applications to Social Philosophy* (1848; London: Parker, Son, and Bourn, 1862), 121, cited in Strober, 2011, 22.

27 See Strober, 2011; see also Susan Knapp, 'Complex Problems Don't Respect Disciplinary Boundaries', *Focus on Faculty: A Supplement to Dartmouth Life*, 1 and 3 December 2006.

28 Lewes to Elma Stuart, 19 February 1873, in Gordon S. Haight (ed.), *The George Eliot Letters*, 7 vols (New Haven and London: Yale University Press, 1954–5), v. 376.

29 Darwin, 1859, 488.

30 *The Descent of Man, and Selection in Relation to Sex*, 2 vols (London: John Murray; 1871), i. 106; see also 71–2.

31 Ekman (ed), 'Preface to the Anniversary Edition', in Darwin, *The Expression of the Emotions in Man and Animals* (1872; London: Harper Collins, 2009), xiii.

32 On Darwin's contribution to psychology see Frank J. Sulloway, *Freud, Biologist of the Mind: Beyond the Psychoanalytic Legend* (New York, Basic Books, 1979), 238–76; James Rowland Angell, 'The Influence of Darwin on Psychology', *Psychological Review*, 16 (1909), 152–69; Robert J. Richards, *Darwin and the Emergence of Evolutionary Theories of Mind and Behavior* (Chicago: University of Chicago Press, 1987); Raymond E. Fancher, *Pioneers of Psychology* (New York: W.W. Norton, 1996); and Paul Griffiths, 'Ethology, Sociobiology, and Evolutionary Psychology', in Sahotra Sarkar and Anya Plutynski (eds), *A Companion to the Philosophy of Biology* (Oxford: Blackwell Publishing, 2008), 393–414. On the continuing relevance of Darwin's ideas in psychology see Ekman, 2009; Ray Crozier, 'Darwin and the Blush', *Wellcome History*, 40 (2009), 16; Jaak Panksepp, *Affective Neuroscience: The Foundations of Human and Animal Emotions* (New York: Oxford University Press, 1998); Alan J. Fridlund, *Human Facial Expression: An Evolutionary View* (New York: Academic Press, 1994); Fridlund, 'Darwin's Anti-Darwinism and *The Expression of the Emotions in Man and Animals*', in K.T. Strongman (ed.), *International Review of Emotion*, 2 (New York: Wiley, 1992), 117–37; N. Frijda, 'The Psychologist's View', in Michael Lewis and Jeannette M. Haviland-Jones (eds), *Handbook of Emotion* (New York: Guilford Press, 2000) 68–87; Ursula Hess and Pascal Thibault, 'Darwin and Emotion Expression', *American Psychologist*, 64 (2009), 120–8; and Brian Parkinson, Agneta H. Fischer, and Antony S.R. Manstead, *Emotion in Social Relations: Cultural, Group, and Interpersonal Processes* (New York and Hove: Psychology Press, 2005). Thomas Dixon argues that Darwin's work on expression provided Freud with alternative evolutionary explanations to natural selection and sexual selection and helps to account for Freud's neo-Lamarkianism. See Dixon, *From Passions to Emotions: The Creation of a Secular Psychological Category* (Cambridge: Cambridge University Press, 2003), 178; and, on Darwin's influence on Freud more generally, see Lucille B. Ritvo, *Darwin's Influence on Freud: A Tale of Two Sciences* (New Haven: Yale University Press, 1990), chapter 12. On Darwin and Lamarck see Peter Bowler, *Evolution: The History of an Idea* (Berkeley and Los Angeles: University of California Press, 1989), 171, 190, 210; and Angelique Richardson, *Love and Eugenics in the Late Nineteenth Century: Rational Reproduction and the New Woman* (Oxford:

Oxford University Press, 2003), 10–12.

33 William James, 'What is an Emotion?', *Mind*, 9 (1884), 188-205 and 'The Physical Basis of Emotion', *Psychological Review*, 1 (1894), 516–29, 529. On parallels between James and Charles Bell six decades earlier see Dixon, '"Emotion": The History of a Keyword in Crisis', *Emotion Review* (2013), doi:10.1177/1754073912445814.

34 See Phillip Prodger, *Darwin's Camera: Art and Photography in the Theory of Evolution* (Oxford: Oxford University Press, 2009).

35 Dixon has recently carried out work with school children on the recognition of emotions, using historical scientific images. The study calls into question the universality of expression: see Dixon, 'Feeling Differently: Using historical images to teach emotional literacy in an East London school', 2011, http://www.qmul.ac.uk/emotions/docs/54589.pdf (accessed 5 July 2012). For other challenges to universality see James M. Carroll and James A. Russell, 'Facial Expressions in Hollywood's Portrayal of Emotion', *Journal of Personality and Social* Psychology, 72 (1997), 164–76; Ruth Leys, 'How Did Fear Become a Scientific Object and What Kind of Object Is It?', *Representations*, 110 (2010), 66–104; Anna Wierzbicka, 'Emotion, Language, and Cultural Scripts', in Shinobu Kitayama and Hazel Rose Markus (eds), *Emotion and Culture: Empirical Studies of Mutual Influence* (Washington: American Psychological Association, 1994), 133–96 ; and, especially, James A. Russell, 'Is there Universal Recognition of Emotion from Facial Expressions? A Review of the Cross-Cultural Studies', *Psychological Bulletin*, 115 (1994), 102–41. For studies which suggest a combination of both universal expressions and cultural variation, see Alan Page Fiske, Shinobu Kitayama, Hazel Rose. Markus and Richard E. Nisbett, 'The Cultural Matrix of Social Psychology', in Susan T. Fiske, Daniel T. Gilbert, and Gardner Lindzey (eds), *The Handbook of Social Psychology*, 2 vols (New York: McGraw-Hill, 1998), ii. 915–81, and Batja Mesquita, Nico H. Frijda, and Klaus R. Scherer, 'Culture and Emotion', in John W. Berry, Pierre R. Dasen and T.S. Saraswathi (eds), *Handbook of Cross-Cultural Psychology*, 2 vols, ii., *Basic Processes and Human Development* (Boston, MA: Allyn & Bacon, 1997), 255–97.

36 See Hess and Thibault, 2009, 123; Nico Frijda, *The Emotions* (Cambridge: Cambridge University Press, 1986); and Klaus R. Scherer, 'On the Nature and Function of Emotion: A Component Process

Approach', in Klaus R. Scherer and Paul Ekman (eds), *Approaches to Emotion* (Hillsdale, NJ: Erlbaum, 1984), 293–318.

37 Antonio Damasio, *Descartes' Error: Emotion, Reason and the Human Brain* (1994; London: Vintage, 2006).

38 Damasio draws on the behaviour of individuals who lost a class of emotions through neurological damage and were unable to make rational decisions; see *The Feeling of What Happens* (1999; London: Vintage, 2000), 41; and *Descartes' Error*, chapter 4.

39 See, for example, in addition to Damasio, the work of Raymond Dolan, Joseph LeDoux, Brian Knutson, and Joshua Greene. On new interdisciplinary attention given to the history of the emotions, see The Centre for the History of the Emotions, founded in 2008 at Queen Mary, University of London. On emotion in the history of science, see *Isis* ('Focus: The Emotional Economy of Science'), 100 (2009). On ways in which Darwin's account of human consciousness lends itself to the concerns of the literary humanities see Patricia Waugh, 'Mind in Modern Fiction: Literary and Philosophical Perspectives', in Nick Saul and Simon J. James (eds), *The Evolution of Literature: Legacies of Darwin in European Cultures* (Amsterdam and New York: Rodopi, 2011), 125–40. Waugh notes the potential of the 'naturalistic' turn to support the kinds of aesthetic experience advocated by many modern creative artists and movements, and helpfully argues that an acknowledgement of a more central role for the sensory and the affective organs in the processes of rational cognition is commensurate with modernist perspectives. On Darwin's sympathetic imagination see Paul White in this collection; see also Levine, *Darwin Loves You: Natural Selection and the Re-Enchantment of the World* (Princeton: Princeton University Press, 2006), chapter 7.

40 See Dolan, 'Speaking about the Other', 2012.

41 Damasio cites the work of Ludvig von Bertalanffy, Kurt Goldstein and Paul Weiss on the idea of an integrated organism, but it was slow to inform the conception of mind and brain.

42 *Descent*, i. 105. On the human–animal relation from a Darwinian perspective, see the work of Marc Bekoff, Jessica Pierce and Frans de Waal. See also Mary Midgely, *Animals and Why They Matter* (1983; Athens, GA: University of Georgia Press, 2006) and *Beast and Man* (1979; London: Routledge, 2002); Donald R. Griffin, *The Question of Animal Awareness: Evolutionary Continuity of Mental Experience* (New York:

Rockefeller University Press, 1978) and *Animal Minds: Beyond Cognition to Consciousness* (University of Chicago Press, 2001); Colin Allen and Marc Bekoff, *Species of Mind: The Philosophy and Biology of Cognitive Ethology* (Cambridge, MA: MIT Press, 1999).

43 *Descent*, i. 154, 155.

44 See, for example, Pawel Fedurek and Katie E. Slocombe, 'Primate Vocal Communication: a Useful Tool for Understanding Human Speech and Language Evolution?', *Human Biology*, 83 (2011), 153–73. Fedurek and Slocombe focus on three main aspects of primate vocal behaviour, functional reference, call combinations, and vocal learning, noting that, important differences notwithstanding, primate vocal communication exhibits some key features characterising human language. They suggest that that comparative research on primate vocal behaviour will deepen understanding of the evolution of human speech and language. On support from comparative neuroimaging work for the notion that the brain regions that process communication signals in the human brain arose from a precursor network of regions present in nonhuman primates, and used for processing species-specific vocalisations, see Benjamin Wilson and Christopher I. Petkov, 'Communication and the Primate Brain: Insights from Neuroimaging Studies in Humans, Chimpanzees and Macaques', *Human Biology*, 83 (2011), 175–89. On human language as a complex reconfiguration of ancestral systems that have been adapted in evolutionarily novel ways see Simon E. Fisher and Gary F. Marcus, 'The Eloquent Ape: Genes, Brains and the Evolution of Language', *Nature Reviews Genetics*, 7 (2006), 9–20. See also C.N. Slobodchikoff, Bianca S. Perla and Jennifer L. Verdolin, *Prairie Dogs: Communication and Community in an Animal Society* (Cambridge, MA: Harvard University Press, 2009). For a recent historical account of the human–animal relation, see Joanna Bourke, *What it Means to be Human: Reflections from 1791 to the Present* (London: Virago, 2011); on primates and language see her chapters two and three.

45 Harriet Ritvo, 'Animal Planet', *Environmental History* 9 (2004), 205.

46 On this, see Marc Bekoff, Foreword, Carol Freeman, Elizabeth Leane and Yyvette Watt, eds, *Considering Animals: Contemporary Studies in Human–Animal Relations* (Farnham, Ashgate, 2011), xv.

47 *Why Animals Matter: Animal Consciousness, Animal Welfare, and Human Well-Being* (Oxford: Oxford University Press, 2012). Dawkins's argument, referred to by Bekoff as Dawkins's dangerous idea in

Psychology Today, http://www.psychologytoday.com/blog/animal-emotions/201205/animal-consciousness-and-science-matter (accessed 5 July 2012), has been received with critical incredulity, as the afterword to this volume notes.

48 The declaration was signed on 7 July 2012 by scientists attending the Francis Crick Memorial Conference on Consciousness in Human and Non-Human Animals, Churchill College, University of Cambridge. See http://fcmconference.org/img/CambridgeDeclarationOnConsciousness.pdf (accessed 10 August 2012); see also http://www.psychologytoday.com/blog/animal-emotions/201208/scientists-finally-conclude-nonhuman-animals-are-conscious-beings (accessed 10 August 2012).

49 Derek C. Penn, Keith J. Holyoak and Daniel J. Povinelli, 'Darwin's Mistake: Explaining the Discontinuity Between Human and Nonhuman Minds', *Behavioral and Brain Sciences*, 31 (2008), 109–30.

50 Gordon Burghardt, 'The Sun Always Rises: Scientists Also Need Semantics', *Behavioral and Brain Sciences*, 31 (2008), 133–4: 134. Elsewhere Burghardt argues for the value of what he terms critical anthropomorphism: see 'Critical Anthropomorphism, Uncritical Anthropocentrism, and Naïve Nominalism', in *Comparative Cognition and Behavior Reviews*, 2 (2007), 136–8. See also Irene M. Pepperberg, 'Difficulties with "Humaniqueness"', *Behavioral and Brain Sciences*, 31 (2008), 143–4. Pepperberg draws on Timothy Q. Gentner, Kimberly M. Fenn, Daniel Margoliash and Howard C. Nusbaum, 'Recursive Syntactic Pattern Learning by Songbirds', *Nature*, 27 April 2006, 1204–7. Gentner et al demonstrate that recursive-style behaviour is not unique to humans and can be observed in, for example, starlings. See Michael Tomasello, Josep Call and Brian Hare, 'Chimpanzees Understand Psychological States – The Question is Which Ones and to What Extent', *Trends in Cognitive Sciences,* 7 (2003), 153–6.

51 Nathan J. Emery and Nicola S. Clayton, 'Imaginative Scrub-Jays, Causal Rooks, and a Liberal Application of Occam's Aftershave', *Behavioral and Brain Sciences*, 31 (2008), 135; the article draws on Sérgio P.C. Correia, Anthony Dickinson and Nicola S. Clayton, 'Western Scrub-Jays Anticipate Future Need States Independently of their Current Motivational State', *Current Biology*, 17 (2007), 856–61. See also N.J. Mulcahy, and J. Call, 'Apes Save Tools for Future Use', *Science,* 312 (2006), 1038–40; C.R. Raby, D.M. Alexis, A. Dickinson, and N.S.

Clayton, 'Planning for the Future by Western Scrub-Jays', *Nature*, 445 (2007), 919–21.

52 Elizabeth V. Hallinan and Valerie A. Kuhlmeier, 'Ontogeny, Phylogeny, and the Relational Reinterpretation Hypothesis', *Behavioral and Brain Sciences*, 31 (2008), 138–9.

53 See, for example, Louis M. Herman, Robert K. Uyeyama, and Adam A. Pack, 'Bottlenose Dolphins Understand Relationships Between Concepts', *Behavioral and Brain Sciences*, 31 (2008), 139–40. See also Louis M. Herman, 'Intelligence and Rational Behavior in the Bottlenosed Dolphin', in Matthew Nudds and Susan Hurley (eds), *Rational Animals?* (Oxford: Oxford University Press, 2006), 439–67, and 'Exploring the Cognitive World of the Bottlenosed Dolphin', in Marc Bekoff, Colin Allen and Gordon Burghardt (eds), *The Cognitive Animal: Empirical and Theoretical Perspectives on Animal Cognition* (Cambridge, MA: MIT Press, 2002), 275–83; and Thomas Suddendorf and Andrew Whiten, 'Mental Evolution and Development: Evidence for Secondary Representation in Children, Great Apes and Other Animals', *Psychological Bulletin*, 127 (2001), 629–50, and 'Reinterpreting the Mentality of Apes', in Kim Sterelny and Julie Fitness, eds, *From Mating to Mentality: Evaluating Evolutionary Psychology* (London and New York: Psychology Press, 2003), 172–96.

54 Michael Siegala and Rosemary Varley, 'If We Could Talk to the Animals', *Behavioral and Brain Sciences*, 31 (2008), 146–7.

55 See also Ray Crozier, ' The Puzzle of Blushing', *The Psychologist*, 23 (2010), 390–3 and 'Darwin and the Blush', *Wellcome History*, 40 (2009), 16.

1

'Love and Hatred are Common to the Whole Sensitive Creation': Animal Feeling in the Century before Darwin[1]

Jane Spencer

Darwin was willing to discern a wide range of emotions in animals: not just those now considered basic in humans – anger, fear, sadness, disgust and enjoyment – but jealousy, pride, shame, ennui, wonder and curiosity. In Part I, chapter iii, of *The Descent of Man* (1871) he writes:

> Most of the more complex emotions are common to the higher animals and ourselves. Every one has seen how jealous a dog is of his master's affection, if lavished on any other creature; and I have observed the same fact with monkeys. This shows that animals not only love, but have desire to be loved. Animals manifestly feel emulation. They love approbation or praise; and a dog carrying a basket for his master exhibits in a high degree self-complacency or pride. There can, I think, be no doubt that a dog feels shame, as distinct from fear, and something very like modesty when begging too often for food. ...
>
> Several observers have stated that monkeys certainly dislike being laughed at ... Dogs show what may be fairly called a sense of humour, as distinct from mere play; if a bit of stick or other such object be thrown to one, he will often carry it away for a short distance; and then squatting down with it on the ground ... will wait until his master comes quite close to take it away. The dog will then seize it and rush away in triumph, repeating the same manoeuvre, and evidently enjoying the practical joke.[2]

A monkey who disliked being laughed at, and a dog who teased his master for fun, would be displaying both self-awareness and awareness of another's perspective, capacities whose ascription to nonhuman animals remains highly controversial today.[3] Darwin's discussion of animal emotion in the *Descent* and in *The Expression of the Emotions in Man and Animals* (1872) influenced later work on the subject, from George Romanes's *Animal Intelligence* (1882) and *Mental Evolution in Animals* (1883) onwards;[4] but it was strongly criticised in the late nineteenth and early twentieth centuries for, among other things, its reliance on anecdote and its anthropomorphism.[5] Darwin 'was convinced and tried to convince his readers that emotions and their expressions were not unique to humans'.[6]

This question was significant for Darwin because similarities between human and animal emotional expression suggested continuity between species, so offering extra support to his evolutionary theory. He ended *Expression* with the cautiously-worded claim that 'the study of the theory of expression confirms to a certain limited extent the conclusion that man is derived from some lower animal form'.[7] That evolutionary conclusion itself has been widely seen as justifying a fundamental re-evaluation of human relations with other animal species. Ousting humanity from its position within Christian tradition as the special creature made in God's image, evolutionary theory places human animals on a continuum with all the others to which they are related. J. Howard Moore, an early-twentieth-century Darwinist and teacher of zoology, used the idea of evolutionary continuity to argue for 'the kinship of all the inhabitants of the planet Earth. . . . they are all related, physically, mentally, morally'.[8] In the late twentieth century James Rachels argued that Darwinism undermined the basis of a traditional morality giving special dignity and consideration to humans. He also emphasises kinship, which he describes in terms of shared feelings and expression: 'Once we see other animals as our kin, we have little choice but to see their condition as analogous to our own. … Darwin stressed that … their nervous systems, their behaviours, their cries, *are* our nervous systems, our behaviours, our cries, with only a little modification. They are our common property because we inherited them from the same ancestors'.[9]

These claims for the far-reaching ethical consequences of Darwinism have recently been challenged by Rod Preece in a history of the status of animals in human thought. He argues that the basis for the ethical consideration of animals had already existed within the long tradition, in many forms of religious and philosophical thought, of a belief in animal–human kinship and in the importance of animal feeling.[10] While much Christian teaching has emphasised a sharp distinction between humans and animals, many Christian thinkers throughout the ages have espoused an alternative tradition in which human–animal similarities are acknowledged and valued. Kinship with animals in this tradition is guaranteed by God's fatherhood of all life, while the attribution of some form of soul to nonhuman animals was the basis for claiming some similarity in attributes of mind between human and animal. Darwinism, Preece argues, did not offer a new account of our responsibilities towards animals, nor did it create a new public awareness of those responsibilities, or promote better treatment of animals.[11] Preece is right that Darwinism offers neither a necessary nor a sufficient condition for reformed ethical attitudes to animals.[12] Continuity, kinship, appreciation of animal abilities and a sense of moral obligation to them had previously been derived in other ways; while there has been no shortage in the years since Darwin of new articulations of human uniqueness compatible with evolutionary theory.[13] What Darwinism did encourage was rather the development of a scientific and secular discourse, rooted in biological and behavioural studies, for articulating concepts of human–animal relationship. While such a discourse of animal–human continuity takes a great deal from Darwin, its roots can be traced well before him – roots of which he was well aware. Always keen to find out whether earlier thinkers had anticipated his ideas – reminding himself at one point to 'Read Aristotle to see whether any of my views are ancient'[14] – Darwin paid particular attention to the natural history and the philosophy of the previous century.

During the eighteenth century, the question of animal feeling was much debated, with several strands of intellectual and scientific discourse having reason to ask about the sensations and passions of brute beasts, and a flourishing tradition of sentimental literature ready to anthropomorphise animals' emotional lives. Descartes's views on animal mechanism, elaborated especially by Malebranche to argue that animals did not feel, were controversial in Britain as elsewhere in Europe, and defences of animal soul were mounted.[15] Early anatomists' discovery of analogous structure in human and ape, especially Edward Tyson's view that the 'orang-outang' had speech organs, encouraged belief in apes' similarity to humans in the matter of mind and feeling as well. Later anatomical researches, such as Petrus Camper's in the 1770s, reasserted human distinction and superiority while adumbrating the growing trend to ground both distinction and likeness in physical causes such as brain size and composition, rather than in a concept of soul.[16] The development and popularity of natural history, particularly with the publication of Buffon's immense *Histoire naturelle* (1749–89), turned attention to the mental as well as physical characteristics of different animal species.[17] The new science of the nerves, tracing sensation to its physiological causes, laid the basis for David Hartley's account of a shared neurophysiological basis for human and animal emotion.[18] David Hume's philosophy of sympathy, elevating the psychological role of the passions as opposed to reason, treated the existence of similar passions in humans and animals as evidence that passion in animals existed on a simple level, and discussed sympathy, or the communication of passion, as a physical process common to human and animal.[19] Francis Hutcheson's moral sense philosophy, acknowledging animals' ability to feel pain and pleasure, laid the groundwork for arguments for animal rights, further developed later in the century.[20] Sentimental poetry, especially that of James Thomson and William Cowper, encouraged readers to sympathise with the affections of beasts, birds and even insects, and were widely cited in later animal welfare campaigns.[21] A complex discourse of animal feeling emerged, in some respects anticipating Darwin's views on the continuity of human and animal emotions.

Eighteenth-century thinkers had two alternative ways of organising the concept of human–animal similarity. One derived from a theist tradition, involved considering the question of the nonhuman animal's soul. Theism, popular in the eighteenth century among many who questioned traditional Christian teachings, is the belief in the existence of a deity, not confined to but including the Christian concept of God. In this tradition, some thinkers base their relatively high estimation of animal capacities on a belief in an immaterial soul which is necessary to move action in animal and human bodies alike. The alternative way of comparing animal and human is an emergent monist and physicalist one, where the similarity of animal and human emotions demonstrates their physical basis; monism is the belief that reality consists in one substance, and therefore mind and matter should not be considered as separate substances. The discourse of animal feeling in the period drew on both of these. In this essay I examine a selection of philosophical, religious and moral texts discussing animal feeling, showing how they variously deploy physicalist and spiritualist explanations for human–animal likeness. I begin with David Hume, whose *Treatise of Human Nature* (1739) offered the most detailed account of animal passions, and whose argument from analogy for crediting nonhuman animals with a mental life has been influential on modern debate; and then discuss David Hartley and Joseph Priestley, whose views of mental continuity between human and animal fed into their monist understanding of body and soul. This physicalist tradition informed the development of Darwin's views on animal feeling, with Hume having the clearest influence on him. I then turn to a number of churchmen[22] who, from the 1750s to the 1790s, wrote sermons or pamphlets in favour of kindness to animals: Richard Dean, James Granger, Humphrey Primatt and Thomas Young. While all these writers hold widely divergent religious and philosophical views, varying between the scepticism of Hume, the Rational Dissent of Priestley and the Anglican Evangelicalism of Dean, they share an emphasis on similarities between animal and human, focused especially on animal capacity for feeling.

The emergent physicalist view of emotion did not simply replace the old paradigm, nor did it by any means dominate the growing literature of animal advocacy.[23] However, as we will see, it became increasingly influential. From the seventeenth century onwards the issue of animal feeling had been intimately tied to religious controversy over the nature of the soul, a controversy which (rather than the questions of kindness and cruelty) had been the main concern of many writers on animals.[24] This intimate tie continues into the period under investigation, but there are signs that it is loosening towards the end of the period. Over the second half of the eighteenth century the question of whether or not brutes possessed an immaterial soul, once considered key to discussion of their minds, became less and less important as animals' mental capacities were increasingly discussed in naturalistic terms.

To draw comparisons between eighteenth-century discussions of animal feeling and Darwin's work on animal emotion is difficult because of the shifting use of feeling terms between the two periods. As Thomas Dixon has shown, the popularity in post-Darwinian psychology of the term *emotion* to refer to feelings arising in bodily impulses and distinguishable through their physiological expression represents the narrowing of a wider discourse of feeling in the eighteenth century, when terms such as *passion*, *sentiment* and *affection* encompassed feelings considered to have spiritual as well as physical aspects.[25] *Emotion* is used relatively little in eighteenth-century philosophy: Dixon traces its first extensive use to David Hume's *Treatise*, and demonstrates its association with a tendency to physicalist views. He also remarks on Hume's loose use of the term relative to later writers: sometimes it means 'something like "feelings" or "movements" or "agitations of the mind", so it "attends" a passion'; but in other places it is used 'just as a kind of synonym for passions, affections, or movement of bodily spirits'.[26] In this essay I discuss eighteenth-century attributions of feeling to animals encompassing both the term *emotion* and the more established *passion*, *sentiment* and *affection*.

Darwin's belief in animal emotion had plenty of precedence in the centuries before him. Despite pervasive beliefs in fundamental differences in kind between human and nonhuman animal life, sentience, at least, had been attributed to animals since ancient times. Aristotle's division of souls into ascending orders of nutritive, sensory, appetitive, locomotive and rational, with only the last of these unique to humanity,[27] influenced the early modern belief that animals, for all their deficiencies in reason, were sensitive to pain and pleasure.[28] The Cartesian explanation of animal motion through pure mechanism, though widely known and debated in the seventeenth and eighteenth centuries, did not command much agreement in Britain.[29] Acceptance of animals' susceptibility to pain and pleasure led easily to a belief in their ability to apprehend these as future states, so animal passions such as desire and fear were also thought possible. As Richard W. Serjeantson has demonstrated, early modern discussions of language reached broad agreement that though animals lacked rational, symbolic language, they had a natural language expressive of their passions.[30] The Italian anatomist Hieronymous Fabricius ab Aquapendente maintained in his *De brutorum loquela* [*On the language of brutes*] (1603) that animals had four main passions: pleasure, desire, annoyance and fear.[31]

A much more ambitious range of animal emotions was described by the Scottish Enlightenment philosopher David Hume. Animal reason and feeling were treated extensively in his *Treatise* (1739). Disappointed by its poor reception, Hume later recast his ideas in more reader-friendly form in *Philosophical Essays Concerning Human Understanding* (1748), later known as *An Enquiry Concerning Human Understanding*. Darwin read the discussions of animals in both works during 1838, while he was beginning to work out the idea of natural selection.[32] In the *Treatise* Hume distinguished between the 'original impressions' of sensation, which included all 'bodily pains and pleasures', and the 'secondary, or reflective impressions' of passion or emotion. Impressions of sensation arise from the constitution of the body, from the 'animal spirits' or from 'the application of objects to the external organs', while passions and emotions proceed either directly from sensations or from the idea of sensations.[33] Animals shared with humans the capacity for secondary passions as well as original sensations. Their passions were presumably more likely to proceed directly from sensations than those of humans, since animals were 'but little susceptible either of the pleasures or pains of the imagination',[34] but even imagination was not ruled out altogether: malice and envy were more common in animals than pity, because less thought and imagination were needed in order to feel them.[35] Hume divided the passions into two kinds: the direct passions arising immediately from pain and pleasure, which he named as 'desire, aversion, grief, joy, hope, fear, despair and security'; and the indirect passions, which were also derived from pain and pleasure but required the addition of more complex circumstances. These included 'pride, humility, ambition, vanity, love, hatred, envy, pity, malice, generosity, with their dependants'.[36] Hume did not even bother to discuss the direct passions in animals, so obvious did he consider it that these were 'of the same nature, and excited by the same causes as in human creatures'.[37] He reserved his attention for the more complex indirect passions, arguing that animals felt pride and humility, envy and malice, and could sense each other's pain and pleasure.[38] While animal pride and humility had simple causes, centred in bodily attributes, and were expressed in bodily postures and movements, they corresponded to the same passions in human beings:

> The very port and gait of a swan, a turkey, or peacock show the high idea he has entertain'd of himself, and his contempt of all others. This is the more remarkable, that in the two last species of animals, the pride always attends the beauty, and is discover'd in the male only. The vanity and emulation of nightingales in singing have been commonly remark'd; as likewise that of horses in swiftness, of hounds in sagacity and smell, of the bull and cock in strength, and of every other animal in his particular excellency. Add to this, that every species of creatures, which approach so often to man, as to familiarize themselves with him, show an evident pride in his approbation, and are pleas'd with his praises and caresses, independent of every other consideration. Nor are they the caresses of every one without distinction, which give them this vanity, but those principally of the persons they know and love; in the same manner as that passion is excited in mankind. All these are evident proofs, that pride and humility are not merely human passions, but extend themselves over the whole animal creation.[39]

Hume's ascription of such complex feelings to nonhuman animals relied on the argument from analogy. Where anatomists had found similar structures in human and animal bodies they concluded that those structures worked in comparable ways, and the same logic, he argued, should be applied to comparisons of animal and human mind: where similar behaviours were observed, we should infer that they shared similar mental causes.[40] Analogy is the keynote of the *Enquiry*'s Section IX, 'Of the Reason of Animals', which opens: 'ALL our reasonings concerning matter of fact are founded on a species of ANALOGY'.[41] Hume's use of analogy in comparing animal and human is seen as a fundamental error by some recent ethologists. Daniel Povinelli, for example, argues that 'identical behaviours may be generated by and/or attended by different psychological representations', and chimpanzees and humans may have arrived at similar behaviour by very different mental routes.[42] The willingness to draw analogies between human and animal which unites Hume and Darwin featured in Enlightenment natural history as well as in the philosophy of sympathy of which Hume was a major exponent. It contributed to one side of the eighteenth-century debate between those who upheld and those who favoured the erosion of a firm dividing line between human and animal.

For Hume the similarity of animal and human passions was important because it indicated the simplicity of the ideas causing them. In his discussion of animal reason in Book I of the *Treatise*, he argued that philosophers tended to explain the mind's workings in too complicated a way, supposing

> such a subtility and refinement of thought, as not only exceeds the capacity of mere animals, but even of children and the common people in our own species, who are notwithstanding susceptible of the same emotions and affections as persons of the most accomplish'd genius and understanding.[43]

A dog, like a man, reasoned by building inferences from past experiences. While humans, and not animals, had the capacity for arguing from their different experiences to a general conclusion, the basic principle of reasoning from an impression on the memory or senses was shared by both.[44] When he turns to the passions in Book II, Hume continues to argue for similarity between human and animal:

> we may observe, not only that love and hatred are common to the whole sensitive creation, but likewise that their causes . . . are of so simple a nature, that they may easily be suppos'd to operate on mere animals. There is no force of reflection or penetration requir'd. Every thing is conducted by springs and principles, which are not peculiar to man, or any one species of animals.[45]

Demolishing the barrier between human reason and animal instinct relied on by many of his contemporaries to guarantee human distinction, Hume announced that 'reason is nothing but a wonderful and intelligible instinct in our souls'.[46] His sensationalist epistemology helped Darwin to work out the ideas expressed in his 1838 Notebooks about the shared basis for simple reasoning in man and other animals.[47] In such Notebook entries as 'Do not our necessary notions follow as consequences on *habitual* or instinctive assent to propositions which are the result of our senses, or our *experience*', and 'Instinct is a modification of bodily structure…& intellect is a modification of instinct', the influence of Humean psychology is evident.[48]

During the eighteenth century the notion that animal and human were governed by similar bodily mechanisms gained ground, and became a basis for challenging belief in distinctions of kind between the two. Darwin knew that the belief he jotted down in 1838, that 'the mind is function of body' had diverse antecedents in the previous century.[49] In France, La Mettrie turned Cartesianism back on itself by arguing that if the actions of beasts could be explained by mechanical causes, so could those of man. His *L'Homme Machine* (1748) was infamous for materialism and atheism.[50] In Britain at around the same time, David Hartley developed a different anti-dualist position in *Observations on Man, his Frame, his Duty, and his Expectations* (1749). The scientist and leading Unitarian theologian Joseph Priestley brought Hartley's work to a wider audience in the 1770s through an abridged edition,[51] ensuring through his mediation that Hartley became a key influence on Unitarian tradition. Priestley followed Hartley in adopting a material theory of the mind, which he maintained (against Hume) to be consistent with religious faith.[52] The unorthodox monism of Hartley and Priestley was a particular influence on Darwin, both through his Unitarian family connections and traditions – Priestley, like both Darwin's grandfathers, Erasmus Darwin and Josiah Wedgwood, belonged to the famous Birmingham Lunar Society – and his reading.[53] His notebooks refer to Priestley and to Hartley's *Observations*, recommended to him through his reading of Priestley and of James Mackintosh's work on seventeenth- and eighteenth-century philosophy.[54]

Hartley traced perception, action and emotion to what he termed 'vibrations' in the nervous system, and combined this early neurophysiological interpretation of psychology with an unorthodox spiritualism: 'Rather than advocating a material universe devoid of spirit, Hartley imagined an active, spiritual cosmos virtually empty of "inert" matter'.[55] Denying the dualism of mind and matter, he considered mind as made up of sensations and ideas that emerge from the organisation of matter.[56] There is thus no need, in Hartley's system, to conceive of nonhuman animals as fundamentally different from human ones in the kind of their souls or in the cause of their mental operations:

> If the Doctrines of Vibrations and Association be found sufficient to solve the Phaenomena of Sensation, Motion, Ideas, and Affections, in

> Men, it will be reasonable to suppose, that they will also be sufficient to solve the analogous Phaenomena in Brutes. And, conversely, it seems probable, that an Endeavour to apply and adapt these Doctrines to Brutes will cast some Light and Evidence upon them, as they take place in Men.[57]

He accounted for brutes' intellectual inferiorities through physical causes: firstly, their smaller brains in proportion to their size, and the 'Imperfection of the Matter of their Brains'.[58] Their lack of verbal language, their greater reliance on instinct and the lesser variety of their mental experiences and impressions, combined to give them different and inferior intelligence.[59] Yet for all his discussion of animal inferiority, Hartley emphasised the basic similarities in mechanism between man and brute. He accepted Cartesian mechanism while applying it equally to man and brute and, crucially, considering it adequate to account for perception and emotion:

> And though I suppose, with *Descartes*, that all [brutes'] Motions are conducted by mere Mechanism; yet I do not suppose them to be destitute of Perception, but that they have this in a manner analogous to that which takes place in us; and that it is subjected to the same mechanical Laws as the Motions. Whether the ideal Vibrations, which take place in the medullary Substances of their Brains, be the Result of former Impressions, or the mere Offspring of their vital and natural Powers … or the compound Effect of both … I always suppose, that corresponding Feelings, and Affections of Mind, attend upon them, just as in us.[60]

Hartley found close relationship and similar feeling between man and brute:

> the Brute Creatures prove their near Relation to us, not only by the general Resemblance of the Body, but by that of the Mind also; inasmuch as many of them have most of the eminent Passions in some imperfect Degree, and there is, perhaps, no Passion belonging to human Nature, which may not be found in some Brute Creature in a considerable Degree.[61]

While brutes could 'scarce ever' achieve the kind of self-consciousness required to 'compare and connect themselves with themselves in different Situations', their 'Reasonings and Affections', though less complex than ours, were similar enough to deserve the same names 'of Sagacity, Cunning, Fear, Love, &c'.[62]

While Hume's belief in animals' feeling did not lead him to advocate any particular tenderness towards them – the example he gives to demonstrate a dog's ability to reason turns rather tellingly on its experience of different beatings[63] – Hartley linked animal feeling to a duty of kindness towards them. A vegetarian himself,[64] he thought that killing animals for food 'does great Violence to the Principles of Benevolence and Compassion'.[65] These principles ought to be applied to animals from religious motives: on the question of kindness to animals Hartley advocated a traditional Christian stewardship. 'We seem to be in the place of God to them, to be his Viceregents, and impowered to receive Homage from them in his Name. And we are obliged by the same Tenure to be their Guardians and Benefactors'.[66] Whereas Hume's views on animal–human continuity led him to argue that the souls of both were mortal,[67] Hartley's on animal–human kinship led in the opposite direction, to speculate that both might be granted eternal life:

> if [animals] should prove to be our brethren and sisters in this higher sense, in immortality as well as mortality, in the permanent principle of our minds, as well as the frail dust of our bodies; if they should be partakers of the same redemption as well as of our fall, and be members of the same mystical body, this would have a particular tendency to increase our tenderness for them.[68]

For Priestley, as for Hartley, a monist view of mind and matter did away with any reason to consider animal and human minds fundamentally different. In his discussions of animals, Priestley belonged firmly to the tradition that placed the question of soul at the centre of the issue of their capabilities and feelings, but his material understanding of the soul significantly shifted the terms of debate. Immortality of the soul, for Priestley, did not consist in the survival of an immaterial substance but on God's eventual resurrection of a body dormant since death. In his *Disquisitions relating to Matter and Spirit* (1777), he offered a history of ancient and modern opinions concerning animal soul. Before Christianity, he claimed, soul was considered to be the same kind of substance whether in man or brute, and thus able to transmigrate between the two: it was Christianity which, by introducing the promise of personal immortality to human souls, 'in such a manner as made it impossible to think that brutes could have any title to it', created 'a great difference between men and brutes':[69] the notion that their souls were of fundamentally different kinds, one immaterial and immortal, the other corporeal and destroyed by death.[70] Priestley evidently thinks this distinction unwarranted, though the opinion he ascribes to Descartes, that the souls of brutes are 'mere *automata*', he finds even 'more extraordinary'.[71] Dismissing Descartes's followers, who denied animal perception,[72] criticising even the 'much more liberal and rational' Cambridge Platonist Ralph Cudworth for considering that brutes lack morality and liberty,[73] and disagreeing with Locke's view that they are unable to form abstract ideas,[74] Priestley concludes, along lines very similar to Hartley's, that though brutes have, 'on account of their wanting a sufficient *quantity of brain*', much simpler ideas and intellects, 'they must necessarily have, *in kind*, every faculty that we are possessed of'.[75] Priestley's argument tends strongly towards the view that human and animal souls are both corporeal and both resurrected into immortal life. If the resurrection of the body is a miraculous event, outside natural laws, it is unlikely to be extended to brutes, since there is no definite Biblical promise to that effect; but if (as Priestley is clearly inclined to believe) resurrection is rather:

> *within the proper course of nature*, extensively considered, and consequently there be something remaining of every organized body that death does not destroy, there will be reason to conclude that they will be benefited by it as well as ourselves. And the great misery to which some of them are exposed in this life, may incline us to think, that a merciful and just God will make them some recompence for it hereafter. He is *their* maker and father as well as *ours*. But with respect to this question, we have no sufficient *data* from which to argue, and must therefore acquiesce in our utter ignorance; satisfied that the Maker and Judge of all will do what is right.[76]

Both Hartley and Priestley, then, mix physicalist accounts of human–

animal similarity with a Christian view of human–animal kinship as based on God's fatherhood of all. Darwin's mind–body materialism and his ideas of continuity between human and animal therefore had strong antecedents in the avant-garde religious traditions of the previous century, traditions which were influential in the development of an ethos of consideration and kindness towards animals.

Sympathy with animal feeling was also recommended by much more orthodox writers. From the 1740s onwards, a number of Anglican churchmen tackled the question of animals' mental and feeling capacities as part of a moral argument for the human duty of kindness to beasts, put forward in tracts and sermons. These writers, unlike Hume, Priestley and Hartley, had no influence on Darwin, whose thought moved in an entirely different direction; but they were important for spreading the idea of human–animal kinship widely among eighteenth-century readers, in a form acceptable to a religiously conservative readership. The first two to be considered here are John Hildrop, a rector in Yorkshire who published *Free Thoughts upon the Brute Creation* in two parts in 1742 and 1743, and Richard Dean, curate of Middleton near Manchester, whose *An Essay on the Future Life of Brutes* appeared 25 years later. Hildrop and Dean, like Hartley and Priestley, emphasised the similarity of animal and human feeling, but they approached the question with a very different agenda. For them, opposition to the Cartesian position on animals was central. Hartley could accept Cartesian mechanism while considering mechanical explanations sufficient to account for perception and feeling in animal and human. Hildrop and Dean, in contrast, shared Descartes's mind–matter dualism, considering an immaterial soul essential to animate a material body, and therefore had to reject his notion of animal as machine in order to argue that animals were capable of experiencing feeling. They both highlighted animals' sensitivity to pain, a sensitivity that they derived ultimately from the possession of an immaterial soul. It was a position shared by John Wesley in his 1763 *Survey of the Wisdom of God in the Creation*: Wesley argued that while both animals and humans may function mechanically in some respects, their 'Sense, Perception and Various Appetites', including the faculties of '*perceiving* or *knowing*' and '*desiring* or *shunning*' derive from a 'superior Principle' or soul which must be immaterial because it has self-motion, while matter is 'wholly and essentially passive'.[77]

Hildrop drew closely on a treatise by the French writer Guillaume-Hyacinth Bougeant, whose *Amusement philosophique sur le langage des bestes* had been translated into English in 1739. He argued for animal sense from the behaviour of dogs (Darwin's later appeal to the evidence offered by pet dogs was to join a long tradition).

> When … I see a Dog hastening to me when I call him, caress me when I stroke him, tremble when I rate him, run away from me when I beat him: nay, further, when I see him reflecting and reasoning upon my Conduct towards him, I must conclude he is acted [upon] by some higher Principle than meer Mechanism.[78]

Animals' language of feeling refuted Cartesianism:

> if by the different Motions and Gestures of their Bodies, or Sound of their Voice, they express their different Sentiments of Joy and Sadness, of Pain or Pleasure, of Fear and Desire, of Love or Hatred; I cannot help concluding from thence, that they have in them some Principle of Knowledge and Sentiment.[79]

Hildrop rejected Bougeant's perhaps facetious explanation that this principle in an animal was the guilty soul of a devil punished by God with confinement in a bestial body,[80] arguing instead that animals had their own immaterial souls,[81] and that they partook without sin in the Fall of Adam and would share in the final redemption.[82]

In his *Essay on the Future Life of Brutes, introduced with Observations upon Evil, its Nature, and Origin* (1767) Richard Dean argued that animal immortality followed logically and morally from animal feeling: logically, because animals' ability to feel pain and pleasure meant they must be directed by a sentient principle, or soul, which in his view entailed immateriality and therefore (unless God chose to annihilate the soul) immortality;[83] morally, because a loving God would not have created animals subject to the pains of this life if he had not intended to compensate their sufferings in a future existence.[84] As Dean's subtitle suggests, for him the question of animal soul was part of the theological problem of the existence of evil, explained within Anglican theology as introduced at the Fall, itself allowed by a loving God in order that man's will should remain free. That animals, though morally blameless, should share in the evil of the Fall without also sharing in the Redemption, seemed to him inconsistent with God's justice and mercy. Unorthodox though the notion of brutes' immortality was, then, it has been suggested that Dean deployed the argument with the intention of defending Anglican tenets of free will from the encroaches of various systems of thought that challenged them: on the one hand, materialist, mechanistic concepts of causation developed after Descartes; on the other, the providentialism of Calvinistic Methodism and Presbyterianism.[85] His *Essay* was part of an Evangelical 'struggle to win back souls from radical dissent and minds from materialist philosophy'.[86]

Dean claimed to 'confute the Doctrine of De Carte [sic], Malebranche [sic], &c.' that 'Brutes are unintelligent Machines'.[87] He stressed that he did not mean to challenge the traditional hierarchy between man and beast, but he criticised the human pride that would increase its distance from animals: 'The Objections against the Futurity of Brutes . . . are not founded in Reason, but in Pride and Envy'.[88] Throughout his essay he emphasises the extent to which humans and nonhuman animals have a shared nature. As creatures with bodies they are necessarily subject to natural evils. While evidently willing to entertain the possibility of complex self-consciousness on the part of beasts, he rests his argument upon their more easily demonstrable simple sensations:

> If brute Animals do not multiply their Sorrows by Reflexion upon what is past, and Fear of what is to come; a Point this perhaps, not quite so clear as to exclude all Doubts and Exceptions; it is however very certain, that they are liable to momentary Sufferings, and transient Evils, they have their dark Hours of Pain and Sickness, and die under the sad Appearances of Agony, like the Beings that are above them.

As 'animated Forms', both humans and animals 'are actuated by Principles of Intelligence peculiar to their Natures . . . [and] all these Souls are immaterial Substances'.[89] Animals cry out in pain; a horse loves to eat sweet hay: 'Does not every day present us with Instances of the Taste of the Brute, and his Power of distinguishing?'[90] Dean's concern is more with animals' sensitivity to physical pain and pleasure than with the passions and affections (though he does say that some brutes show affection and fidelity, and even that they like to go to church).[91] It is this bodily sensitivity which he calls on to demonstrate both animals' ensoulment and the human duty to treat them well. A reader convinced of his arguments will, he hopes, 'consider, that as Brutes have Sensibility, they are capable of Pain, feel every Bang, and Cut, and Stab, as much as he himself does, some of them perhaps more, and therefore he must not treat them as Stocks, or Stones, or Things that cannot feel'.[92]

In the last thirty years of the century, members of the clergy continued to produce works of animal advocacy, but there was a shift in their emphasis. They came to depend less on the argument for the immateriality and immortality of brute souls. The physical sensibility of animals, recognised by Dean, became a more central part of the case for treating them well; and the influence of both Hume and Priestley can be detected in some of their discussions of animal feeling.

James Granger, vicar of Shiplake, Oxford, writing a sermon against cruelty to animals a few years after the publication of Dean's essay, appeals to the notion of animal sensitivity without deriving it from an immaterial soul. It is the likeness between human and animal sensitivity which argues for compassion: 'the great *law* of *humanity* … comprehends every kind of being that hath the same acute sense of pain, which [man] finds in his own frame'.[93] In this short sermon Granger is more concerned to preach a message of kindness than to venture onto the controversial territory of the soul.[94] His advice is framed within the concept of Christian stewardship: like David Hartley, he considers that man is the 'substitute or vice-roy of the Almighty, with respect to the animal creation', and therefore accountable to God for how he treats it.[95] Granger is explicit about human kinship with animals, based on the notion of the chain of being linking all life, and on scriptural texts. A righteous man 'extends his benevolence to the brute creation; to which he thinks himself allied, as well as to the higher orders of being, well knowing, that as he can say to an Angel, *Thou art my brother*; so he can say to the worm, *Thou art my sister*'.[96]

Humphrey Primatt's *Dissertation on the Duty of Mercy and Sin of Cruelty to Brute Animals* (1776), also appeals to the chain of being and to human kinship with animals, arguing on these bases for a widening circle of compassion: 'our Love and Mercy . . . are to be extended to every object of the Love and Mercy of GOD *the universal Parent*'.[97] While acknowledging man's place at the top of the scale of terrestrial animals,[98] Primatt emphasises what men and beasts have in common: constituted of breathing dust, both are living souls,[99] and most importantly, both feel pain: 'Superiority of rank or station exempts no creature from the sensibility of pain, nor does inferiority render the feelings thereof the less exquisite. Pain is pain, whether it be inflicted on man or beast'.[100] Primatt, like Dean and Hildrop, broaches the question of the nature of the animal soul, but comes to no clear conclusion. At one point he says that man only has an immortal soul,[101] but at another suggests that this may be a mistaken belief born of human presumption and pride.[102] He adopts Joseph Priestley's view that if animals have no future life to recompense them it is all the more important to see to their happiness in this world.[103] For Primatt, however, the question of the materiality or immateriality of the animal soul is finally irrelevant. Whatever a being's physical shape, and whether its soul is that of a brute or a man, 'the substance or matter, of which the creature is composed, would be equally susceptible of feeling'.[104] His argument for animal sensitivity is the Humean one from the analogy between their bodies:

> A *Brute* is an animal no less sensible of pain than a Man. He has similar nerves and organs of sensation; and his cries and groans, in case of violent impressions upon his body, though he cannot utter his complaints by speech or human voice, are as strong indications to us of his sensibility of pain, as the cries and groans of a *human* being, whose language we do not understand.[105]

In Primatt's work we see the incorporation of physical analogies derived from Enlightenment science into a Christian argument for recognising animal feeling.

With Granger, and more substantially in Primatt, we see that clerical animal advocacy has moved away from the concerns of Hildrop and Dean. The question of animal capacity for feeling is no longer closely tied up with and subordinate to the debate about the nature of the soul. Animal feeling is significant primarily for the argument about how animals should be treated. Primatt's contribution to this growing discussion was significant, and his *Dissertation*, reprinted and redacted later in the century, influenced the early nineteenth-century campaigners for animal welfare legislation. By the time Thomas Young wrote his *Essay on Humanity to Animals* in 1798, the debate about the treatment of animals had shifted again. Young (who later became a rector in Yorkshire) still finds a fundamentally Christian justification for his doctrine of kindness: God's will for all his creatures' happiness is the foundation for animal rights, and of the various reasons he gives for acting kindly to animals the one he considers 'the strongest, perhaps', is scriptural;[106] a large proportion of the text is devoted to discussion of scriptural passages. However, the emphasis of Young's tract is practical. Writing at a time when the idea of legal protection for animals had been mooted by Jeremy Bentham and Thomas Lawrence,[107] Young describes specific abuses that ought to be tackled, and writes to rouse indignation against them. He ignores the Cartesian debate and the whole question of the soul, except for brief remarks on the smallest 'mite, worm, or fly', which 'exceeds the finest piece of man-made mechanism'; such creatures, more importantly, have within them 'the principle of life'.[108] He relies instead on the tradition of sentimental writing invoking sympathy for the pains and pleasures of animals. Drawing on this sentimental tradition, Young is able to ascribe complex and refined affections to nonhuman animals. The affection of a mother bird for her young, for example, is described with reference to passages from Thomson's *Seasons*.[109]

During the latter part of the eighteenth century, then, animals' capacity for feeling was made the basis for increasingly confident and specific calls for reformed social attitudes to them. The question of the nature of animals' souls, once central to the debate about their treatment, lost its pivotal significance, though it did not disappear altogether.[110] Animal advocacy retained, for the most part,[111] its explicitly Christian framework and a fondness for Biblical testimony, while incorporating to some degree those physiological explanations for both human and animal passions and affections that were later to attract Darwin. The result was a fertile soil for the investigation of animal and human emotion carried out in the following century.

This essay forms part of my AHRC-funded project, 'Representing Animals in Late Eighteenth-Century Britain: Natural History, Narrative Sympathy, and Human and Animal Rights'. I am grateful to the AHRC for its generosity offered through its Research Leave scheme.

Notes

1 David Hume, *A Treatise of Human Nature*, ed. by L.A. Selby-Bigge, rev. by P.H. Nidditch (1739; Oxford: Clarendon Press,1978), Book II, Part ii, Section 12, 397.

2 Charles Darwin, *The Descent of Man, and Selection in Relation to Sex* (1871; 2nd edition. London: John Murray, 1874), 71.

3 Research on these abilities has concentrated especially on primates: for an introduction to the late twentieth-century shift in studies of animals from a behaviourist paradigm to a cognitive approach more willing to ascribe mental states, see Michael Tomasello and Josep Call, *Primate Cognition* (Oxford: Oxford University Press, 1997). For an introduction to recent debates on anthropomorphism see Sandra D. Mitchell, 'Anthropomorphism and Cross-Species Modeling', in Lorraine Daston and Gregg Mitman (eds), *Thinking with Animals: New Perspectives on Anthropomorphism* (New York: Columbia University Press, 2004), 100–17.

4 Tomasello and Call, 1997, 4–5.

5 Paul Ekman, 'Introduction to the Third Edition', in Darwin, *The Expression of the Emotions in Man and Animals* (1872; London: HarperCollins, 1998), xxix–xxxiv.

6 Ekman, 1998, xxix.

7 Darwin, 1872, 360.

8 J. Howard Moore, *The Universal Kinship* (London: G. Bell and Sons, 1906), vii.

9 James Rachels, *Created from Animals: The Moral Implications of Darwinism* (Oxford and New York: Oxford University Press, 1990), 131, emphasis in original.

10 Rod Preece, *Brute Souls, Happy Beasts, and Evolution: The Historical Status of Animals* (Toronto: University of British Columbia Press, 2005). For an article that summarises his main arguments see his 'Thoughts out of Season on the History of Animal Ethics', *Society and Animals*, 15 (2007), 365–78. For his argument that early Darwinians failed to take a pro-animal stance in the Victorian debate about vivisection see his 'Darwinism, Christianity, and the Great Vivisection Debate', *Journal of the History of Ideas*, 64 (2003), 399–419.

11 Preece, 2005, 336.

12 On this see also Mark Rowlands, 'Philosophy and Animals in the Age of Empire', in Kathleen Kete (ed.), *A Cultural History of Animals in the Age of Empire* (Oxford and New York: Berg, 2007), 135–52. Discussing Darwin's comments in 1837 on animals as 'our fellow brethren', who 'may partake of our origin in one common ancestor – we may all be melded together', Rowlands points out that while the evolutionary justification for the view of animals as brethren, sharing important similarities with humans, is new, the sentiment itself is not, 139.

13 As one example see the 'Personal Footnote' in Martin Kemp, *The Human Animal in Western Art and Science* (Chicago and London: University of Chicago Press, 2007), 267–74, which reinstates the notion of human difference in kind, not merely degree, from other animals within an evolutionary framework in which developing complexity of human consciousness eventually resulted in a decisive change: 'the conjunction of mental and bodily powers we have developed has crucially granted us a capacity to think and act differently in kind from animals', 273.

14 Darwin, 'Darwin's Reading Notebooks', in *The Correspondence of Charles Darwin, Volume 4, 1847–1850*, ed. by Frederick Burkhardt and Sydney Smith (Cambridge: Cambridge University Press, 1988), Appendix IV, 438.

15 For a discussion of Cartesian views on animal sentience in Britain see Keith Thomas, *Man and the Natural World: Changing Attitudes in England 1500–1800* (1983; London: Penguin, 1984), 30–6; Preece, 2005, 144–53. For a discussion of the reception of Cartesian views on animal mechanism in France see Hester Hastings, *Man and Beast in French Thought of the Eighteenth Century* (Baltimore: Johns Hopkins Press, 1936), 20–63.

16 On Tyson and Camper, and for a discussion of the eighteenth-century debate about the relation between human and ape, see Carl Niekerk, 'Man and Orangutan in Eighteenth-Century Thinking: Retracing the History of Dutch and German Anthropology', *Monatshafte*, 96 (2004), 477–502.

17 On Buffon see Jacques Rogier, *Buffon: A Life in Natural History*, trans. by Sarah Lucille Bonnefoi, ed. by L. Pearce Williams (Ithaca: Cornell University Press, 1997).

18 On Hartley see Richard C. Allen, *David Hartley on Human Nature* (Albany: SUNY Press, 1999).

19 On Hume's 'ontological naturalism', according to which human beings are 'part of the *seamless natural order* of causes and effects in which there is no categorical divide that distinguishes human beings from the rest of nature', see Paul Russell, *The Riddle of Hume's* Treatise: *Skepticism, Naturalism and Irreligion* (Oxford: Oxford University Press, 2008), 202–3, emphasis in original.

20 Aaron Garrett, 'Francis Hutcheson and the Origin of Animal Rights', *Journal of the History of Philosophy*, 45 (2007), 143–265.

21 See David Perkins, *Romanticism and Animal Rights* (Cambridge: Cambridge University Press, 2003), for discussions of the contribution of sentimental poetry to the development of thought on animal rights. Thomson is discussed especially in chapter 4, on hunting, and Cowper in chapter 3. Perkins considers that 'no writer had more effect on transforming attitudes to animals than Cowper', 44.

22 Richard Dean was curate of Middleton, near Manchester; James Granger was vicar of Shiplake, Oxfordshire, where he preached a sermon against cruelty to animals in 1772; Humphrey Primatt was a vicar and a Doctor of Divinity; Thomas Young was a fellow of Trinity College, Cambridge, and later a rector in Yorkshire. See Aaron Garrett (ed.), *Animal Rights and Souls in the Eighteenth Century* (Bristol: Thoemmes Press, 2000), vol.1, for an introduction to these and other writers on animal rights, i–xxv. Dean's work is reprinted in vol. 2 and Primatt's in vol. 3 of this 6-volume collection.

23 See Andreas-Holger Maehle, 'Cruelty and Kindness to the "Brute Creation": Stability and Change in the Ethics of the Man–Animal Relationship, 1600–1850', in Aubrey Manning and James Serpell (eds), *Animal and Human Society: Changing Perspectives* (London and New York: Routledge, 1994), 81–105, who shows that the growing European literature against cruelty to animals from the seventeenth century onwards generally used three main arguments: Biblical examples; God's mercy to all creatures; and the danger to humans of being brutalised by their own cruelty, 83.

24 Hastings, 1936, 14–15.

25 Thomas Dixon, *From Passions to Emotions: The Creation of a Secular Psychological Category* (Cambridge: Cambridge University Press, 2003).

26 Dixon, 2003, 105.

27 Gary Steiner, *Anthropomorphism and its Discontents: The Moral Status of Animals in the History of Western Philosophy* (Pittsburgh: University of

Pittsburgh Press, 2005), 76.

28 Thomas, 1984, 30–6.

29 Thomas, 1984, 35. Hastings concludes that by the middle of the eighteenth century the consensus in France was a rejection of the Cartesian view in favour of the belief that animals possessed feeling and some form of soul, 63.

30 Richard W. Serjeantson, 'The Passions and Animal Language, 1540–1700', *Journal of the History of Ideas*, 62 (2001), 425–44.

31 Serjeantson, 2001, 440.

32 In the N Notebook, begun October 1838, Darwin records: 'Hume has section (IX) on the Reason of Animals, Essays Vol. 2' [ie. the *Enquiry Concerning Human Understanding*, which has a section IX so titled, and which went through several editions as part of a collected *Essays and Treatises on Several Subjects*] (N Notebook, 101); and later notes down 'Humes Dissertation on the Passions' [ie. a recast version of Book II of the *Treatise*] (N Notebook, 184). Paul H. Barrett (ed.), *Early Writings of Charles Darwin*, Book 2 of Howard E. Gruber, *Darwin on Man: A Psychological Study of Scientific Creativity together with Darwin's Early and Unpublished Notebooks Transcribed and annotated by Paul H Barrett* (London: Wildwood House, 1974), 348, 351. On the relation between Hume's *Treatise* and *Enquiry*, and their publication history, see Tom L. Beauchamp, 'Editor's Introduction' to Hume, *An Enquiry concerning Human Understanding* (1748; Oxford: Oxford University Press, 1999), 7–9.

33 Hume, 1739, Book II, Part i, Section 1, 275.

34 *Ibid.*, Book II, Part ii, Section 17, 397.

35 *Ibid.*, Book II, Part ii, Section 16, 298.

36 *Ibid.*, Book II, Part i, Section 1, 276–7.

37 *Ibid.*, Book II, Part iii, Section 9, 448.

38 *Ibid.*, Book II, Part i, Section 12, 326; Book II, Part ii, Section 12, 398.

39 *Ibid.*, Book II, Part i, Section 12, 326.

40 *Ibid.*, Book I, Part iii, Section 16, 176–7; Book II, Part I, Section 12, 325–6.

41 *Ibid.*, 165, emphasis in original.

42 Daniel J. Povinelli, *Folk Physics for Apes: The Chimpanzee's Theory of how the World Works* (Oxford: Oxford University Press, 2000), 50. Povinelli criticises Hume's and Darwin's use of the argument from analogy to support belief in comparable mental operations in humans and

nonhuman animals, 2–11.

43 Hume, 1739, Book I, Part iii, Section 16, 177.

44 *Ibid.*, Book I, Part iii, Section 16, 177–9.

45 *Ibid.*, Book II, Part ii, Section 17, 397.

46 *Ibid.*, Book I, Part iii, Section 16, 179.

47 Robert J. Richards, 'Darwin on Mind, Morals and Emotions', *The Cambridge Companion to Darwin,* ed. by Jonathan Hodge and Geoffrey Radick, 2nd edn (Cambridge: Cambridge University Press, 2009), 96–119: 99.

48 N Notebook, 17, 48, emphasis in original; Barrett, 1974, 333, 339.

49 N Notebook, 5; Barrett, 1974, 331.

50 Hastings, 1936, 94–108; see also Robert Boakes, *From Darwin to Behaviourism: Psychology and the Minds of Animals* (Cambridge: Cambridge University Press, 1984), 90–2.

51 Allen, 1999, 376.

52 Robert E. Schofield, *The Enlightened Joseph Priestley: A Study of his Life and Work from 1773 to 1804* (University Park: Pennsylvania State University Press, 2004), 37–42.

53 Adrian Desmond and Jim Moore, *Darwin* (1991; New York and London: W.W. Norton and Co., 1994), 7–9.

54 'Hartley's. observ. on man. Recommended by Mackintosh and Priestly', 'Darwin's Reading Notebooks', Burkhardt and Smith, 1988, iv. 444. Darwin knew James Mackintosh's *Dissertation on the Progress of Ethical Philosophy, Chiefly during the Seventeenth and Eighteenth Centuries*: see Burkhardt and Smith, 1988, iv. 540. See also Notebook N, 184: '"Hartley" I should think well worth studying –', Barrett, 1974, 351.

55 Allen, 1999, 382–3.

56 Allen, 1999, 204.

57 David Hartley, *Observations on Man, his Frame, his Duty and his Expectations,* 2 vols (London: S. Richardson, 1749), i. 404.

58 *Ibid.*, i. 405.

59 *Ibid.*, i. 405–13.

60 *Ibid.*, i. 413.

61 *Ibid.*, i. 413–14.

62 *Ibid.*, i. 414.

63 'As you vary [your dog's] experience, he varies his reasoning. Make a beating follow upon one sign or motion for some time, and afterwards upon another; and he will successively draw different conclusions,

according to his most recent experience'. Hume, 1739, Book I, Part iii, Section 16, 178.

64 Allen, 1999, 429.

65 Hartley, 1749, ii. 222.

66 *Ibid.,* i.,415.

67 Russell, 2008, 201.

68 Hartley, 1749, ii. 223.

69 Joseph Priestley, *Disquisitions relating to Matter and Spirit. To which is added, the history of the philosophical doctrine concerning the origin of the soul, and the nature of matter; with its influence on Christianity, especially with respect to the doctrine of the pre-existence of Christ* (London: J. Johnson, 1777), 234.

70 *Ibid.*, 235.

71 *Ibid.*, 1777, 235.

72 *Ibid.*, 1777, 235.

73 *Ibid.*, 1777, 236–7.

74 *Ibid.*, 1777, 238–9.

75 *Ibid.*, 1777, 238, emphasis in original.

76 *Ibid.*, 1777, 239–40. In his *Institutes of Natural and Revealed Religion*, 2nd edn (Birmingham: Pearson and Rollason, for J. Johnson, 1782), Priestley is somewhat more cautious about the ascription of immortality to brutes, but still hints that it is possible, and certainly in moral and compassionate terms desirable: 'Brute creatures too have faculties similar to ours, since they differ from us in *degree* more than in *kind.* ... But then the difference is so great ... and man is so evidently the most distinguished of all the creatures of God upon the face of the earth, that there seems to be foundation enough for our expecting a preference [in respect of immortality]. Or, if the brute creation should be interested in a future life, we shall certainly have more reason to rejoice in it, than to be offended at it; and many of them seem to have more pain than pleasure in this', 144, emphasis in original.

77 John Wesley, *A Survey of the Wisdom of God in the Creation: or a Compendium of Natural Philosophy*, 2 vols (1763; Bristol: W. Pine, 1770), i. 107, emphasis in original.

78 John Hildrop, *Free Thoughts upon the Brute-Creation: or, an Examination of Father Bougeant's Philosophical Amusement, &c. In Two Letters to a Lady. Letter I* (London: R. Minors, 1742), 7. Compare Guillaume-Hyacinthe Bougeant, *A Philosophical Amusement upon the Language of Beasts. Written Originally in French by Father Bougeant* (London: T. Cooper, 1739), 5: 'I see

a Dog hastening to me when I call him, caress me when I stroke him, tremble and run away when I rate him, obey me when I command him, and give all the outward Signs of many different Sentiments: of Joy and Sadness, of Grief and Pain, of Fear and Desire, of Passions, of Love and Hatred. I immediately conclude from thence, that a Dog has in him a Principle of Knowledge and Sentiment . . . '.

79 Hildrop, 1742, 8.

80 See Bougeant, 1739, 21–2.

81 John Hildrop, *Free Thoughts upon the Brute-Creation: or, an Examination of Father Bougeant's Philosophical Amusement, &c. In Two Letters to a Lady. Letter II* (London: R. Minors, 1743), 60.

82 Hildrop, 1743, 77–86.

83 Richard Dean, *An Essay on the Future Life of Brutes, introduced with Observations upon Evil, its Nature, and its Origin*, 2 vols (Manchester: J. Harrop, 1767), ii. 49–69.

84 *Ibid.*, ii. 69–87.

85 David Allan, 'An Eighteenth-Century Lancashire Minister on Animal Sentience: Richard Dean's "Essay on the Future Life of Brute Creatures"', *Transactions of the Lancashire and Cheshire Antiquarian Society*, 94 (1998), 31–63.

86 Allan, 1998, 61.

87 Dean, 1767, i. xv.

88 *Ibid.*, ii. 93.

89 *Ibid.*, ii. vii–viii.

90 *Ibid.*, ii. 54.

91 *Ibid.*, ii. 71–2.

92 *Ibid.*, ii. 106–7.

93 James Granger, *An Apology for the Brute Creation, or Abuse of Animals Censured; in a Sermon on Proverbs xii.10. Preached in the Parish Church of Shiplake, in Oxfordshire, October 18, 1772* (London: T. Davies, 1772), 8, emphasis in original.

94 The only mention of the soul is a reference in the Dedication to transmigration, when the author warns the countryman 'Tom' that his soul after death may 'inhabit the fore-horse of a dray, and suffer all the pain that guilt and whiplash can give'. Granger, 1772, Dedication, n.pag.

95 *Ibid.*, 8.

96 *Ibid.*, 6, emphasis in original.

97 Humphrey Primatt, *A Dissertation on the Duty of Mercy and Sin of Cruelty to Brute Animals* (London: R. Hett, 1776), rept. in Garrett, 2000, iii. iii–iv, emphasis in original.

98 *Ibid.*, 6.

99 *Ibid.*, 96, 99.

100 *Ibid.*, 7.

101 *Ibid.*, 101.

102 'An *hereafter* for a Brute, a recompence for a Beast has a strange sound in the ears of a man. We cannot bear the thought of it . . . in the pride of our heart we rather say, – Let Man be happy, though all creation groan'. Primatt, 1776, 42.

103 Primatt, 1776, 66–7. Primatt refers in a footnote to Priestley's *Institutes of Natural and Revealed Religion.*

104 Primatt, 1776, 16. In a footnote he adds that the question of whether the soul is inseparable from the body or a separate, spiritual substance is not relevant to the point at issue, that of animal feeling.

105 Primatt, 1776, 13–14, emphasis in original.

106 Thomas Young, *An Essay on Humanity to Animals* (London: T. Cadell, jun. and W. Davies, 1798), 9.

107 On the early campaign for legal reform on animals see Mike Radford, *Animal Welfare Law in Britain: Regulation and Responsibility* (Oxford: Oxford University Press, 2001), 15–31.

108 Young, 1798, 193–4.

109 *Ibid.*, 47–8.

110 In 1790 Catherine Macaulay made an argument for animals' immortal souls in *Letters on Education* (London: C. Dilly, 1790), 1–6.

111 An interesting exception is John Oswald, who drew on Hindu thinking for his *The Cry of Nature: or, an Appeal to Mercy and to Justice, on behalf of persecuted Animals* (London: J. Johnson, 1791). See Garrett, 2000, i. xxii–xxv.

2

'The Book of the Season': The Conception and Reception of Darwin's *Expression*

Angelique Richardson

In the summer of 1838, when Victoria had been queen for a year, the Chartists were demanding parliamentary reform, and Dickens published *Oliver Twist,* Darwin was keeping a private notebook on the expression of emotion. Bound in red leather, marked M and 'Expression', it was succeeded in the autumn by a red-rust notebook N, also marked 'Expression' and 'private'. These were pivotal months for Darwin. As he filled up the M and N books he was also gathering data on the transmutation of species and its causes, and while the notebooks are concerned largely with expression they are far-reaching in scope, embracing human thought, habit and instinct, heredity, evolutionary origins and human–animal continuities.[1]

Darwin continued to think through the ideas sketched out in these notebooks which, more than thirty years later, would become *The Expression of the Emotions in Man and Animals*. Writing to the geologist Charles Lyell shortly after he published the *Origin of Species*, he remarked that he had 'collected a good many facts' and 'speculated' on expression of countenance. He added 'I do not suppose I shall ever publish; but it is an uncommonly curious subject'.[2] It was, he wrote soon after to his botanist friend J.D. Hooker, one of his 'hobby-horses'.[3] The *Expression,* the most human-centred of Darwin's works, received a striking degree of attention, and remains important in psychology and evolutionary biology, though it is now generally less widely discussed than the *Origin* and *The Descent of Man, and Selection in Relation to Sex* (1871). Appearing immediately after the *Descent,* in which it was originally going to be included – until the *Descent* grew too big – it told the story of the origins of day-to-day, hourly expressions as Darwin set out to demonstrate that emotional expressions have evolved and are, or have been, adaptive. Human–animal kinship, and the physical basis of the emotions, were crucial to Darwin's theory of emotion. Questions about the relation of mind and body underpinned his research, allowing him to further integrate humans into the natural world as he developed and extended his idea of evolution.

This chapter explores Darwin's ideas on expression from notebook to publication, charting his methods, sources (including novels) and historical antecedents, and his research on both the role of expression in social relations and the evolutionary relation between the social instincts and the moral sense. Contextualising his ideas within the shift towards a physical understanding of life in the second half of the nineteenth century, it considers medical and popular contemporary responses to the *Expression*, its dialogue with the novel and other art forms, and the extent to which it became part of everyday culture. As Darwin's science challenged boundaries between genus, *Expression* pulled at distinctions of genre, providing an important example of the deep reciprocity of science and culture.[4]

Expressemotions: Darwin's Notebooks

'What is Emotion' wrote Darwin on 21st September 1838 in the M notebook. He underlined Emotion in blue crayon, and continued: 'analysis of expression of desire – is there not a protrusion of chin, like bulls & horses'.[5] This deft and easy move between humans and animals was characteristic. Pondering whether human language, which he would describe in the *Descent* as 'half-art and half-instinct', began with singing, he muses on frogs chirping in harmony.[6]

The meaning of emotion was both complex and shifting. As Thomas Dixon notes, during the eighteenth century it carried a physical sense, coming increasingly to refer to 'the bodily stirrings accompanying mental feelings', but from the middle of that century it moved from the bodily to the mental domain.[7] Bodily stirrings remained constitutive of emotion, though. Darwin's ideas were informed by the physiological and developmental turns taken in moral philosophy, notably by Hume, in natural theology, and, later, in psychology by Alexander Bain and Herbert Spencer.[8] Dixon notes that Hume's use of the word 'emotions' in his *Treatise of Human Nature* (1739–40) was similar to its present-day usage, and credits the Edinburgh professor of moral philosophy Thomas Brown with introducing the term as a psychological category in his lectures on mind in 1820. He observes that in these lectures the term took on 'a newly systematic theoretical role in the science of the mind'.[9]

Darwin's 'Old & Useless Notes', useless in the sense that they had been fully exploited (he added the title some years after they were written, probably after he completed the *Descent*) contain the statement: 'Those emotions which are strongest in man, are common to other animals & therefore to progenitor far back'.[10] This commonality evidenced the common kinship of humans and animals. Early in his note-taking he observed continuity in the actions, emotions and minds of humans and animals, pointing out that they responded to the world through the same five senses. He remarked that it 'would indeed be wonderful' (the word here carrying the sense of strange) 'if mind of animal was not closely allied to that of men, when the *five senses* were the same – In its action – emotions –'.[11]

In the second of his transmutation notebooks Darwin wrote 'Study Bell on Expression', adding 'if [Bell] shows that a man grinning is to exposes [sic] his canine teeth. no doubt a habit gained by formerly being a baboon with great canine teeth'. 'This way of viewing the subject important'.[12] Charles Bell, a Scottish surgeon and anatomist, and author of the fourth of the *Bridgewater Treatises*, *The Hand: Its Mechanism and Vital Endowments as Evincing Design* (1833), was the most influential theorist of emotion before Darwin. He had argued for the physical basis of emotions in his *Essays on The Anatomy of Expression in Painting* (1806).[13] But Darwin would take this further, bring the expression of emotions into the story of evolution. He later returned to the passage, adding '(This may be made capital argument if man does move muscles for uncovering canines) –. Blend this argument with his having canine teeth at all'.

In Notebook N, Darwin wrote under 'convulsive actions' 'But why does joy, & *OTHER EMOTIONS* make grown up people cry. – What is emotion?'[14] He then paused briefly on Burke's *Philosophical Inquiry into the Origin of Our Ideas of the Sublime and the Beautiful*, and Wordsworth's Preface to *Lyrical Ballads*. Ten days later he was puzzling over the relation between emotions: 'love being an emotion does it regard <<is it influenced by >> other emotions?'[15] The questions continued. Weeks before marrying his cousin Emma Wedgwood, he remarked: 'What passes in a man's mind when he says he loves a person – do not the features pass before him marked, with the habitual expressemotions, which make us love him, or her'.[16] The compressed term 'expressemotions' is striking, conflating expression and emotion. The extent to which emotions were constituted by their expression was central. 'the mind is function of body' he declared in this notebook.[17] Bringing humans firmly into the physical world, the statement offered a naturalistic understanding of mind, challenging dualistic thought. The body is given a primary role, though whether it is entirely causative, or associative, is left ambiguous: 'Emotions are the heredetary [sic] effects on the mind, accompanying certain bodily actions'.[18] Emotions were expressed physically because they resulted from a physical disturbance: 'Expression, is an heredetary [sic] habitual movement consequent on some action, which the progenitor did, when excited or disturbed by the same cause, which <<now>> excites the expression.–'[19] However, questions remained: 'but what first caused this bodily action. if the emotion was not first felt?' He then added 'without <<slight>> flush, acceleration of pulse. or rigidity of muscles. – man cannot be said to be *angry*.'[20] Expression here precedes, or occurs at the point of, emotion, attesting to its physical basis.

In the same notebook Darwin pondered 'the *Blushing* of Camelion & Octopus'. Here he suggested such blushing 'an excitement of surface *under the will?* of the animal'.[21] The role of the will in the expression of emotion was a central question in Darwin's research, as his hesitancy here signifies, and he would increasingly argue for the involuntariness of expression. He concluded that the blush was uniquely human, the result of a more developed self-consciousness: 'animals. not being such thinking people. do not blush' (the phrase confidently, and even humorously, expressing his views on continuity).[22] But while blushing provided an instance of thought causing expression, it was an involuntary process, and intimately connected to the body, as he would elaborate in *Expression*.

Darwin's research bears interesting relation to the existing practices of physiognomy and phrenology. He was concerned to distinguish expression from physiognomy, which, notwithstanding the attention it gave to the body, sidestepped questions of the material basis of life, and was increasingly viewed as dualistic. Franz Joseph Gall, founder of the more materialist phrenology, would question the emphasis physiognomy placed on the soul rather than the material organs.[23] But an early conversation with the Scottish philosopher Sir James Mackintosh also expunged what little faith Darwin had in phrenology. Aged twenty he wrote to his cousin William Darwin Fox 'had some talk with [Sir J. Mackintosh] about Phrenology, & he has entirely battered down the very little belief of it that I picked up at Osmaston. He says, as long as Education is supposed to have any effect in decreasing the power of any organ of the brain, he cannot see how it ever can be proved true'.[24] By contrast, Darwin's theory of expression would offer a much more fluid conception of character in which individual expression developed through a welter of habits, instincts and, crucially, social relations. Nonetheless, elements of the language of physiognomy can be traced in Darwin's work, as it pervaded Victorian culture, permeating novels.[25] Darwin took notes from the Swiss physiognomist Johann Kaspar Lavater: 'every man is born with a portion of physiognomical sensation', adding 'I think that this cannot be disputed anymore in men. than in animals'.[26] This sensation was the ability to *read* others. As Darwin developed his understanding of emotion he saw expression as constituted in part through community and, for humans in particular, carrying with it a sense of perception by others. Man, he declared, was a 'socialist' (by which he meant 'Social animal') and knew 'other men . . . by looks'. 'No one', he remarked, 'can say expression was invented to conceal one's thought'.[27] The point, once made, seemed both obvious and irrefutable.

Social Instincts and the Moral Sense

Darwin was establishing that expression was a way of communicating without verbal language. This also brought humans and animals into new proximity, revealing shared methods of communication. His early notes show the centrality of sociality to expression across species. In his transmutation notebooks continuity of expression attested to the mutability of species '– Laughing modified barking., smiling modified laughing', offering new insight into communication: 'Barking to tell other animals in associated kinds of good news'. 'Under this point of view', he concluded, 'Expression' (he inserted 'of all animals') 'becomes very curious'.[28] While some expressions may be innate, experience was often required for them to be recognised. Darwin recorded his eldest sister noting: 'very young children, express the greatest surprise at emotions in her countenance – before they can have learnt by experience, that movements of face are more expressive than movements of fingers'.[29] Characteristically, he added a comparison with animals: '– like Kitten with mice. –'. Later, he wrote:

> The distinction <<as often said >> of language in man is very great from all animals – but do not overrate – animals communicate to each other. – Lonsdale's story of Snails, Fox of cows, & many of insects – they likewise must understand each other expressions, sounds, & signal movements. – some say dogs understand expression of man's face.[30]

Urging a striking degree of proximity between humans and animals, these remarks highlighted their shared sociality, and the dependence of that sociality on expressions. They were necessary – and had to be understood – in order for communities to work.[31] It was becoming increasingly clear to Darwin that the moral sense developed from the social instincts and was not, thus, uniquely human. Taking notes on the moral sense from Mackintosh's recently published *Dissertation on the Progress of Ethical Philosophy* (1837), he referred to the observation by the seventeenth-century Dutch philosopher Hugo Grotius that 'even inferior animals under the powerful though transient impulse of parental love, prefer their young to their own safety or life'.[32] He added 'we see in other animals they consist in such active sympathy that the individual forgets itself, & aids & defends & acts for others at its own expense'.

First publishing on emotions in the *Descent* Darwin observed there that it was 'well-established' that the lower animals are excited by the same emotions as ourselves. He gave illustrations of terror, rage and love, stating that 'complex emotions' were common to humans and the higher animals.[33] On primates he wrote that they have 'the same senses, intuitions and sensations – similar passions, affections, and emotions, even the more complex ones', and focused on 'the all-important emotion of sympathy',[34] arguing that it had become an instinct, strengthening the mutual love of a community. Remarking that he was the first to approach the question of the moral sense 'exclusively from the side of natural history', Darwin observed, 'Besides love and sympathy, animals exhibit other qualities which in us would be called moral'. The moral sense developed out of the social instincts. Both related at first, he noted, 'exclusively to the community'.[35] He concluded his discussion of the moral sense with the statement 'the difference in mind between man and the higher animals, great as it is, is certainly one of degree and not of kind'.[36]

Darwin's Global Network

Studying his infants and animals exhaustively, Darwin needed to go further afield for evidence of uniformity of expression. Though he read travel books,[37] and began studying photographs, it was his global network of correspondents that would prove vital. They ranged from friends, neighbours, relations and fellow men of science to missionaries, entrepreneurs and government colonials.[38] In the 1860s he formulated a list of queries as the basis of his method of inquiry, sending this out with numerous letters (see Figure 2.1).[39] The sixteen questions, reprinted in the *Expression*, ranged from 'Is astonishment expressed by the eyes & mouth being opened wide, & by the eyebrows being raised?' to 'Do the children when sulky pout or greatly protrude the lips?' He stated at the foot of the questions 'Observations on natives who have had little communication with Europeans would be of course the most valuable, though those made on any natives would be of *much* interest to me' (16, emphasis in original). He was after detail, anxious for specificity: 'General remarks on expression are of comparatively little value'. He reminded his assistants not to rely on memory, 'so deceptive in subjects like these' (16). But he knew the very conditions of the expression of emotions, their sociality, might hinder their effective study. It would be difficult for the observer not to be drawn in to the communicativeness of the encounter and thus lose objectivity. Writing to the geologist and mining engineer Thomas Belt for information on expression in 'Negroes or American Indians' in Nicaragua, he cautioned: 'You would have to read the queries several times over, and you will find it *curiously* difficult to observe; for any emotion excites some sympathy in observer's mind and this

Figure 2.1: Darwin's printed queries, *The Expression of the Emotions in Man and Animals* (London: John Murray, 1872), 15-16.

(1.) Is astonishment expressed by the eyes and mouth being opened wide, and by the eyebrows being raised?

(2.) Does shame excite a blush when the colour of the skin allows it to be visible? and especially how low down the body does the blush extend?

(3.) When a man is indignant or defiant does he frown, hold his body and head erect, square his shoulders and clench his fists?

(4.) When considering deeply on any subject, or trying to understand any puzzle, does he frown, or wrinkle the skin beneath the lower eyelids?

(5.) When in low spirits, are the corners of the mouth depressed, and the inner corner of the eyebrows raised by that muscle which the French call the "Grief muscle"? The eyebrow in this state becomes slightly oblique, with a little swelling at the inner end; and the forehead is transversely wrinkled in the middle part, but not across the whole breadth, as when the eyebrows are raised in surprise.

(6.) When in good spirits do the eyes sparkle, with the skin a little wrinkled round and under them, and with the mouth a little drawn back at the corners?

(7.) When a man sneers or snarls at another, is the corner of the upper lip over the canine or eye tooth raised on the side facing the man whom he addresses?

(8.) Can a dogged or obstinate expression be recognized, which is chiefly shown by the mouth being firmly closed, a lowering brow and a slight frown?

(9.) Is contempt expressed by a slight protrusion of the lips and by turning up the nose, with a slight expiration?

(10.) Is disgust shown by the lower lip being turned down, the upper lip slightly raised, with a sudden expiration, something like incipient vomiting, or like something spit out of the mouth?

(11.) Is extreme fear expressed in the same general manner as with Europeans?

(12.) Is laughter ever carried to such an extreme as to bring tears into the eyes?

(13.) When a man wishes to show that he cannot prevent something being done, or cannot himself do something, does he shrug his shoulders, turn inwards his elbows, extend outwards his hands and open the palms; with the eyebrows raised?

(14.) Do the children when sulky, pout or greatly protrude the lips?

(15.) Can guilty, or sly, or jealous expressions be recognized? though I know not how these can be defined.

(16.) Is the head nodded vertically in affirmation, and shaken laterally in negation?

Observations on natives who have had little communication with Europeans would be of course the most valuable, though those made on any natives would be of much interest to me. General remarks on expression are of comparatively little value; and memory is so deceptive that I earnestly beg it may not be trusted. A definite description of the countenance under any emotion or frame of mind, with a statement of the circumstances under which it occurred, would possess much value.

---

at the moment makes the point to be forgotten'.[40] Expressions were elaborately connected to community.

From Down to Nicuragua, from babies to garden birds, Darwin turned the world into a laboratory for his study of expression. To his niece Lucy Wedgwood he wrote 'I hear that your Dog is a barker: please observe for me whether the (upper) lips are at all retracted or everted when he barks & just before he barks'. His own dog, Polly, provided no enlightenment on the subject: 'Our Polly will not bark except as she rushes away to some supposed enemy in forest'. Defensive barking would not do: 'a joyful bark or bark of good spirits w^{d}. be best'. He also asked Lucy, who was such a good listener – 'I know I can trust your conclusion' – if she would 'Think of any fact about expression of any emotion in any of your birds'.[41]

In 1868 Darwin asked T.H. Huxley, the famously vociferous public defender of evolution, if Mrs Huxley would respond on query 5 – 'When in low spirits are the corners of the mouth depressed & the inner corner or angle of the eyebrows raised & contracted by that muscle which the French call the grief muscle?' – 'when one of her children is struggling & just going to burst out crying'.[42] He told Huxley how, in order to assist his researches, one young lady near Downe had made her young child cry until she saw 'the eyebrows for a second or two beautifully oblique, just before the torrent of tears began'.

This was the year that Huxley argued in his Edinburgh public lecture, 'On the Physical basis of Life', for a unity pervading the whole living world, from 'the great Finner whale' to 'invisible animalcules'; from the flower in a girl's hair to the blood in her youthful veins.[43] Likewise, he related all human activities to the body: 'Even those manifestations of intellect, of feeling, and of will, which we rightly name the higher faculties', were 'transitory changes in the relative positions of parts of the body'.[44] Published in the *Fortnightly Review* the following year it caused a sensation.[45] Two years later, in his Presidential address to the Belfast meeting of the British Association for the Advancement of Science, the physicist John Tyndall would declare: 'Every meal we eat, and every cup we drink, illustrates the mysterious control of Mind by Matter'.[46] Darwin's research emerged in this new materialist context.

The Expression of the Emotions

The *Expression* was published on 26 November 1872. By this time, emotion had become established as a psychological category and a subject of study which grouped together passions and affections. Darwin recorded 'my book sold largely; 5267 copies having been disposed of on the day of publication'.[47] With seven heliotype plates it was one of the first books to include photographs, and these ranged from babies to actors and (courtesy of James Crichton-Browne, Medical Superintendent of the West Riding Asylum), lunatics.[48] Priced at 12 shillings, it was translated into Dutch, French, German, Italian, Polish and Russian within his lifetime.[49] The first edition listed in its front pages the first cheap, sixth, edition of the Origin at 7s. 6 d., also by John Murray, published in February of the same year. His work was circulating at a new rate (ii).

Darwin emphasised that to appreciate the significance of his expression research it was necessary to accept common descent: 'He who admits, on general grounds, that the structure and habits of all animals have been gradually evolved, will look at the whole subject of Expression in a new and interesting light' (12). The work persuasively underlines the unconscious, instinctive, nature of expression. The involuntariness of expression was cogent evidence of the embodied nature of emotion, and shed new light on the relation of mind, emotion and the body. Darwin observed 'Most of our emotions are so closely connected with their expression, that they hardly exist if the body remains passive – the nature of the expression depending in chief part on the nature of the actions which have been habitually performed under this particular state of the mind' (239).

Figure 2.2: Small dog watching a cat on a table. From a photograph by Oscar Rejlander, *Expression of the Emotions in Man and Animals* (London: John Murray, 1872), 43.

The Darwin of the *Expression* is not a detached observer. He addresses his readers directly, offering anecdotes from friends, expresses wonder, amusement and affection, as he observes his children, his pets, and species in the Zoological Gardens. He recalls his play-biting terrier giving wags of his tail 'which seems to say "Never mind, it is all fun"' (63). The physical expressions of a dog greeting his 'beloved master' are analogous to 'the beaming eyes and smiling cheeks of a man when he meets an old friend' (11). Through such absorbing details Darwin revealed a striking degree of human–animal continuity. Expression in humans, though, had become more muted and complex, as, for example, the perceived link between the snarl and the more covert sneer suggested (249-53). There was an analogy to be drawn between the tickled mind and the tickled body; as children laugh and anthropoid apes utter a comparable sound when tickled unexpectedly (though not by a stranger, which would produce fear), so with the mind when a 'novel or incongruous idea' breaks through a habitual train of thought, producing laughter at the ludicrous (202). Physical actions accompanying – and to varying degrees constituting, emotion, are colourfully detailed. Astonished monkeys keep their eyebrows permanently raised (143). Rabbits stamp loudly on the ground to signal to their comrades (93). Porcupines 'rattle their quills and vibrate their tails when angered' and 'storks, when excited, make a loud clattering noise with their beaks' (93, 94). An angry lion erects his mane; the hair on a fearful cat, especially on its tail, bristles (96). Kittens, puppies, young pigs, 'and probably many other young animals', alternately push with their fore-feet against the mammary glands of their mothers, to excite a free secretion of milk', this action, initially developed for survival, now accompanies other physical pleasure: seen, for example, in a cat having its back scratched, so that it 'has almost become the expression of a pleasurable sensation' (46-7).[50]

The *Expression* is packed to the brim with stories, ranging from Darwin's entertaining visits to the Zoological Gardens to a young man 'of strongly nervous temperament' who, on hearing that a fortune had been bequeathed him, showed all the signs of drunkenness.[51] Material explanations are offered for coughing or sneezing, for differences in laughter between men and women, for expressions of love: 'we long to clasp in our arms those whom we tenderly love. We probably owe this desire to inherited habit, in association with the nursing and tending of our children, and with the mutual caresses of lovers'. The work bustles with detail about the mental and physical lives of animals, from affectionate monkeys, yawning baboons, fighting or curious deer, impatient horses, excited rattlesnakes, moving to the seemingly anthropomorphic – dogs have ennui, just as Darwin remarks that all animals, 'feel Wonder' (30-40; 42). Mammals jostle with humans – frowning babies, playing children – all are fully integrated into the naturalistic economy of emotion.

Darwin's research was organised around three principles of expression, each resulting in expressions that were independent of will. The first, the principle of serviceable habits, posits that useful expressive habits are transmitted to the next generation, for example raising eyebrows in surprise was useful as the field of vision was increased, and the eyes were able to move easily in any direction. Many expressions carried with them evidence of their origin. For example, eyebrows were raised, Darwin observed, 'when we earnestly desire to remember something; acting as if we endeavoured to see it' (228). A dog paying close attention would often double up one of its fore-legs, an action which originated with the cautious steps of approaching prey, and was readily illustrated by Darwin's fox terrier Polly (see Figure 2.2). Darwin also considered reflex actions, relating how he put his face to the glass-plate in front of a puff-adder in the Zoological

Gardens, to see if he could resist the reflex of jumping if the snake struck at him. He jumped backwards as the snake made a move, finding his will and reason powerless against imagined danger (38).

The second principle was one of antithesis; unconscious, communicative expressions developed as opposites to an earlier more basic one. Shoulders shrugged as opposed to kept raised in preparedness for aggression, for example, indicated impotence or an apology (63) and, most memorably, the hot-house face of Bob, Darwin's retriever, whose head and ears would droop into 'hopeless dejection' when he saw they were headed not for a long walk but to see some of Darwin's other species in his hot house (57, 60, 64).[52] The third was the principle of 'direct action of the excited nervous system on the body'; trembling was one example, and 'was of no service, often of much disservice, and cannot have been at first acquired through the will, and then rendered habitual in association with any emotion' (67). Another example was joy, which quickens the circulation, stimulates the brain and excites the sensorium, which in turn reacts on the whole body: 'We see this in our young children, in their loud laughter, clapping of hands, and jumping for joy' (76). The principle could be seen at work in the ease with which an unexpected event could lead a young child just beginning to cry instead to laugh, expending its superfluous energy equally well (200-1). It wasn't always possible, though, Darwin noted, to decide how much weight to attribute to a principle in any given expression (82).

On the opening page of *Expression* Darwin credited Bell with having laid the foundations of expression as a branch of science. His study was intended, though, to challenge Bell's claim that certain muscles had been created especially for expression, as he set out to show that the habit of expressing feelings by certain movements 'though now rendered innate' had been 'in some manner gradually acquired', offering a naturalistic explanation (19). But, their differences notwithstanding, Darwin's work was also informed by Bell and other natural theological writers such as Thomas Burgess.[53] While Darwin contested Bell's argument that humans had special muscles to express certain feelings, he nonetheless took his main principle of expression, namely the idea of 'serviceable associated habits', from the third edition of Bell's *Essays* (1844).[54] He drew also on new developments in psychology. The empiricist Alexander Bain published *The Senses and the*

Intellect in 1855, emphasising the role of experience and the coalition of mind and body, and psychological and physiological processes. He wrote: 'I look upon the expression so-called as part and parcel of the feeling. I believe it to be a general law of the mind that, along with the fact of inward feeling or consciousness, there is a diffusive action or excitement over the bodily members'.[55] Darwin quoted this passage directly in *Expression* (9), drawing also on *The Emotions and the Will* (1859), in which Bain had stated 'emotion is the name here used to comprehend all that is understood by feelings, states of feeling, pleasures, pains, passions, sentiments, affections'.[56] Spencer's work on the bodily nature of expression also informed Darwin's thinking; he quoted from his *Principles of Psychology* (1855) on the bodily symptoms of fear – 'palpitations and tremblings' – and referred to his later work on the physiology of laughter, in which Spencer insisted on 'the general law' that feeling, passing a certain pitch, 'habitually vents itself in bodily action'.[57] While Darwin did not define emotions in *Expression* he observed in the opening chapter Spencer's distinction between emotions and sensations, with sensations being 'generated in our corporeal framework'.[58] Spencer had defined emotion as the kind of feeling that was not a direct result of any action on the organism but either an indirect result of such action, or arising quite apart from such action, generated not in the body but in consciousness, and objected to Bain's ignoring of any such distinction.[59]

For Darwin, though, emotions had a clear physical correlation and he linked them both to environment and to the body. They were caused by states of affairs outside the body, as opposed to sensations such as hunger, pleasure and pain, but experienced in the body, and constituted by bodily actions.[60] The expressive body responded to external stimuli, and the affected body in turn affected the mind. Weeping, Darwin argued, was itself caused by an unconscious and habitual chain of events; prolonged screaming in children leads inevitably to:

> the gorging of the blood-vessels of the eye; and this will have led, at first consciously and at last habitually, to the contraction of the muscles round the eyes in order to protect them. At the same time the spasmodic pressure on the surface of the eye, and the distension of the vessels within the eye, without necessarily entailing any conscious sensation, will have affected, through reflex action, the lacrymal glands.[61]

While Darwin also gave examples of thought causing expression, these, again, demonstrated a close relation between mind and body, a simultaneity, even: 'as soon as some melancholy thought passes through the brain, there occurs a just perceptible drawing down of the corners of the mouth, or a slight raising up of the inner ends of the eyebrows, or both movements combined, and immediately afterwards a slight suffusion of tears'.[62] In a chapter devoted to blushing, Darwin explored the direct effect of attention on the body: 'The hypothesis which appears to me the most probable, though it may at first seem rash, is that attention closely directed to any part of the body tends to interfere with the ordinary and tonic contraction of the small arteries of that part.'[63] He moved from this instance of physical attention to the perception of such attention and, from there, to perceived thoughts about character – each caused the same affect on the body:

> Whenever we believe that others are depreciating or even considering our personal appearance, our attention is vividly directed to the outer and visible parts of our bodies; and of all such parts we are most sensitive about our faces, as no doubt has been the case during many past generations.

What was initially a physical response to physical attention – 'assuming for the moment that the capillary vessels can be acted on by close attention, those of the face will have become eminently susceptible' – is complicated by self-conscious thought, but the response still registers physically: 'Through the force of association, the same effects will tend to follow whenever we think that others are considering or censuring our actions or character'.[64] Darwin returned in *Expression* to his earlier insistence on the vital communicative function of expression. Expressions that were useless in the sense of having no direct and immediate survival value through, for example, protection of the body, still had an important social role to play, as, as he has shown in the *Descent*, characteristics such as ornamental appendages which had no direct use for survival led, through sexual selection, to survival through progeny:

> The movements of expression in the face and body, whatever their origin may have been, are in themselves of much importance for our

> welfare. They serve as the first means of communication between the mother and her infant; she smiles approval, and thus encourages her child on the right path, or frowns disapproval. We readily perceive sympathy in others by their expression; our sufferings are thus mitigated and our pleasures increased; and mutual good feeling is thus strengthened. The movements of expression give vividness and energy to our spoken words. They reveal the thoughts and intentions of others more truly than do words, which may be falsified.[65]

As physical expression might be more truthful than words, so Darwin pointed out that the imagination could serve as an impediment to the accurate representation of expression: 'how easily we may be misguided by our imagination';[66] likewise, as he had cautioned correspondents responding to his questionnaire, sympathy could compromise accuracy. 'When we witness any deep emotion, our sympathy is so strongly excited that close observation is forgotten or rendered almost impossible'.[67]

These effects had implications for the reliability of Darwin's most ready historical sources. The study of the expressions required a close and careful dialogue between science and art. Drawing extensively on the realm of art, intimately connected with the emotions, Darwin looked to visual, plastic and narrative art forms for examples of expression, but acknowledged that they were unlikely to represent it with scientific accuracy. He observed that the fine arts tended to avoid or to misrepresent the fleeting and indefinite nature of the emotions, while narrative art was also prone to misrepresent them, misled by sympathy, the imagination or convention.[68] While his sources included books of illustrations he had turned to photography as potentially more reliable, corresponding with photographers such as Adolphe Kindermann, Dr George Wallich, and Oscar Rejlander, and collecting about a hundred images.[69]

Nonetheless, examples from art and from classical and modern literature were vital to Darwin's study, and proliferate in the *Expression*. Equally, Darwin brought his powers of acute observation and analysis to the ordinary, to subjects traditionally the province of the novelist and painter, at times adopting a narrative voice common to the novelist, at others quoting directly from contemporary novels in ways which highlighted the interfusion of science and culture. In *Oliver*

Twist (1838) Darwin found a graphic example of the sneer:

> Dickens, in speaking of an atrocious murderer who had just been caught, and was surrounded by a furious mob, describes 'the people as jumping up one behind another, snarling with their teeth, and making at him like wild beasts'. Every one who has had much to do with young children must have seen how naturally they take to biting, when in a passion. It seems as instinctive in them as in young crocodiles, who snap their little jaws as soon as they emerge from the egg.[70]

Mrs Gaskell's description of a baby in *Mary Barton* (1848) refusing porridge – 'it made its mouth like a square, and let the porridge run out at all four corners' – forms the basis of his discussion of a baby's imperfect control of 'the depressor muscles of the angles of the mouth' (151); Jack Brownlow in Margaret Oliphant's *The Brownlows* (1868) provides an accurate illustration of resolute determination – 'As shrugging the shoulders generally implies "I cannot do this or that," so by a slight change, it sometimes implies "I won't do it"'; 'He thrust his hands deep down into his pockets, and set up his shoulders to his ears, which was a good warning that, come right or wrong, this rock should fly from its firm base as soon as Jack would; and that any remonstrance on the subject was purely futile'. As soon as the son got his own way, he 'put his shoulders into their natural position' (270). Likewise, *Wenderholme* (1869), a novel by P.G. Hamerton (the *Saturday Review* art critic), published the following year, provided Darwin with an example of the whistling that could accompany surprise through a modification of the open mouth of astonishment – 'if the mouth is not much opened, whilst the lips are considerably protruded' (285). He remarked 'We Europeans often whistle as a sign of surprise; thus, in a recent novel it is said, "here the man expressed his astonishment and disapprobation by a prolonged whistle"' (286). These popular works of fiction would underscore the wider resonance of his research, demonstrating that he was bringing new scientific understanding to facts partially known but unexplained. In turn, writers were fascinated by the *Expression*. George Eliot and Lewis Carroll read it as soon as it appeared; Eliot took notes,[71] and Carroll sent Darwin a photograph of a young girl, which he headed 'No Lessons Today', hoping Darwin would include it in subsequent editions as an example of happiness through absence of work.[72]

Darwin also turned to earlier sources. To Seneca and Macrobius he looked for information on the expression of shame, and Aesop's 'Ox and the Frog' supplied details of a frog who became so swollen from vanity and envy that it burst (this illustrated the tendency of amphibians and reptiles to enlarge themselves to excite fear in an enemy, and also to prevent themselves from being eaten by a snake) (104). Language too revealed human-animal continuity, with the word toad 'expressing in all the languages of Europe the habit of swelling' (104). Shakespeare was a frequent source in *Expression,* affording examples of rage, disgust, jealousy, shame and astonishment. Arguing for the uniformity of the expression of disgust Darwin moved from Shakespeare's Duke of Norfolk declaring 'I spit at him – call him a slanderous coward and a villain' – to one of his babies at five months ejecting his first mouthful of water (261). A month later the baby, full of surprise, pushed out a piece of ripe cherry with his tongue. Was this, Darwin wondered, why lolling out the tongue served everywhere as a sign of contempt? (261). But, he pointed out, Shakespeare wasn't always right. He erred in thinking Juliet didn't blush because it was dark – blushing could take place both in solitude and darkness. Indeed, this was evidence of its involuntariness (365). Critics would observe that the theatre had much to gain from the *Expression*: 'To artists and actors this work must become a sort of text-book on the facts of facial and bodily expression'.[73]

Darwin's motives found clear expression in his conclusions, as he sought to establish common descent and the unity of races. In the final paragraph of the *Expression* he reiterated his earlier observation that '[t]he different races of man express their emotions and sensations with remarkable uniformity throughout the world' (139; 367), pointing out that the study of the theory of expression confirmed both animal descent and the unity of races. James Moore and Adrian Desmond have recently revealed the extent to which Darwin was driven in his work by his antipathy towards slavery, and any research which might further his commitment to monogenism would have been doubly welcome.[74] Darwin added to his concluding remarks on uniformity of expression 'as far as my judgment serves, such confirmation was hardly needed' (367). The passage was one of the most frequently cited in reviews.[75]

Darwin's Readers

The audience of the *Expression* was broad and varied. It ranged from philosophers and men of science to ordinary men and women seeking erudition, entertainment and, as the popular press would have it, the romance of novels. Newspapers and scientific and literary journals alike clamoured to give the book attention. For the *Telegraph* it was 'one of the most eagerly anticipated books of the season'.[76] The phrase 'book of the season', more usually used of fiction than scientific treatises, reflected the hybrid nature both of the *Expression* and its audience. The colonial newspaper the *Argus* echoed the praise: it 'is probably destined to be the book of the season'.[77] Its subject material suggested it would generate even more interest than the *Origin*, even if readers often took away only a partial rendering of Darwin's ideas, accompanied by a scepticism or outright rejection of evolution. 'The book is at once the most popular, and yet the most profound in its suggestiveness, of any of the volumes Mr Darwin has written', declared the *New York Tribune*.

Reviewers stressed that it would be read for entertainment as well as by those in search of new knowledge, suggesting readers need not – or would not – be bound by Darwin's intentions. The *Daily News* remarked:

> Darwin has given us in this volume a book which will, perhaps, be read with equal interest by two distinct and almost antagonistic classes of readers – those who seek only enlightenment, and those who seek only amusement. The child who reads 'Gulliver's Travels' as a story can hardly be more different from the scholar who studies it as a wonderful piece of satirical art, than one set of Mr Darwin's readers, so far as this volume is concerned, will probably be from another. For it is, apart from all graver considerations, one of the most entertaining books we have read for a long time. If it were merely a collection of miscellaneous observations as to the method of expressing the various emotions in all races of man and in the lower animals, it would still be found full of a peculiar and an absorbing interest.[78]

Expression touched everyone – this was one factor in accounting for the interest it excited. 'There is nothing more wonderful in nature than the expressions of the human face, nor anything which excites a wider sympathy than the expression of the emotions', remarked *The Times*.[79] Some reviewers suggested that Darwin's observations would be of value in domestic life no less than in the advancement of science. *The Times* wrote 'we should recommend to mothers a series of photographs of sulky, screaming, and smiling children, which might enable them, in the midst of some of their less agreeable experiences, to find relief in scientific observation'.[80] In the words of *The Globe*: 'his treatises are universally said to be more interesting than novels, and his books are, in fact, in as much request at the circulating libraries as the regulation three volumes of romance'.[81]

Several reviewers stressed that the subject required detailed explication to be understood, ascribing to Darwin the role of detective in a narrative punctuated with homely detail. This combination of familiarity, novelty and strangeness served to increase the popular appeal of the work. Emphasising the need for expert guidance, the *Lancet* remarked:

> The expression of the emotions commences so early and is so constantly before us, we are so accustomed to read the passing thought in the play of feature and the gestures of the body, that we scarcely realise the complexity of the muscular actions by which they are effected, or the difficulties that present themselves in attempting to frame a rational explanation of their cause.[82]

The weekly reform newspaper the *Examiner* followed suit, observing: 'We are all familiar with the language of emotion', but that this needed explanation to be properly understood: 'few are able to specify distinctly what those signs are. To determine with precision what the facts are, is, therefore, no easy task, and Mr Darwin has resorted to various methods of ascertaining them'.[83]

But, with Darwin's shaping philosophy and elucidation, the expression of the emotions were made newly accessible, and acclaimed as such. The *New York Tribune* observed of the *Expression* 'No one can read it without understanding it, nor can any read it without learning something'.[84] Darwin's writing as much as the subject matter held the attention of readers. 'Every page abounds with illustration and anecdote', declared the *Telegraph*. The *Athenaeum* remarked 'Whatever view of the matter the reader may take, he will not fail to be interested by Mr Darwin's exposition, and amused by his numerous illustrations'.[85] The *Daily News* similarly picked up on the richness of 'apposite anecdote and curious scraps of description', and the *Examiner* referred to Darwin's 'felicity of style', his 'extraordinary range of knowledge', and 'a happy, if occasionally somewhat bold, spirit of speculation'.[86] 'Not one of Mr Darwin's dissertations is tame', it declared: 'all of them, on the expression of slyness, shrugging the shoulders, 'uncovering the canine', nodding and shaking the head, are extremely interesting'. Numerous reviews relayed Darwin's puff-adder story. [87] Intrigued by the idea that 'Fine music causes a shiver to run down the backs of some persons',[88] the *English Independent* delighted in the story of the nervous young man who gained a fortune and appeared drunk, concluding 'We did not know until now that one could so literally be made drunk with joy'. The press was fascinated by Darwin's discovery of the physical explanation of tears, with several reviewers quoting the passage in full.[89]

The originality of *Expression* was acknowledged and seen as striking. The natural history journal *Land and Water* remarked: 'his present subject is yet so fresh in itself, so entirely new in its mode of treatment, that this last work, like its two predecessors, brings us at once into an unknown train of thought, and fresh series of ideas ... In fact, we are much mistaken if Mr Darwin has not invented a new science, or at least a new development of an old one'.[90] The *Argus* was confident not only that *Expression* would be extensively read, but also that it attracted new readers to Darwin's theory of evolution, 'whom the technical and scientific character of preceding volumes has hitherto deterred from following out the speculations which they contain – speculations which, within the space of a few years, have revolutionised our ideas of natural history, and even of man himself'.[91]

The most insightful reviews emphasised the overriding interest in causes and origins. The *Athenaeum* reminded its readers always to bear in mind that Darwin's observations referred 'not so much to the manifestation of emotion on the faces and limbs of living creatures as to the causes or motive powers of those manifestations, or, to speak more strictly, the media between the emotions and the manifestations'.[92] Anton Dohrn, a prominent German Darwinist whose zoological station in Naples, the first in the world, was about to open, wrote at once to tell him that, like his other works, *Expression* contained the material and principles of a new science.[93] Subsequently reviewing it in the *Academy: A Monthly Record of Literature, Learning, Science and Art*, which prided itself on scholarly impartiality, Dohrn credited Darwin with bringing together morphology, concerned with the form and structure of living beings, and physiology, traditionally concerned with their function, praising him for bringing order and philosophy to facts. He stressed that Darwin not only connected 'by new ties man and his mental world to animal life and its bodily constitution', but also traced 'a new and highly important line of inquiry with respect to the origin of functions in showing how in many cases the function of expressing emotions has its origin in other functions'. In doing so, he declared, Darwin had led us to that point in physiology 'where life itself becomes but a function of matter'. He emphasised that Darwin's subject was 'not the anatomy nor the physiology of expression of which he treats', but that his object was rather 'to trace its origin in the ancestors of man, and to give an exposition of the principles involved in the different modes of expression in man as well as in other animals'. Dohrn concluded 'From a biological point of view' the main interest of the book was 'its successful attempt to trace the origin of special functions', introducing the theory of evolution 'into the domain of physiology'.[94] In this way, the press both popularised Darwin's new work and offered guidance on how to read it. The *Expression* was, the *Argus* observed, an important link in the development of Darwin's theory. Previous writers, it noted, had appeared 'convinced that species, man of course included, came into existence in their present condition'.[95]

The reviews ensured that Darwin's methods, as well as the reaches of his research, received wide public attention. 'It is not a new subject to him, for his notes go so far back as 1838, and for many years he has kept it in view and been adding to his stock of observations', wrote the *Examiner*. In the same piece, Darwin's questions, and the global reach of his correspondents, were noted and acclaimed: 'Thirty-six sets of answers were returned, and these are most valuable in determining the universality of certain expressions among the different races of men'.[96] The questions were of especial interest to overseas papers such as the *Argus*.[97] The *Daily News* reproduced the questionnaire in full, but it was circumspect, predicting that the *Expression* 'will of course give birth to much controversy; many of the facts – to say nothing of the principles – will be disputed'.[98] The *Standard* stressed the originality of Darwin's methods: the subject 'is handled in so original and scientific a manner, it contains so large an amount of information gathered from so large a variety of sources, that it is quite a work *sui generis*',[99] and the *Examiner* drew attention to the domestic setting of his research: 'He has himself closely watched the early days of his own children, and his observations form one of the most valuable portions of his book'.[100] Reviewers were also struck by how Darwin had made his method the subject of his research; *The Times* remarked that he had turned 'his favourite method of inquiry' – his observation of humans and animals – 'into a subject of great popular interest'.[101]

Others noted the limitations of Bell's work on the emotions. The *Examiner* observed: 'Sir Charles Bell stopped short at the facts, without attempting any philosophical explanation', and praised Darwin for beginning where Bell had ended and seeking, through extensive observation, to explain expression by the theory of descent.[102] Some critics made connections with physiognomy. Darwin had conceded 'whatever amount of truth the so-called science of physiognomy may contain' appeared to depend on people bringing into frequent use different facial muscles, according to their dispositions; thus possibly increasing the development of these muscles and deepening lines or furrows on the face; 'the free expression by outward signs of an emotion intensifies it'. Dohrn remarked that 'The three principles with the many instances by which they are illustrated in Mr Darwin's book are of interest not only as teaching physiognomy, but, as we have already indicated, in attempting to introduce the theory of Evolution into our knowledge of the functions of man and other animals'.[103]

But the idea that Darwin's work on the expressions constituted a break from physiognomy generated more interest. The *Athenaeum* commented that *L'Art de Connaître des Hommes par la Physionomie* (1835) by Lavater 'contains a great deal of twaddle, and, owing to the writer's prodigious industry and wonderful complacency, combines more blunders with more knowledge than any other book of the kind'.[104] *Land and Water*, whose founder, Francis Buckland, opposed Darwinian evolution, remarked that Lavater's 'curious and clever book' had 'proved the text-book for all such as care to inquire into the subject' and was 'a record of shrewd observation, and considerable inductive skill', but noted it had not entered upon 'the proximate causes of expression, much less upon their first origin'.[105]

Newspapers and magazines not only brought the *Expression* to wide public attention, they also reiterated Darwin's theory of descent. Bringing it more firmly into circulation, they drew the attention of new audiences and contributed to the ongoing ferment of ideas on evolution. Darwin had stressed in the *Origin* that the idea of the independent creation of species was an obstacle to understanding relations between them, remarking now that it was 'as pernicious with respect to Expression as to every other branch of natural history' (12). A number of journals quoted this remark directly, with the *Telegraph* reproducing his argument that 'Many writers consider the whole subject of expression as inexplicable . . . No doubt as long as man and all other animals are viewed as independent creations, an effectual stop is put to our natural desire to investigate as far as possible the causes of Expression'.[106] The *Argus* noted that while sceptics might accept similarity of passions in men and animals but still be 'a long way from accepting the Theory of Evolution', Darwin had, with this work, 'forc[ed] Expression into the service of the Evolution Theory'. This it declared to be a 'brilliant' notion, 'worthy of the acumen and ingenuity of its author'.[107] The *Examiner* echoed the *Argus*, reporting 'another province is added to the realm of natural history as re-arranged by natural selection'.[108] Natural history had been revolutionised, and the *Expression* was part of this revolution.

While some acclaimed the work in full, others were more guarded, or hostile. Although *Land and Water* was not without praise for Darwin's latest research, the review concluded satirically: 'Our own usually placid physiognomy wears "a derisive and sardonic smile" when we read that "our male semi-human progenitors possessed great canine teeth" .... with all deference to Mr Darwin, we must decline to receive the "early progenitor" who fought with his teeth, moved his ears, and did not blush, into our family'.[109] The *Athenaeum* was sceptical of the idea of animal morality, couching its objections in humour: 'As to a conscience in dogs in general, no man of experience and common sense believes in a canine conscience any more than in a legal conscience'.[110] It conceded: 'It is a significant fact that the more the habits of any particular animal are studied by a naturalist, the more he attributes to reason and the less to unlearnt instincts', but under-

Figure 2.3: Robert Braithwaite Martineau, 'Kit's Writing Lesson'. Reproduced courtesy of the Tate Gallery.

scored the line that Darwin was seeking to erase: 'it is in all that relates to the development by evolution, or by natural selection, of moral and intellectual faculties and emotions that he is manifestly feeblest. He does his best, but he rows against wind and tide'. The following year *The Times* praised Darwin as an observer of natural history, but on continuity of mental powers in humans and animals and on the moral sense declared 'Mr Darwin appears quite out of his element'.[111] Remarking 'Our characteristic objects and motives are invisible, mental, and ultimate; theirs are visible, corporeal, and immediate', it insisted on 'an absolute difference in kind' between humans and animals. By contrast the *Lancet* welcomed Darwin's theory of common descent, and met his anthropomorphism with a fascinated approval – 'Dogs and the higher monkeys he thinks undoubtedly express pleasure by grinning and incipient smiles; chuckling and a kind of laughter can be induced in chimpanzees and young orangs by tickling or presentation of favourite food'.[112]

The reception of *Expression* was characterised by the dialogue between literature and science that was part of its make-up. The *Scientific Review* quoted Darwin quoting Shakespeare and the *Literary World* concluded that *Expression* afforded 'pleasant reading', offering entertainment that was 'intellectual and improving'. The *Telegraph* remarked:

> An especially notable feature in the book is the way in which the author lays the whole field of literature under contribution to illustrate his descriptions of various expression: Homer, Spenser, Shakespeare, Tennyson, and Dickens, ancient classics and modern travellers, are all in turn made to lend a light to the scientific record of symptoms, and a fresh charm to the fascinating page.[113]

There was a general consensus that the study of the expressions inspired dialogue between science and art. The *Lancet* pointed out that it was in reading fiction that expressions were most likely to be seen:

> the habitual repression of the expression of the emotions, in the educated classes at least, causes the full action of the several muscles to be rarely witnessed but most people must have experienced the remains of these associated actions...in reading an affecting passage in a good writer aloud.[114]

The subject matter of the *Expression* drew frequent literary comparison. The *Evening Herald*, in an upbeat review headed 'Social Notes by a Plain Man', was particularly taken with the illustrations. Focusing on the portrait of 'a young lady who is supposed to be tearing up the photograph of a despised lover', whom he likened to Tennyson's Lynette, he saw it as ' a good sample of the more interesting – where all are interesting – parts of the book, and as showing how wide and familiar a range of topics the author has introduced'.[115] The critics were intrigued by Darwin's observations that due to imitation 'Children learning to write often twist about their tongues as their fingers move, in a ridiculous fashion'. Reporting on 'The habit which many youngsters have while they are learning to write, of rolling their tongues in accord with the motions of their hands', the *Athenaeum* noted that this was 'probably due to imitation' and 'was amusingly depicted in the late Mr Martineau's picture "Kit's Writing Lesson" – an illustration of "The Old Curiosity Shop"' (see Figure 2.3).[116] The *Argus* recalled the same passage 'He gives some curious instances at length. Some actions appear independent of habit, and due rather to sympathy, as when an awkward writer twists his jaws and tongue about in an uncouth manner while trying to form his letters'. It added an example of its own:

> One may remark here that singular instances of an analogous nature may be observed any night in a theatre. Some persons will follow the facial movements and distortions of the clown and pantaloon, seeming unable to keep from imitating them; and children especially exhibit the strongest sympathy with what is going on on the stage, throwing up their arms, suspending their laughter, and kicking violently with their feet, as if waiting for and repeating the movements of the performers.[117]

Some reviewers, echoing Darwin's more general reservations on the limitations of art in representing expressions accurately, saw his research as a corrective to the depiction of expression in fiction. The *Daily News* remarked:

> Most of us...have got into conventional ways of assuming certain expressions of face to follow certain emotion. Our authors have done a good deal to perpetuate error in this way, for only a very few writers really take the trouble to do more, when describing emotion, than to copy a description of somebody else.

While blushing and blanching were staple ingredients of the novel, the form was insufficiently empirical to be of reliable use in the study of expression. The reviewer continued:

> Nothing is more common in books than to hear of some heroine whose colour went and came, who turned from pale to red, and from red to pale several times in instant succession, although we venture to think that no one in real life ever witnessed such a phenomenon. We have at present in our recollection the heroine of a really eminent author who blushed so that the colour gradually overspread her whole frame; although the authority of the most experienced physicians has satisfied Mr Darwin that only in very extraordinary instances indeed does the crimson colour spread below the upper part of the chest. People are not always correct as to the expression on their own faces,

> even when requested to consider the subject. A lady mentioned in this volume was convinced that she had blushed, while those around her saw that she had really turned pale.

The wide appeal of *Expression*, circulating as fast as a bestselling romance, intensified anxieties for its opponents as they braced themselves for an unprecedented popularisation of the idea of evolution. They sought to use the qualities that had secured its popularity to undermine its seriousness. The *English Independent* remarked that Darwin's work was entertaining rather than philosophically valuable.[118] *The Globe* warned against 'Drawing-Room Darwinism' (the title of the review), seeking to disarm it through irony, and pushing it towards comedy:

> Society pricked up its ears literally when it was informed that we have a reminiscence of our monkey stage of existence in certain formations of the organ of hearing, and raised its eyebrows on being told of the connection between a nictitating action and a similar movement constantly indulged in by Jacko. But we imagine the new work on expression must be still more attractive to polite circles. Here, for instance, we are instructed in the profound doctrine of The Sulk, considered from a very different stand point from that from which it has been hitherto regarded. It may be well said we knew not indeed what we did when we pouted.[119]

Opposed to the idea of common descent, the reviewer reflected on his mockery:

> We are not, as it may be perceived, treating Mr Darwin's volume seriously. It is beyond question that to thousands of readers it will appear in a somewhat absurd light, and so far we confess we believe no harm can be done by it. But a little learning, especially a very little, is a dangerous thing, and drawing-room Darwinism occasionally degenerates into materialistic slipshod, of a contemptible instead of a comic description.

Remarking 'might we not enter thus early a mild protest against Darwin at dinner and Darwin at tea', it warned 'It is now the rage amongst the people in town to talk of this wonderful book, and a wonderful book is made of it when its facts and theories are imperfectly recollected and incoherently stated'. It urged for Bell's work to be given equal attention:

> If Mr Darwin is to be a boudoir favourite, so ought his opponent. Both sides should have a hearing. Under such circumstances a universal course of anatomy and physiology should be undertaken. Without an acquaintance with these studies the terminology of Mr Darwin must often appear obscure, if not unintelligible, not only to most ladies but to a great many gentlemen. And it must, moreover, be said that the mere 'lay' taste for a work like the 'Descent of Man' has in it a savour of vulgar curiosity rather than traces of a sincere desire to search after truth.[120]

There was particular anxiety over women readers:

> We dare not say what ladies ought to read or ought not to read now-a-days, but we may venture to hint that they should not be seen floundering beyond their depth . . . Young ladies are able to talk a little on the descent of man, and, although the peculiar philosophy of the origin of species might be too much for them, they contrive at any rate to get the amusing facts and illustrations of the theory into their heads for conversational use. It is after all only natural that so universal a concern should be experienced about the productions of a *savan* who is so picturesque and vivid in enforcing his views.[121]

Such warnings went unheeded. An immediate sensation, bringing evolution into physiology and psychology in the most captivating of ways, the ideas of the *Expression* were quick to permeate the wider culture, and also to change fundamentally Victorian understanding of emotion, reshaping perceptions of the place of humans in nature. And it was perhaps in the novel, where the emotions had long received detailed, if not always empirically accurate, attention, that this would be most fully in evidence.

I am grateful to the Syndics of Cambridge University Library for permission to quote from the Darwin archive.

Notes

1 See Paul H. Barrett, Peter J. Gautrey, Sandra Herbert, David Kohn, Sydney Smith (eds), *Charles Darwin's Notebooks* (1836–1844; London and Cambridge: Natural History Museum and Cambridge University Press, 1987), 561. Unless otherwise stated, notebook page references are to this volume.

2 Darwin to Charles Lyell, 10 January 1860, Darwin Correspondence Database, http://www.darwinproject.ac.uk/entry-2647 (accessed 17 February 2011).

3 Darwin to J.D. Hooker, 24 December 1862, Darwin Correspondence Database, http://www.darwinproject.ac.uk/entry-3875 (accessed 17 February 2011).

4 On Darwin's writing see Gillian Beer, *Darwin's Plots: Evolutionary Narrative in Darwin, George Eliot and Nineteenth-Century Fiction*, 3rd edn (1983; Cambridge: Cambridge University Press, 2009), and George Levine, *Darwin the Writer* (Oxford: Oxford University Press, 2011).

5 Notebook M, entry 146e, 556. << indicates Darwin's insertions.

6 Notebook N, entry 18, 19 October 1838, 568; 2 October 1838, 609.

7 Thomas Dixon, '"Emotion": The History of a Keyword in Crisis', *Emotion Review* (2013), doi:10.1177/1754073912445814. See also Dixon (ed.), *Thomas Brown: Selected Philosophical Writings* (Exeter: Imprint Academic, 2010), 1–30.

8 See Dixon, *From Passions to Emotions: The Creation of a Secular Psychological Category* (Cambridge: Cambridge University Press, 2003), especially chapter four; and chapter one in this volume.

9 Thomas Brown, *Lectures on the Philosophy of the Human Mind* (Edinburgh: Tait, 1820). See Dixon, 2003, 104, 109; see also Dixon, 2013, and Dixon (ed.), 2010.

10 'Old & Useless Notes about the Moral Sense & Some Metaphysical Points Written about the Year 1837 & Earlier', in Barrett et al, 1987, 597; entry 29, 2 October 1838, 609.

11 'Old & Useless Notes', entry 8a, 600.

12 Notebook C, entry 243, 315. Stéphanie Dupouy argues for a shift from continuity in the notebooks to distinction in the *Expression*, noting the construction of the books, which separates the expressions of animals and men (Dupouy, 'The Naturalist and The Nuances: Sentimentalism, Moral Values, and Emotional Expression in Darwin and The

Anatomists', *Journal of the History of the Behavioural Sciences*, 47 (2011), 351), but in *Expression* Darwin was writing, possibly defensively, for probably his broadest audience yet and the more careful approach he adopted here does not necessarily reflect a shift of opinion.

13 On the popularity of Bell's book and its editions, see Lucy Hartley, *Physiognomy and the Meaning of Expression in Nineteenth-Century Culture* (Cambridge: Cambridge University Press, 2001), 79, 198, note 11. On conceptions before Darwin of the biological or physiological origins of expression see Dupouy, 2011, 341; on Bell more specifically see Paul White, 'The Face of Physiology', *19: Interdisciplinary Studies in the Long Nineteenth Century*, 7 (2008), 1–22.

14 Notebook N, entry 57, 27 December 1838, 578.

15 Notebook N, entry 59, 6 January 1839, 579.

16 *Ibid.*

17 Notebook N, entry 5, 3 October 1838, 564.

18 Notebook N, entry 5, 3 October 1838, 564; Notebook N, entry 66, 16 March 1839, 581.

19 Notebook M, entry 107, 545.

20 Notebook N, entry 66, 582.

21 Notebook N, entry 60, 579, emphasis in original.

22 Notebook N, entry 52, 577.

23 See François Joseph Gall, *On the Functions of the Brain*, trans. by Winslow Lewis, 6 vols (Boston: Marsh, Capen and Lyon, 1835), v. 265; cited in Sally Shuttleworth, *Charlotte Brontë and Victorian Psychology* (Cambridge: Cambridge University Press, 1996), 61. For an early rejection of physiognomy as unscientific, see Georges Louis Leclerc, Comte de Buffon, *Natural History, General and Particular*, trans. by William Smellie, 3rd edn, 9 vols (1749; London: W. Strahan and T. Cadell, 1791), ii. 451–2.

24 Darwin to William Darwin Fox, 3 January 1830, Darwin Correspondence Project, http://www.darwinproject.ac.uk/entry-75 (accessed 17 February 2011).

25 Physiognomy continued to have cultural resonance through the nineteenth century; on physiognomy and the *Expression* see Rosemary Jann, 'Evolutionary Physiognomy and Darwin's *Expression of the Emotions*', *Victorian Review*, 18 (1992), 1–27; and on physiognomical discourse in phrenology see Richard Twine, 'Physiognomy, Phrenology and the Temporality of the Body', *Body and Society*, 8 (2002), 67–88.

On the reliance of genres from medical writing to etiquette books on physiognomy see Emma Spary, 'Political, Natural and Bodily Economies', in Nicholas Jardine, James A. Secord and Emma Spary (eds), *Cultures of Natural History* (Cambridge: Cambridge University Press, 1996), 178–96. On physiognomy and the *Expression,* see Hartley, 2001; Graeme Tytler, *Physiognomy in the European Novel: Faces and Fortunes* (New Jersey: Princeton University Press, 1982); and Melissa Percival and Graeme Tytler (eds), *Physiognomy in Profile: Lavater's Impact on European Culture* (Newark: University of Delaware Press, 2005); Twine, 2002, and chapter 5 in this volume. For an excellent study of physiognomy in the novels of Charlotte Brontë see Shuttleworth, 1996, and on physiognomy in visual culture see Mary Cowling, *The Artist as Anthropologist: The Representation of Type and Character in Victorian Art* (Cambridge: Cambridge University Press, 1989).

26 Johann Kaspar Lavater, *Essays on Physiognomy* (London, C. Whittingham, 1804) 4 vols, i. 86; Notebook M, 23 September 1838, entry 145e, 556.

27 Notebook N, entry 109, 593; entry 35, 30 October 1838, 572.

28 Notebook C, entry 243, 315.

29 Notebook N, entry 37, 20 November, 573.

30 Notebook M, entry 97, 542.

31 For recent studies of emotions and sociality, see Ursula Hess, Rainer Banse and Arvid Kappas, 'The Intensity of Facial Expressions is Determined by Underlying Affective State and Social Situation', *Journal of Personality and Social* Psychology, 69 (1995), 280–8; and James A. Russell and José Miguel Fernández-Dols (eds), *The Psychology of Facial Expression: Studies in Emotion and Social Interaction* (New York: Cambridge University Press, 1997).

32 'Old & Useless Notes', entry 42, 5 May 1839, 619.

33 Darwin, *The Descent of Man, and Selection in Relation to Sex*, 2 vols (London: John Murray; 1871), i. 39, 40–1

34 *Ibid.*, i. 48

35 *Ibid.*, i. 71, 78, 97.

36 *Ibid.*, i. 105.

37 Darwin, *The Expression of the Emotions in Man and Animals* (London: John Murray, 1872), 22. Where subsequent references to this work are contextually clear, page numbers will be given parenthetically in the text.

38 See Adrian Desmond and James Moore, *Darwin* (Harmondsworth: Penguin, 1992), 593.

39 R.B. Freeman and P.J. Gautrey note that Darwin was asking these questions as early as 1860: 'Charles Darwin's Queries About Expression', *Bulletin of the British Museum (Natural History) Historical Series*, 4 (1972), 217.

40 Darwin to Thomas Belt, 6 January 1868, Darwin Correspondence Database, http://www.darwinproject.ac.uk/entry-6546 (accessed 17 February 2011), emphasis in original.

41 Darwin to Lucy Wedgwood, 8 June [1867–72?], Darwin Correspondence Database, http://www.darwinproject.ac.uk/entry-7223 (accessed 17 February 2011).

42 Darwin to T.H. Huxley, 30 January 1868, Darwin Correspondence Database, http://www.darwinproject.ac.uk/entry-5817 (accessed 17 February 2011).

43 T.H. Huxley, 'On the Physical basis of Life', delivered 8 November 1868, published in the *Fortnightly Review*, 5 (1869), 129–45 and, in the United States, as 'New Theory of Life' in the New York newspaper *The World*, 18 February 1969. Reprinted in Huxley, *Collected Essays*, 9 vols (London: Macmillan, 1904), i. 132,133–4.

44 Huxley, 1904, i. 134.

45 See John Morley, *The Works of Lord Morley* (London: Macmillan, 1921), 82. On Huxley's significance in historical perspective, see G. Rickey Welch and James S. Clegg, 'From Protoplasmic Theory to Cellular Systems Biology: A 150-Year Reflection', *American Journal of Cell Physiology*, 298 (2010), 1280–90.

46 John Tyndall, *Address Delivered Before the British Association Assembled at Belfast* (London: Longmans, Green, and Co, 1874), 54.

47 Francis Darwin (ed.), *The Life and Letters of Charles Darwin, including an Autobiographical Chapter*, 3 vols (London: John Murray, 1887), i. 48.

48 See Phillip Prodger, *Darwin's Camera: Art and Photography in the Theory of Evolution* (Oxford: Oxford University Press, 2009).

49 See http://darwin-online.org.uk/EditorialIntroductions/Freeman_OntheOriginofSpecies.html (accessed 21 February 2011).

50 This passage was quoted in full in the *Examiner* (First Notice), 23 November 1872, 1155.

51 James Crichton-Browne, *Medical Mirror* (1865), cited in Darwin, 1872, 76 note 10.

52 The hot-house face dog was a favourite with critics; see, for example, *Examiner* (Second Notice), 7 December 1872, 1205.

53 Dixon, 2003, 170–1. Dixon notes that Darwin adopted several

examples on blushing from Burgess. See also Dupouy, 2011, on continuity with earlier thought on expression.

54 Darwin, 1872, 10; see Dixon, 2013 and 2003. Dupouy argues that by the time of *Expression* Darwin was emphasising the attenuated nature of human expressions, compared with those of animals, and thus the ability of humans to control and inhibit expressions (Dupouy, 2011, 337). She draws on Anne Vincent-Buffault, *Histoire des Larmes* (Paris: Payot , 1986, repr. 2001); Roger Smith, *Inhibition: History and Meaning in the Sciences of Mind and Brain* (Berkeley, CA: University of California Press, 1992) and Lorraine Daston and Peter Galison, *Objectivity* (Cambridge: Zone Books, 2007), suggesting that in *Expression* Darwin tended to see emotions as softened by culture, within a framework of the development of the will and learning (353). Dixon argues that through challenging Bell, Darwin came to endorse in places the view that expressions were vestiges of a time when human beings had a greater voluntary control over their bodies, which was the opposite of his main argument; see Dixon, 2003, 174–76, and, on utility and the *Expression*, see 168–77.

55 Bain, 1864, 96.

56 Bain, *The Emotions and the Will* (London: Parker, 1859), 3, in Dixon, 2013. Herbert Spencer complained that Bain ignored the distinction between emotions and sensations, including sensations under the umbrella of emotions. He considered this retrograde and suggested they both be embraced by the term feeling. See Spencer, *Illustrations of Universal Progress: A Series of Discussions* (1865; D. Appleton and Co, 1870), 320. For discussion of Bain see Rick Rylance, *Victorian Psychology and British Culture 1850–1888* (Oxford: Oxford University Press, 2000), 150–202; and Dixon, 2003, 150–5. Dixon points out that a number of leading Victorian psychologists endorsed the separation of intellectual thoughts from emotional feelings; he references Bain, *The Senses and the Intellect* (London: Parker, 1855; 2nd edn, 1864); and 1859; J. McCosh *Psychology: The Cognitive Powers* (London: Macmillan, 1886); and *Psychology: The Motive Powers; Emotions, Conscience, Will* (London: Macmillan, 1887); J.M. Baldwin, *Handbook of Psychology I: The Senses and the Intellect; II: Feeling and Will* (London: Macmillan,1891); and James Sully, *The Human Mind: A Textbook of Psychology*, 2 vols (London: Longmans, Green, and Co., 1892). See also Annette Baier, 'What Emotions Are About', *Philosophical Perspectives*, 4 (1990), 1–29; Richard Wollheim, *On the Emotions*

(New Haven: Yale University Press, 1999); and Dylan Evans, *Emotion: The Science of Sentiment* (Oxford: Oxford University Press, 2001) on historical accounts of emotions.

57 Darwin, 1872, 9. Darwin quotes Spencer, *Principles of Psychology* (London: Longman, Brown, Green and Longmans, 1855), and Spencer, *Essays: Scientific, Political and Speculative, Second Series* (London: Williams & Norgate, 1863), 111.

58 Spencer, *Essays*, 1863, 138; quoted in Darwin, 1872, 27.

59 Spencer, *Essays*, 1863, 138.

60 See Paul Griffiths, 'Emotions', in Stephen P. Stich and Ted A. Warfield (eds), *The Blackwell Guide to Philosophy of Mind* (Oxford: Blackwell, 2002), 288–308; and Dixon, 2013. Dixon demonstrates that while Charles Bell conceived of emotion as a movement of the mind he was the first to give a constitutive role to bodily moments. On emotions as action patterns and adaptive responses to particular events or context see Lewis in this volume.

61 Darwin, 1872, 176. On Darwin and tears see Paul White, 'Darwin Wept: Science and the Sentimental Subject', *Journal of Victorian Culture*, 16 (2011), 226–41; and Thomas Dixon, 'The Tears of Mr Justice Willes', *Journal of Victorian Culture*, 17 (2012): 1–23.

62 Darwin, 1872, 197.

63 Darwin, 1872, 338–9.

64 Darwin, 1872, 339.

65 Darwin, 1872, 365–6. While Dixon underlines the emphasis in *Expression* on the uselessness of expression, pointing out that the most useful emotions and forms of expression, i.e. moral sentiments and language, respectively, were dealt with in the *Descent* (Dixon, 2003, 175), in the expression notebooks Darwin considers the sociality (and hence utility) of expression at length. On pre-Darwinian ideas of expression as useful in Louis-Jacques Moreau de la Sarthe, Bell and Louis-Pierre Gratiolet see Dupouy, 2011, 342; and on uselessness in the *Expression* see Dupouy 2011, 348. For contemporary views on the question of utility in *Expression* see Dohrn, 1873, 210.

66 Darwin, 1872, 14.

67 Darwin, 1872, 13.

68 Darwin, 1872, 13, 14, 184-5.

69 See Janet Browne, *Charles Darwin: The Power of Place* (New York: Knopf, 2002). Browne notes that in 1866 he paid £14 for photographs. On

photography in *Expression* and Darwin's relationship with Rejlander see Prodger, 2009.

70 Darwin, 1872, 243.

71 GHL Diary, ms diaries for 1869-1876, Beinecke Rare Book and Manuscript Library, Yale University, cited in Eliot, *George Eliot's Daniel Deronda Notebooks*, ed. by Jane Irwin (Cambridge, Cambridge University Press, 1996), 242, note 2.

72 See Darwin to Charles Lutwidge Dodgson, 10 December 1872 and 14 December 1872, cited in Christoph Heyl, 'The Posthumous Evolution of the Dodo: From Charles Darwin to Lewis Carroll and Beyond', Darwin our Contemporary conference, Siegen University, 2009.

73 Unsigned review, 'The Expression of the Emotions', *The Times*, 13 December 1872, 4. On Darwin and the stage see Tiffany Watt-Smith, 'Darwin's Flinch: Sensation Theatre and Scientific Looking in 1872', *Journal of Victorian Culture*, 15 (2010), 101–18; Jonathan Smith, *Charles Darwin and Victorian Visual Culture* (Cambridge: Cambridge University Press, 2009); Kirsten Shepherd-Barr, 'Darwin on Stage: Evolutionary Theory in the Theatre', *Interdisciplinary Science Reviews*, 33 (2008), 107–15; and Jane R. Goodall, *Performance and Evolution in the Age of Darwin: Out of the Natural Order* (London: Routledge, 2002). See also chapter five, 147, in this volume.

74 Desmond and Moore, *Darwin's Sacred Cause: Race, Slavery and the Quest for Human Origins* (Harmondsworth: Penguin, 2009). See also Greg Radick, 'Race and Language in the Darwinian Tradition (and what Darwin's Language–Species Parallels have to do with it)', *Studies in History and Philosophy of Biological and Biomedical Sciences*, 39 (2008), 359–70, and 'Did Darwin change his Mind about the Fuegians?', *Endeavour*, 34 (2010), 50–4.

75 See, for example, unsigned review, 'Mr Darwin's New Book. The Expression of the Emotions in Man and Animals', *Daily News*, 5 November 1872.

76 Unsigned review, 'Mr Darwin on Expression', *Telegraph,* 5 November 1872.

77 Unsigned review, 'Expression in Man and Animals' (First Notice), *Argus*, 23 January 1873.

78 *Daily News*, 5 November 1872.

79 *The Times*, 13 December 1872, 4.

80 *Ibid.*

81 'Darwin in the Drawing Room', *The Globe*, 7 November 1872.

82 Unsigned review, 'The Expression of the Emotions in Man and Animals' (First Notice), *Lancet*, 7 December 1872, 817.

83 *Examiner* (First Notice), 23 November 1872, 1155.

84 *New York Tribune*, 14 January 1873.

85 Unsigned review, 'The Expression of the Emotions in Man and Animals' (Second Notice), *Athenaeum*, 16 November 1872, 632.

86 *Daily News*, 5 November 1872; *Examiner* (First Notice), 23 November 1872, 1155.

87 See *Examiner* (Second Notice), 7 December 1872, 1205; *New York Tribune*, 14 January 1873; unsigned review, 'The Expression of Emotions in Man and Animals', *Scientific Review*, 2 December 1872, 176–8.

88 Darwin, 1872, 68, cited in unsigned review, 'Darwin on the Emotions', *English Independent*, 14 November 1872, 1209. See chapter three in this volume.

89 See unsigned review, 'Mr Darwin on The Expression of the Emotions' (Second Notice), *Examiner*, 7 December 1872, 1206.

90 Unsigned review, 'The Expression of the Emotions in Man and Animals', *Land and Water*, 9 February 1873.

91 *Argus*, 23 January 1873.

92 *Athenaeum*, 16 November 1872, 591.

93 Anton Dohrn to Darwin, 13 November 1872, http://www.darwinproject.ac.uk/entry-8624 (accessed 17 May 2011).

94 Anton Dohrn, 'The Expression of the Emotions in Man and Animals', *Academy*, 4 (1873), 209. See Darwin's scrapbook of reviews, DAR226, Darwin Archive, Cambridge University Library, 211.

95 *Argus*, 23 January 1873.

96 *Examiner* (First Notice), 1872, 1155.

97 *Argus*, 23 January 1873.

98 *Daily News*, 5 November 1872.

99 'Mr Darwin's New Book', *Standard*, 8 November 1872, 5–6.

100 *Examiner* (First Notice), 23 November 1872, 1156.

101 *The Times*, 13 December 1872, 4.

102 Unsigned review, 'Mr Darwin on The Expression of the Emotions' (First Notice), *Examiner*, 23 November 1872, 1155.

103 Dohrn, 1873, 210.

104 Unsigned review, 'The Expression of the Emotions in Man and Animals', *Athenaeum*, 9 November 1872, 591.

105 *Land and Water*, 9 February 1873.

106 Darwin 1872, 12; *Telegraph*, 5 November 1872. See also, for example, *Argus*, 23 January 1873, on Darwin's break with the idea that all species had come into existence in their present condition.

107 *Argus,* 23 January 1873.

108 *Examiner* (Second Notice), 7 December 1872, 1205–7.

109 *Land and Water*, 9 February 1873.

110 Unsigned review, 'Descent of Man and Selection in Relation to Sex', *Athenaeum,* 4 March 1871, 276.

111 Unsigned review, 'The Expression of the Emotions', *The Times*, 13 December 1872, 4.

112 Unsigned review, 'The Expression of the Emotions in Man and Animals' (First Notice), *Lancet*, 7 December 1872, 818.

113 *Telegraph,* 5 November 1872.

114 Unsigned review of *Expression* (Second Notice), *Lancet*, 14 December 1872, 852.

115 Reference to *Idylls of the King: Gareth and Lynette,* i. 574. See *Darwin,* 1872, 255.

116 *Athenaeum*, 16 November 1872, 591.

117 *Argus*, 23 January 1873. The reference is to Darwin, 1872, 36.

118 Unsigned review, 'Darwin on the Emotions', *English Independent,* 14 November 1872, 1209.

119 'Darwin in the Drawing Room', *The Globe*, 7 November 1872.

120 On these questions see Gowan Dawson, *Darwin, Literature, and Victorian Respectability* (Cambridge: Cambridge University Press, 2007).

121 'Darwin in the Drawing Room', *The Globe*, 7 November 1872. On anxieties over reading women see Kate Flint, *The Woman Reader 1837–1914* (Oxford: Clarendon Press, 1993) and Angelique Richardson, *Love and Eugenics in the Late Nineteenth Century: Rational Reproduction and the New Woman* (Oxford: Oxford University Press, 2003), chapter four.

3

The Backbone Shiver: Darwin and the Arts

Gillian Beer

> But why does joy, & *other emotions* make grown up people cry What is emotion? At end of Burke's essay on the sublime and Beautiful there are some notes, & likewise in Wordsworth's dissertations on Poetry.
>
> Charles Darwin, in the N Notebook[1]

Darwin and Aesthetic Emotion

In his private Notebooks, kept during the late 1830s when he was first formulating his theories of evolutionary descent, Darwin repeatedly draws on his intense experiences in response to music, fine art, theatre, and literature, as he does also in his early letters and in the *Voyage of the Beagle* in 1839.[2] What was the place of the arts in Darwin's creativity? What happened to this eager emotional and sensory responsiveness later in his life? In this essay I first explore his responses and then ask whether, or in what ways, these experiences enter his later theories. But there is a preliminary difficulty in thinking about Darwin and aesthetic experience and it comes from his own late commentary.

Late in his life, in the reminiscences he composed for his family, Darwin looked back with a sense of loss on the intensity of his earlier pleasure in the arts:

> I have said that in one respect my mind has changed during the last twenty or thirty years. Up to the age of thirty, or beyond it, poetry of many kinds, such as the works of Milton, Gray, Byron, Wordsworth, Coleridge, and Shelley, gave me great pleasure, and even as a schoolboy I took intense delight in Shakespeare, especially in the historical plays. I have also said that formerly pictures gave me considerable,

> and music very great delight. But now for many years I cannot endure to read a line of poetry; I have tried lately to read Shakespeare, and found it so intolerably dull that it nauseated me. I have also almost lost my taste for pictures or music. Music generally sets me thinking too energetically on what I have been at work on, instead of giving me pleasure. I retain some taste for fine scenery, but it does not cause me the exquisite delight which it formerly did.[3]

That passage in his *Autobiography* has been read by some as implying that he never had a strong aesthetic sense. The contrary is manifestly true, and is vouched for by the deep regret, as well as the rueful amusement, with which he puzzles over his loss (intense delight, exquisite delight, great pleasure). The words delight and pleasure run through the passage evoking the telling ghosts of that former experience, which he can just recall. Music now turns his mind back obsessionally on his own thoughts instead of releasing them into music's counter-creativity (though it remains powerful). He exempts only one form of art, the novel, and that as a concession to what he sees as his own lowered taste:

> On the other hand, novels which are works of the imagination, though not of a very high order, have been for years a wonderful relief and pleasure to me, and I often bless all novelists. A surprising number have been read aloud to me, and I like all if moderately good, and if they do not end unhappily – against which a law ought to be passed. A novel, according to my taste, does not come into the first class unless it contains some person whom one can thoroughly love, and if it be a pretty woman all the better.[4]

As one might expect, he attempts to understand the physical source of this attrition and to generalise from it.

> My mind seems to have become a kind of machine for grinding general laws out of large collections of facts, but why this should have caused the atrophy of that part of the brain alone, on which the higher tastes depend, I cannot conceive. A man with a mind more highly organised or better constituted than mine, would not I suppose have thus suffered; and if I had to live my life again I would have

> made a rule to read some poetry and listen to some music at least once every week; for perhaps the parts of my brain now atrophied could thus have been kept active through use. The loss of these tastes is a loss of happiness, and may possibly be injurious to the intellect, and more probably to the moral character, by enfeebling the emotional part of our nature.[5]

He fears that the loss of aesthetic pleasure may enfeeble the emotions and perhaps the intellect. He deliberately frames the sentence with a generalising plural – 'the emotional part of *our* nature' – but it tells a story of himself. The exquisite spontaneity of response to the arts has long been lost to him though he imagines a more orderly programme that might sustain such tastes.

What happened to this somatic responsiveness? Did it simply die down or was it transformed and marshalled into his theory? Had these pleasures silted through into the soil of his thinking or was there a void where they had been? I hope to demonstrate that his early responses to music, literature, and visual art first subside and then are gathered, to differing degrees, into the implications and arguments of his late works.[6]

First, let's note their importance to him, as he dates it, '*up to the age of thirty, or beyond it*' (my italics): that is, not only through his childhood and youth and the formative years as a student in Edinburgh and then Cambridge, but through the much *more* formative five years of the *Beagle* voyage, the return to London life, and the composition of the Notebooks, where in the late 1830s, when he was still in his twenties, the first thronging ideas for his theory gushed forth. These same notebooks proved to be a great resource for his later work in the 1870s when he first placed the human prominently among other species in his argument, having deliberately avoided discussion of humankind, for what he called 'diplomatic' reasons, in the *Origin of Species* (1859).[7]

It is striking that in his description of what most delighted him in his youth, music and poetry are the most prominent. His reference to high art and paintings is somewhat guarded and his pleasure is most freely expressed in landscape. Of his taste on return from the *Beagle* voyage he writes:

> About this time I took much delight in Wordsworth's and Coleridge's poetry, and can boast that I read the *Excursion* twice through. Formerly Milton's *Paradise Lost* had been my chief favourite, and in my excursions during the voyage of the *Beagle*, when I could take only a single small volume, I always chose Milton.[8]

He was fourteen when his sensibility awoke to landscape: 'In 1822 a vivid delight in scenery was first awakened in my mind, during a riding tour on the borders of Wales, and this has lasted longer than any other aesthetic pleasure'.[9] That pleasure remained, and it was one he shared with Ruskin, though on evolution they differed bitterly. Darwin's response to landscape was much expanded by his land journeys from the *Beagle* where he encountered the extraordinary variety of the earth, in some cases (in contrast to much of the British landscape), primordial growth unmixed with the arts of human design. He describes these natural scenes, often with rapture, but also quite frequently by an unexpected comparison with Western arts.

More than once the comparison is with engravings:

> The view was striking: it may aptly be compared to a framed engraving, where the frame represents the breakers, the marginal paper the smooth lagoon, and the drawing the island itself'[10]

– not altogether easy to visualise!

Outside Rio he entered 'a noble forest':

> At this elevation the landscape attains its most brilliant tint; and every form, every shade, so completely surpasses in magnificence all that the European has ever beheld in his own country, that he knows not how to express his feelings. The general effect frequently recalled to my mind the gayest scenery of the opera house or the great theatres.[11]

High on a mountain in Chile he is for a time alone:

> The atmosphere resplendently clear, the sky an intense blue; the profound valleys; the wild broken forms; the heaps of ruins, piled up during the lapse of ages; the bright-coloured rocks, contrasted with the quiet mountains of snow; all these together produced a scene

> no one could have imagined. Neither plant nor bird, excepting a few condors wheeling around the higher pinnacles, distracted my attention from the inanimate mass. I felt glad that I was alone; it was like watching a thunderstorm, or hearing in full orchestra a chorus of the Messiah.[12]

These synaesthetic responses – sight conjuring music in particular – move freely across natural and cultural forms (thunderstorms and Handel's *Messiah*, forest and theatre). They also cross sound and silence (the quiet mountains of snow...watching a thunderstorm). Darwin strains to communicate to the reader who has not shared his journey what he also hyperbolically asserts cannot be imagined. To do so, he merges senses, tumbles culture and nature together, and poises all these materials in a written style that expresses with great clarity the role of the silent observer.

Fuegians and Freischutz

Sometimes Darwin's invocation of cultural material that is remote from the actual scene carries much more personal significance than appears on the surface. At his first encounter with the Fuegians, Darwin gives a closely observed account of their facial ornamentation:

> The old man had a fillet of white feathers tied round his head, which partly confined his black, coarse, and entangled hair. His face was crossed by two broad transverse bars; one, painted bright red, reached from ear to ear and included the upper lip; the other, white like chalk, extended above and parallel to the first, so that even his eyelids were thus coloured. The other two men were ornamented by streaks of black powder, made of charcoal. The party altogether closely resembled the devils which come on the stage in plays like *Der Freischutz*.[13]

That derisive remark at the end of the paragraph carries freight that is not at first apparent. His struggles to familiarise the unknown can sometimes give us insights into Darwin's unuttered formative experiences – experiences both intimately personal and leading into his scientific practices and beliefs. Let me illustrate that by pursuing this example.

Darwin gropes for a comparison to communicate with his English audience at home and finds it in the gothic opera *Der Freischutz* (roughly translated, the Free Shooter). *Der Freischutz*, Carl Maria von Weber's opera, is based on a German folk tale full of menace and dread. By selling his soul to the devil the huntsman obtains seven unerring bullets; release from this terrible bargain can only be obtained by providing another victim within the deadline. The villain tries to entrap the hero by offering him the magic bullets. (In his youth and young manhood Darwin himself was a skilled and enthusiastic shot.) In the famous scene in Wolf's Glen the huntsman invokes the devil with rites and magic spells and their horrible ingredients. Samiel, the devil himself, finally appears, accompanied by his host of lesser devils. The surging and tumultuous music of this scene powerfully evokes a scene of primitive passions and sorcery. Hearing the music makes you understand the intensity of dread and suspense that Darwin is recalling. The reference to this story expressed as sound evokes Darwin's own dread of entrapment during this first encounter with Fuegian men. The ridicule is defensive.

Darwin's invocation of the opera with its visual extremes – the painted devils alongside *sturm und drang* music – also implicitly acknowledges that there is no simple upward trajectory from 'primitive' life to modern European life, with the older cultural and emotional phases safely sloughed off. The same dreads, and even the same practices, are active in them both. European art toys with powerful folk sources, and in performance is overwhelmed by their reality. For Darwin and his first readers *Der Freischutz* was a more or less contemporary opera; it was first produced in London, in English, in July 1824 when Darwin was sixteen and had been shown a number of times after that. Darwin was writing the *Voyage of the Beagle* just a dozen years after the first London production so that it summoned a theatrical experience shared with many of his contemporary readers.

Perhaps it is not stretching context too far to observe that at the height of the hero's terror, in the opera's climactic Wolf's Glen scene to which Darwin refers here, the ghost of the hero's mother appears to him in her grave clothes, urging him to retreat. Darwin's mother, dead since he was eight years old, may not have entered his conscious mind as he evoked this scene. But he is remembering his first encounter with the devastatingly unfamiliar inhabitants of Tierra del Fuego whose reputation for violence was formidable indeed. That violence even seemed to be potentially enacted in the elaborate aggression of the old man's face-painting. Darwin's dread and the desire to retreat express themselves in his evocation of that moment of the opera with its silent presence of the phantasmal mother. The devils enter the stage as she vanishes.

All this hallucinatory recollection is compacted immediately after his very careful description of the Fuegian's painted ornamentation – ornamentation that is itself performance. So what may appear at first as just a throwaway joke at the Fuegian's expense *also* harbours, and seeks to ward off, Darwin's intense emotions that must be subdued: fear of what is strange, dread of sorcery and unknown powers, and – as he acknowledges – outright terror at the possibility of hostilities – emotions almost certainly shared by the Fuegians in the encounter! Opera and folk-tale together give a language for such feelings. These responses he will much later, in the *Descent of Man* and even more in *Expression of the Emotions,* investigate as *universal,* not only within humankind but across the animal kingdom.

The Backbone Shiver

A shiver that may be dread or delight is a typical response of the young Darwin: they mingle in his reaction to the imaginative experience of the arts. In a memorable phrase in his early 1838 autobiographical fragment he remembers how his own lies, or 'story-telling' fabulations as a child, gave him 'pleasure, like a tragedy'.[14]

Music is the art that when young thrills him most directly, in the body. 'How's your backbone?' he asks his friend J.M. Herbert after they have listened together to an anthem. In the *Autobiography* he remembers the music at Cambridge, particularly the voices of King's College choir: 'I acquired a strong taste for music, and used very often to time my walks so as to hear on weekdays the anthem in King's College Chapel. This gave me intense pleasure, so that my backbone would sometimes shiver'.[15] This phenomenon is borne out by the comments of his friends: for example, Herbert, writing to him while Darwin was on the *Beagle* in 1832, mingles Darwin's love of poetry and music in his recollection of him. (The mention of 'your old friend Keats' is striking here.) Herbert writes:

> I was last night at a Quarterly Meeting of the Choral, which you used to patronize to such an extent; and if your old friend Keats was right you will, as you used to, feel a thrill through your backbone at the very mention of some of your old favourites: for 'Heard melodies are sweet, But those unheard are sweeter- '.[16]

Somatic experience revives through memory. Herbert remembers the intensity of Darwin's response to harmony, as do other of his friends such as Jonathan Cameron. (Simple melody apparently left him unaffected.) But in a later letter Herbert teases Darwin for his odd mixture of intense response and inexpert knowledge:

> yr. want of musical memory, the coexistence of which with yr. musical taste & longings has been often the subject of my most curious speculations...At times I imagine yr. fondness for music must be the result of reasoning, tho' the intensity of yr. feeling it, as evinced by yr. *back-bone* criticisms on Handel's Choruses, wd lead one to infer a very large development [sic] of the organ of Tune.[17]

Punningly, Herbert turns to phrenology (the *organ* of Tune) to solve the mystery of Darwin's absorbing music into his body. Darwin much later self-deprecatingly puzzled over the same thing: 'I am so utterly destitute of an ear, that I cannot perceive a discord, or keep time and hum a tune correctly; and it is a mystery how I could possibly have derived pleasure from music'.[18] But in the N Notebook, closer to his time at Cambridge, Darwin acknowledges through an anecdote the 'primitive' sensory pleasures of music as compared with the fine arts:

> Old man at Cambridge observed the ignorant merely looked at picture as works of imitation – Hence pleasure in the beautiful (distinct from sexual pleasure) is acquired taste. – Whilst music extremely primitive. – almost like tastes of mouth and smell.[19]

Strikingly, he returns to this comparison in the *Descent*: 'But if it be further asked why musical tones in a certain order and rhythm give man and other animals pleasure, we can no more give the reason than for the pleasantness of certain tastes and smells'.[20]

Ralph Colp describes Darwin's responsiveness as showing the characteristics 'of a labile individual: who because of a reactive or labile automatic nervous system, responded to a wide range of unpleasant, and/or pleasant, mental stimuli by experiencing brief episodes of trembling, chills, shivering, and gastro-intestinal upsets'.[21]

Whereas Darwin's taste for music was somatic, appetitive, his response to high art was much more learnt: he was, as he put it, '"inoculated" with a taste for pictures and good engravings' which was 'not natural to me'. (He sold his Wedgwood copy of the Portland Vase in 1858, perhaps because he feared for it in the midst of family life, or simply because he had no special pleasure from ceramics.) In the *Autobiography* he grades his pleasure in the different arts: 'I have also said that formerly pictures gave me considerable, and music very great delight'.

> I was very intimate with Whitley, who was afterwards Senior Wrangler, and we used continually to take long walks together. He inoculated me with a taste for pictures and good engravings, of which I bought some. I frequently went to the Fitzwilliam Gallery, and my taste must have been fairly good, for I certainly admired the best pictures, which I discussed with the old curator. I read also with much interest Sir Joshua Reynolds' book. This taste, though not natural to me, lasted for several years and many of the pictures in the National Gallery in London gave me much pleasure; that of Sebastian del Piombo exciting in me a sense of sublimity [perhaps 'The Raising of Lazarus' with its tumult of figures around the central event].[22]

There is an implicit comparison here: Darwin acknowledges his more conventional 'learnt' response to paintings –notice the defensiveness of: 'my taste must have been fairly good, for I certainly admired the best pictures, which I discussed with the old curator' (*that's* all right then). His descriptions of his musical experience are much more eager.

Sex played a part in his young musical tastes too. Hearing the soprano/mezzo-soprano Malibran (who he spells Malibrand) at 'the music meeting' he is overwhelmed by her:

> words cannot praise her enough, she is the most charming person I ever saw...I like Madame Blasis very much but nothing will do after

> Malibrand, who sung some comic songs, a persons heart must be made of stone not to have lost it to her.[23]

Such experience much later on would feed his understanding of how powerfully song acts in sexual selection, demonstrating how many species *sing* in passion and pleasure (gibbons, seals, and birds are among his examples).

Art, Observation and Writing

When, on the *Beagle*, he writes to Whitley longing for their old pleasures, he particularly invokes a Titian picture of Venus in the Fitzwilliam Museum in Cambridge. The exciting nature of this picture was signalled by its being kept behind a curtain, not to offend female eyes, and incidentally to give an extra frisson to male spectators as the curtain was drawn back, theatrically, to reveal her naked figure. (This picture may have been Titian's 'Venus with a Lute-Player' or else the 'Sleeping Venus' by Padovanino, now labelled 'After Titian' but then accepted as his.) Darwin writes to Whitley, expressing both estrangement and nostalgia, a kind of libidinal mourning for the lost past of friendship, of culture, and even of self. (I have published an extensive analysis of the deep concerns underlying this particular invocation in an essay 'Four Bodies on the *Beagle*', in my book *Open Fields*.[24]) Darwin moves from the 'glory and luxuriance' of the tropics, to his astonishment on first seeing 'a naked Fuegian his long hair blowing about, his face besmeared with paint'.[25] The unintelligible cries and 'inconceivably wild' figure of the Fuegian occupy a paragraph. The 'inconceivable' figure is also Darwin's double:

> As Darwin recollects it here, they seem to be alone together. Revealingly, closer to the event, he writes of a *group* of men on the mountain. In this dream-like and desolated scanning for meaning, an intimate, alienated pair of men emerges: Darwin and the naked Fuegian.[26]

He then swerves to a recollection of his Cambridge past:

> When I return to England you must take me in hand with respect to the fine arts. I yet recollect there was a man called Raffaelle Sanctus.

> How delightful it will be to see once more in the Fitzwilliam Titian's Venus; how much more than delightful to go to some good concert or fine opera. These recollections will not do. I shall not be able tomorrow to pick out the entrails of some small animal, with half my usual gusto.[27]

That last wrench, towards the body of the 'small animal' tugs in its wake the body of the Venus and of the Fuegian. High art and animal bodies are held in the same gaze. Only the Fuegian looks back at him.

> The topographical and cultural distance Darwin supposes between himself and the Fuegians (though disagreeable to us now) is valuable to him in reaching a new theoretical position. It allows Darwin to extend the spectrum of humankind to such an extent that it begins to overlap the distance between humankind and other animals: instead of the severe disjunction between the human and other species enjoined by creationist theory the whole measure can be encompassed, 'we are all netted together'.[28]

Most fine art in the great galleries and museums gave him only temperate gratification (delightful) compared with his visceral response to music (more than delightful). But although his pleasure in fine art was somewhat conventional, *the visual* permeated his practice, his materials, and his pleasures. It escaped the bounds of art from the start. Observation, his primary scientific method, demands sight and sustained attention; he relished both the wide sweep of a landscape and the scrutiny of a flower or a beetle. Indeed, he believed that this capacity for responsiveness extended to other creatures too: in an early notebook he writes 'I cannot help. thinking horses admire a wide prospect'.[29] He learnt from illustrations by botanic artists (though he was wary of their tendency to generalise or epitomise a so-called 'standard' plant).[30] His own concern was with the slight variations between specific plants that enable evolutionary change, so nothing to him was trivial and every detail mattered. He learnt also by studying the plant samples of men like Henslow and Joseph Hooker.[31] He practised, though without great expertise, the dissection of dead creatures (never of living ones). He could not draw, he said. He wasn't much good at dissection, he said. Writing was his daily medium of thought and of communication and his writing was ballasted with his reading in Wordsworth and Thomson and Scott, those great authors of the natural world in its detail and sweep. His reading lists are staggering in their scope and show him at ease with poetry from Dante, Shakespeare, and Milton, to his near contemporaries such as Byron and Wordsworth, alongside travel literature, breeding manuals, and philosophers such as Hume.

He delighted in intricate verbal description, so much so that his children teased him and said that he sounded as if he were writing an advertisement:

> In writing, he sometimes showed the same tendency to strong expressions that he did in conversation. Thus in the 'Origin', p.440, there is a description of a larval cirripede, 'with six pairs of beautifully constructed natatory legs, a pair of magnificent compound eyes, and extremely complex antennae'. We used to laugh at him for this sentence, which we compared to an advertisement. This tendency to give himself up to the enthusiastic turn of his thought, without fear of being ludicrous appeared elsewhere in his writings.[32]

He sought to evoke in every detail a visual presence for the object described. He watched with supreme attention the searching tendrils of climbing plants and the agitation of worms placed close to the bass notes of a grand piano.

Learning the Arts of Others

But there remains the question of what 'the arts' included for him. Did he recognise indigenous arts on his voyage? That remark about the Fuegian's face 'smeared with paint' may suggest not. But it was not his only reaction and may indeed be conjuring up a *past* reaction he can expect that he *would* have shared with his conservative friend Whitley at that earlier period he is here nostalgically evoking. He reports in Tierra del Fuego how surprised the Fuegians were by 'our' dancing and singing. This is a good reminder that his contact with indigenous arts was not only with the arts of exotic tribes but with the arts of hornpipe and sea-shanty, those of sailors on shipboard and on shore. (Most of the Fuegians simply watched in amazement but one young

Fuegian man did consent to try the waltz.) Did Darwin dance? His description sounds as if he might have done (he was a supernumerary not an officer on board ship so not completely bound by naval hierarchy), and he is certainly identifying with the dancers. He writes again about it, very conscious that the English are the exhibition now and the Fuegians the audience:

> The women worked hard, while the men lounged about all day, watching us. ...They asked for everything they saw, and stole what they could. They were delighted at our dancing and singing, and were particularly interested in seeing us wash in a neighbouring brook.[33]

In barter, Darwin observes the droll and levelling circumstance that both sides equally think they are getting the better of the other:

> Both parties laughing, wondering, gaping at each other; we pitying them, for giving us good fish and crabs for rags, etc.; they grasping at the chance of finding people so foolish as to exchange such splendid ornaments for a good supper.[34]

The sailors seek food; the Fuegians beauty.

In particular, Darwin was struck by the sensory immediacy of the Fuegians' pleasures: they were far more interested in 'the beauty of scarlet cloth or blue beads', he wrote, than in the complexity of the ship on which the travellers arrived. And he remembered the observation of the French explorer Louis-Antoine de Bougainville, that the Pacific islanders treated the masterpieces of human industry exactly as they treated the laws and phenomena of nature.[35] In one sense, then, art seemed to have no separate category in Fuegian life, but – as Darwin more and more recognised – that did not mean that art had no place in their culture. The contrary was true. The 'arts' were assimilated into natural law and into the performance of daily life despite dreadful physical conditions: persistent rain, damp cold, choking smoke. Later, Darwin affirmed that the intricate aesthetics of the everyday are held across a spectrum stretching from humankind through other life-forms, including wasps and worms.

Darwin learnt to relish the bodily arts in which the person becomes the canvas for display. When they reached Tahiti, he observed:

> The common people, when working, keep the upper part of their bodies quite naked; and it is then that the Tahitians are seen to advantage. They are very tall, broad-shouldered, athletic, and well-proportioned. It has been remarked, that it requires little habit to make a dark skin more pleasing and natural to the eye of an European than his own colour. A white man bathing by the side of a Tahitian, was like a plant bleached by the gardener's art compared with a fine dark green one growing vigorously in the open fields. Most of the men are tattooed, and the ornaments follow the curvature of the body so gracefully, that they have a very elegant effect. One common pattern, varying in its details, is somewhat like the crown of a palm-tree. It springs from the central line of the back, and gracefully curls round both sides. The simile may be a fanciful one, but I thought the body of a man thus ornamented was like the trunk of a noble tree embraced by a delicate creeper.[36]

His similes here affirm the accord of the Tahitians with the natural world, in bodily skin colour (as if growing vigorously in the open fields rather than bleached by the gardener's art). The distinction offers a striking look forward as metaphor towards the concept of natural as opposed to artificial selection. And there is already a glimpse of co-evolution in the simile of the man as 'a noble tree embraced by a delicate creeper'. Most immediately, there is physical pleasure in the beauty of the Tahitian man's body.

The visual world is not outside only. It is inside, in the memory. Darwin's writing in the first part of his career immediately on and after the *Beagle* voyage shows the traveller's longing to retain and communicate unrepeatable experiences. It demonstrates also Darwin's need to retain details that will allow him to pursue differences: *variability* becomes crucial in the formulation of natural selection. Superabundance and variability entwine and cut across each other. Darwin is moved by variety.

'Spots of Time'

Toward the end of his life in the *Autobiography* he returns specifically to the emotional effects of memory. Here one can discern the effects of Wordsworth's evocations of place as condensed emotion and moments of inner change, the 'visionary dreariness' he experienced as a young boy . As a young boy at his first school Darwin recalls:

> I remember clearly only one other incident during this year whilst I was at Mr. Case's daily school – namely, the burial of a dragoon soldier; and it is surprising how clearly I can still see the horse with the man's empty boots and carbine suspended to the saddle, and the firing over the grave. This scene deeply stirred whatever poetic fancy there was in me.[37]

On St. Jago Darwin first realised some of his possible powers and that favourite word 'thrill' (the backbone shiver) occurs again as he recalls both the sensation and the scene fixed in his memory:

> It then first dawned on me that I might perhaps write a book on the geology of the various countries visited, and this made me thrill with delight. That was a memorable hour to me, and how distinctly I can call to mind the low cliff of lava beneath which I rested, with the sun glaring hot, a few strange desert plants growing near, and with living corals in the tidal pools at my feet.[38]

That capacity for vivid local recall stood Darwin in good stead in his scientific work (curiously, he found that while he could learn poetry with ease he forgot it again within forty-eight hours).

Observation relies on the present moment of close scrutiny but, in his earlier life, during the land voyages from the *Beagle* especially, he was always relinquishing that present and travelling onwards. So he perforce augmented his visual encounters by writing his field notes, then at large his Journal, and letters, in which he recorded his recollections: as he put it 'I took much pains in describing carefully and vividly all that I had seen'.[39] 'Carefully and vividly': the defence against fading.

Later, when settled at Down House, recurrence formed an essential part of his often extended experimental work in the greenhouses and garden. And despite his feeling of having been abandoned by, or having abandoned the arts, each day made space for novel-reading (often two, sometimes three, sessions a day from Francis Darwin's account) and for listening to Emma playing the piano. Emma had taken some lessons with Chopin and she enjoyed playing Mozart and

Mendelssohn, as well as the then popular and distinguished pianist and composer Moscheles who was Mendelssohn's teacher. Her style, apparently, was disciplined and restrained, very *unromantic,* at the opposite extreme from *Der Freischutz*.

So though he had lost the sublime emotions generated by his earlier experiences of music and literature and landscape, music and literature continued to feed his domestic emotional life, levelled now with the everyday but totally integrated into it. In the steady rhythm of his daily life the presence of novel reading cannot have failed to be a resource to his thinking, emphasising as it does empathy, scepticism, questions of belief and of gesture, the expression of the emotions, and above all difference, as well as strong persistent stories common across time. Together he and Emma read and read again the novels of Walter Scott, Jane Austen, and Elizabeth Gaskell, as well as many more ephemeral or unacknowledged works.

In the *Descent* he cites Schopenhauer on the subject of romances and love affairs:

> The final aim of all love intrigues, be they comic or tragic, is really of more importance than all other ends in human life. What it all turns upon is nothing less than the composition of the next generation...it is not the weal or woe of any one individual, but that of the human race to come, which is here at stake.[40]

Arts Animal and Human

It is in the works of the late 1860s and 1870s that questions concerning human experience are given their full force, alongside the experience of other organisms. In the *Descent* and *Expression* humans are foregrounded, but alongside them go continuing enquiries into climbing plants and variability, and last of all his volume on *The Formation of Vegetable Mould, through the action of worms, with observations on their habits* (1881).

In the *Descent,* his early delight in music bears fruit in his theory of how language developed:

> We must suppose that the rhythms, and cadences of oratory are derived from previously developed musical powers. We can thus

> understand how it is that music, dancing, song, and poetry are such very ancient arts. We may go even further than this and...believe that musical sounds afforded one of the bases for the development of language.[41]

Language, which he twice calls 'half-art, half-instinct' is in his view still in the process of evolution and thus not a secure separate mediator of actuality. Its origins, he argues, are to be found by analogy with bird-song:

> The slight natural differences of song in the same species inhabiting different districts may be appositely compared, as Barrington remarks, 'to provincial dialects'; and the songs of allied, though distinct species may be compared with the languages of distinct races of man. I have given the foregoing details to show that an instinctive tendency to acquire an art is not peculiar to man.[42]

'An instinctive tendency to acquire an art is not peculiar to man': that surely is one of his most radical statements.

The arts in his argument have become fundamental, 'instinctive', to human and other kinds. (Indeed in his recollections of the Fuegians in the conclusion of the *Descent* he typifies them as 'barbarians' who 'possessed hardly any arts' and for this and other reasons not to be preferred to other more humble creatures.) He continues, on the same page, to argue from bird-song that 'Language owes its origin to the imitation and modification of various natural sounds, the voices of other animals, and man's own instinctive cries, aided by signs and gestures'.[43] Rather than music being the outcome of the tones of conversation, as Herbert Spencer argued, for Darwin music is originary, prior to language, and intimately bound in to the process of sexual selection:

> All these facts with respect to music and impassioned speech become intelligible to a certain extent, if we may assume that musical tones and rhythm were used by our half-human ancestors, during the season of courtship, when animals of all kinds are excited not only by love, but by the strong passions of jealousy, rivalry, and triumph.[44]

The backbone shiver that Darwin experienced in his youth as he listened to young voices in chorus, or combined in opera with orchestra and drama, goes back directly into the primeval and beyond, goes back before language, into arts shared instinctively by the human and many other kinds. Darwin's own bodily experience of music's power informs the development of his theory here. And, reinforcing his association of humankind with all other organic life, he argues that: 'Whether we can or not give any reason for the pleasure thus derived from vision and hearing, yet man and many of the lower animals are alike pleased by the same colours, graceful shading and forms, and the same sounds'.[45]

Of Beauty

So the question of beauty comes to the fore in his later thinking. Indeed, it had provoked him already on the *Beagle* voyage. Gazing at the play of creatures in the sea he records in his diary his puzzlement over why God had made such beauty when there was no one to behold it (no wonder Coleridge's *The Ancient Mariner* spoke so strongly to him). That puzzlement derived from two assumptions that he later shed: one assumed that God had created the world for man's pleasure; the other that human beings possess the sole criterion for judging beauty. As he worked he more and more realised that beauty is variable, that each group or kind prepares its own criteria and that it is within each group that beauty is realised. The rapture of attentiveness that characterised his powers of observation was in accord with the activity of other occupants of the natural world, especially during the period of courtship. His theory of sexual selection paid attention to the ways in which beauty became the driver for survival.

In the fourth edition of the *Origin,* published in 1866, and again in the fifth published in 1869, the years during which he was preparing the *Descent,* Darwin added several pages on the idea of beauty, seeking in particular to refute Ruskin's insistence on the centrality of the human in its significance. Ruskin had been arguing for that centrality, and confirmed it in *Fors Clavigera* thus:

> For leaf, and root, and fruit exist, all of them, only – that there may be flowers. He disregarded the life and passion of the creature, which

> were its essence....Now in exactly the sense that modern Science declares that there is no such thing as a Flower, it has declared that there is no such thing as a Man, but only a transitional form of Ascidians and apes. It may or may not be true – it is not of the smallest consequence whether it be or not. The real fact is, that, rightly seen with human eyes, there is nothing else but man; that all animals and things beside him are only made that they may change into him; that the world truly exists only in the presence of Man, acts only in the passion of Man.[46]

Ruskin claims passion as the central value. Darwin cools such heated rhetoric and thinks instead about the claims of all life forms to their own validity, unencumbered by human preferences:

> If beautiful objects had been created for man's gratification, it ought to be shown that there was less beauty on the face of the earth before man appeared than since he came on the stage. Were the beautiful volute and cone shells of the Eocene epoch, and the gracefully sculptured ammonites of the Secondary period, created that man might ages afterwards admire them in his cabinet?

'No', is the necessary answer. Darwin addresses the question of flowers head-on, immediately after this passage: 'Flowers rank among the most beautiful productions of nature; and they have become through natural selection beautiful,...that they might be easily observed and visited by insects, so that their fertilisation might be favoured'.[47] Beauty is formed *by means of* natural selection – and in this latter phase of his thinking, through sexual selection too. A peacock can't entice an ant with the glory of his tail – it's strictly for the pea-hen: and of course any passing mortal dazzled by irrelevant beauty. Indeed, the functionalism implicit in Darwin's insistence on the selective process pays rather little attention to the aesthetic pleasures still felt by those who have no wish to couple with a peacock. That is his necessary recoil from Ruskin's argument, but it is one that admits and admires the gorgeous in natural life for its own energy, rather than for human gratification. His emotion is focused on all the life-forms that so inventively and magnificently survive, and embraces a deep appreciation of those that survived in the past but are no longer present.

An early review of the *Origin* in *All the Year Round* (Dickens's weekly journal) sneered at Darwin's ideas as will-o-the-wisps. Perhaps there was something in it, perhaps someone else would come along to make sense of the maze, the reviewer conceded, but so far as Darwin was concerned the reviewer held out the prospect of a long life mortified by error without outcome.[48] Had that proved to be the case, for all the charms of Darwin's person and programme, we would not still be analysing the implications of his experience for his ideas. But his sturdy observation and experiment, and the accrued and sometimes transformative work of others set in train by his extraordinary prescience, make him still present all around us now. That achievement was sprung on his responsiveness to the world around him. He relished the aesthetic choices of worms, the energy and discrimination of climbing plants, the erotic ingenuity of orchids, the reasoning gaze of infants. He heard primordial chatter in bird-song as well as the snarl or silence of subjugation as species declined.

Darwin considered the arts to be prior to science, indeed the basis for science, and fundamental to species life:

> All Science is reason acting /systematizing /on principles, which even animals practically know (art precedes science – art is experience and observation) in balancing a body & an ass knows one side of triangle is shorter than two.[49]

He provided half of what's needed for the arts to flourish. He was the audience to the organic world: engrossed, responsive, and observant. He watched and listened. He lay on his front to scrutinise the soil and its creatures, and on his back to gaze into the canopy of leaves, birds, and the sky. All the materials of the arts were present in his mind, not as discrete events, but in the rush and tangle, the intricacy and precision, of the natural world and in the laws he drew from those 'endless forms'. Darwin's theories of natural and sexual selection are bedded in preoccupations with emotion and experience, with what is held in common and what is diverse across the whole span of organic life, across all extant creatures and plant life – and fossilised life too.

Notes

1 Paul H. Barrett (ed.), *Metaphysics, Materialism, Evolution of Mind: Early Writings of Charles Darwin* (Chicago: University of Chicago Press, 1980), 81.

2 The work now known as *The Voyage of the Beagle* was first published in 1839 as *Journal and Remarks 1832–36,* the third and final volume of Captain Fitzroy's *Narrative of the Surveying Voyages of His Majesty's Ships, Adventure and Beagle,* the official account of the voyage.

3 Charles Darwin, *The Autobiography of Charles Darwin and Selected Letters*, ed. by Francis Darwin (1892; New York: Dover Books, 1958), 53. First published in an edited version in 1887; written in the late 1870s.

4 *Ibid.*, 54.

5 *Ibid.*, 54.

6 I have written about the literary connections at length in *Darwin's Plots: Evolutionary Narrative in Darwin, George Eliot and Nineteenth-Century Fiction*, 3rd edn (Cambridge: Cambridge University Press, 2009); in 'Darwin's Reading and the Fictions of Development', in D. Kohn (ed.), *The Darwinian Heritage* (Princeton: Princeton University Press, 1985); and in 'Darwin and Romanticism', *Wordsworth Circle*, 41 (2010), 3–9. In the present essay I am more concerned with vision and music.

7 Darwin, *The Descent of Man, and Selection in Relation to Sex*, ed. by Jim Moore and Adrian Desmond (1871; 2nd edn, 1874; London: Penguin, 2004); *The Expression of the Emotions in Man and Animals,* ed. by Paul Ekman (1872, 2nd edn 1889; London: Harper Collins, 1998).

8 Darwin, 1892, 33; my recent essay, 'Darwin and Romanticism' (2010) sets out from this same passage in the *Autobiography* and initially pursues some of the same argument. See also the new chapter, 'Darwin and the Consciousness of Others' in *Darwin's Plots,* 2009, 242–55; and Beer, 1983.

9 Darwin, 1892, 10.

10 Darwin, *The Voyage of the Beagle*, ed. by David Amigoni (1839; Ware: Wordsworth Editions Ltd, 1997), 386.

11 *Ibid.*, 33.

12 *Ibid.*, 307.

13 *Ibid.*, 196; see also Gillian Beer, *Open Fields: Science in Cultural Encounter* (Oxford: Oxford University Press, 1996).

14 Darwin, *Autobiographies*, ed. by Michael Neve and Sharon Messenger

(London: Penguin, 2002), 3.

15 Darwin, 1892, 20.

16 *The Correspondence of Charles Darwin*, ed. by Frederick Burkhardt and Sydney Smith, 19 vols (Cambridge: Cambridge University Press, 1985), i. 223.

17 *Ibid*, i. 523.

18 Darwin, 1892, 20.

19 Barrett, 1980, 74.

20 Darwin, *Descent*, 2nd edn (1874; 2004), 635.

21 Ralph Colp, 'More on Darwin's illness', *History of Science*, 38 (2000), 219.

22 Darwin, *Autobiography*, 1892, 20.

23 Burkhardt and Smith, 1985, i. 94.

24 Beer, 1996, 13–30.

25 Burkhardt and Smith, 1985, i. 397.

26 Beer, 1996, 25–6.

27 Beer, 1996, 26.

28 Beer, 1996, 26. A commentator on this essay sees the cultural distance of the Fuegians as problematic for Darwin's theories of human unity, but as I have argued above, the extended spectrum of the human is invaluable to him in re-imagining kinship among humans and other animal species. For further discussion of the Fuegians who Darwin encountered see: 'Travelling the Other Way: Travel Narratives and Truth Claims', Beer, 1996, 55–71. See also, for recent commentary on Darwin's assessment of the Fuegians, Greg Radick, 'Did Darwin change his mind about the Fuegians?', *Endeavour*, 34 (2010), 50–4. See also wider discussion in David Amigoni, *Colonies, Cults and Evolution: Literature, Science and Culture in Nineteenth Century Writing* (Cambridge: Cambridge University Press, 2007).

29 26 August 1838, Notebook M, in Paul H. Barrett, Peter J. Gautrey, Sandra Herbert, David Kohn and Sydney Smith (eds), *Charles Darwin's Notebooks, 1836–1844* (London and Cambridge: Natural History Museum and Cambridge University Press, 1987), 546.

30 Jim Endersby, *Imperial Nature: Joseph Hooker and the Practices of Victorian Science* (Chicago: Chicago University Press, 2008), 137–69.

31 Endersby, 2008.

32 Darwin, 1892, 105.

33 Darwin, 1839, 97, 212, 217.

34 *Ibid.*, 217.
35 *Ibid.*, 217.
36 *Ibid.*, 384.
37 *Ibid.*, 8.
38 Darwin, 1892, 31.
39 *Ibid.*, 29.
40 Darwin, *Descent*, 2nd edn (1874; 2004), 653.
41 *Ibid.*, 131.
42 *Ibid.*, 109.
43 *Ibid.*, 109.
44 *Ibid.*, 638.
45 *Ibid.*, 115.
46 John Ruskin, *Fors Clavigera: Letters to the Workmen and Labourers of Great Britain*, 8 vols (1871; New York: John Wiley, 1881), v. 85–6. For extensive and often brilliant discussion of Darwin's relation to visual culture see Jonathan Smith, *Charles Darwin and Victorian Visual Culture* (Cambridge: Cambridge University Press), 2006.
47 Darwin, *The Origin of Species: A Variorum Text*, ed. by Morse Peckham (1859; Philadelphia: University of Pennsylvania Press, 1959), 370.
48 *All the Year Round*, 3 (1860), 299.
49 Barrett, 1980, 73.

4

Becoming an Animal: Darwin and the Evolution of Sympathy

Paul White

In his discussion of the moral sense in *Descent of Man*, Darwin locates human virtue in the social instincts that we share with, and have inherited from, animals, instincts that compel us to act for the community good. Scholars have rightly sought the roots of Darwin's moral theory in the eighteenth-century philosophy of Adam Smith, David Hume, and James Mackintosh, who based human society on the power of sympathy. Darwin's moral theory has an important and peculiar place in the long history of sympathy, for it rests unusually on the study of animal instincts and the physiology of emotions. For eighteenth-century theorists, sympathy was not, strictly speaking, an emotion, but rather a process of registering and communicating feelings between persons, and between and within bodies; a process that could be sudden and automatic, or extraordinarily refined and deliberative. For Darwin, sympathy was a more emotional affair. It was closely linked to love and the tender affections. Its evolution was associated with the extension and softening of the emotions that attends the growth of mental powers; but emotions, in turn, were conceived as reflex mechanisms, instinctive bodily responses that helped organisms to survive, adapt, and reproduce, and communities to grow and flourish.

Darwinian sympathy also conjoined the human and the animal in novel ways. Darwin's moral theory was not based on a generalised notion of the human and animal, but rather on a broad study of human groups and extensive observations of animal species, many of which would seem quite remote from humans, thus ants, bees, and birds, as well as primates and domestic animals. Some of these (especially dogs) were his companions and friends. This proximity to, and community with, animals in all of their variety, idiosyncrasy, and personality, had profound implications for the way in which human–animal boundaries were configured by Darwin, and for the nature of the sympathy that lay at the heart of human and animal communities. Darwin's theory of instinctive sympathy challenged prevailing divisions and hierarchies between the human and the animal, and within the human frame, between mind and body, and reason or will, and emotion. For Darwin the moral sense was literally that – a physical sensation, a feeling of pleasure in exercising one's social instincts, or of pain when those instincts went unheeded. It was also an emotional bond of love, but a love composed of bodily affections and attachments that were shared by many animals. Was it better to be an insect, a dog, or a monkey, to experience the flow of sympathy without hesitation, rather than a philosopher, burdened with ceaseless deliberations and pangs of conscience?

Shortly after the *Descent of Man* was published, Darwin received a letter from a stranger remarking on his striking resemblance to a 'venerable old Ape'.[1] The letter was inspired by a portrait that had appeared in the *Illustrated London News*, a typical engraving of the naturalist with deep set eyes, prominent brow, and bearded face. Soon other likenesses would appear, showing Darwin literally transformed, with monkey hands and tail, swinging from a tree, an incarnation of his theory of evolution.[2] Darwin collected such ephemera avidly, and was not offended. Through Darwin's surviving manuscripts and correspondence, as well as his publications and reviews, we can trace his reflections on animals and emotions over nearly four decades; the interweaving of sympathy with the tender affections in the domestic sphere of his work, his web of family, friends and collaborators, his devoted and dutiful pets; his efforts to build sympathy with readers and critics, incorporating and challenging them, with all of their marked differences, in a vision of community embracing all of nature. Darwinian sympathy was not just an emotional bond with another, be it human or animal. It was potentially more transformative – a process of becoming another, especially if that other was animal.

Animal Studies

Darwin's earliest memory was of a cow. It gave him a start as it ran suddenly past the window while he sat on his sister's knee. In his autobiographical notes and reflections, Darwin described his childhood delight in the company of animals, fishing for newts in a quarry pool, collecting insects, taking poetic pleasure in the gulls and cormorants in flight by the sea. This childhood enthusiasm led him to conclude: 'I was born a naturalist'.[3] Inspired by Gilbert White's *Natural History of Selbourne*, he began to study the habits of birds. He also found excitement in the rural sports of the gentry, for whom animals were game to be stalked, chased, and shot. In his university years at Edinburgh and Cambridge, he learned other approaches to animals: anatomy as the basis for comparative zoology and taxonomy, systematic collecting and observation in the field. While continuing his passion for hunting, he pursued a specialist interest in coleoptera, amassing an impressive range of beetle specimens, carefully pinned, labeled, and boxed: 'No pursuit at Cambridge was followed with nearly so much eagerness or gave me so much pleasure'.[4]

The affective field of animal study was greatly enlarged on the *Beagle* voyage. From the deck of the ship, Darwin watched as aeronautic spiders shot silk threads like rays of light into the currents of warm air. Crabs used tiny serrated and flattened claws like sucker cups to cling to the bodies of medusa. In his diary of the voyage, painstaking observation and collection were punctuated by shooting, eating, and cavorting: a rare species of ostrich inadvertently consumed around a campfire with gauchos, a giant Galapagos tortoise ridden like a horse.[5] The rich particularity of animal life and many joyful encounters unfolded in an experience of Nature as an affecting whole. Animals were not only observed and collected as specimens. They were touchstones for more powerful emotional bonds forged in a landscape of beauty and rarity, wonder and sublimity:

> If the eye attempts to follow the flight of a gaudy butter-fly, it is arrested by some strange tree or fruit; if watching an insect one forgets it in the stranger flower it is crawling over. – if turning to admire the splendour of the scenery, the individual character of the foreground fixes the attention. The mind is a chaos of delight, out of which a world of future & more quiet pleasure will arise. –[6]

The landscape of pleasure and abundance was disturbed, however, by the presence of humans, the native and enslaved peoples who evoked for Darwin a more alien and debased form of animal life. The Chilean miners were worked and fed like draft horses. The slaves in Brazil were lowered to a condition of helpless brutes. The savages of Tierra del Fuego huddled naked in the sleet, ate raw fish, and howled like wild beasts.[7] Human cruelty or harsh conditions animalised man. When identified with the savage or slave, the animal became a degraded form of life. Darwin's emotional response to these encounters was interestingly varied. His deep aversion to slavery is well known, and his *Beagle* diary and letters home display great fervour for the abolitionist cause.[8] Darwin extended fellow-feeling to slaves when they expressed particular longings attesting to their human dignity: to return to their families in Africa, to be allowed to work for wages rather than under duress. But when witnessed in an 'animal' state, they were revolting. Darwin tried to communicate with one. Having difficulty, he began to gesture, moving his hands near the man's face. Suddenly the slave was reduced to cowering, sure he was about to be beaten for some unknown offence, and afraid even to protect himself. At this point, Darwin was overcome with feelings of 'disgust and shame', disgust at seeing 'a powerful man...trained to a degradation lower than the slavery of the most helpless animal', and shame at seeing himself in the person of a slave master. It is curious that Darwin does not express sympathy for the poor man, but rather a sense of revulsion, forced upon him by the wretched feelings of another.[9] The slave is abhorrent because he has been degraded morally and politically, but why should such degradation signify animality? Why has the animal become derogatory when it is civilised men who have debased their fellow creatures?

Darwin's sentiments for the Yhagan people of Tierra del Fuego were similarly divided. Working in the well-established tradition of the natural history of man, he regarded the native peoples as objects of study. Because of their extreme wildness, the 'Fuegians', like the flora and fauna of the tropical forest, inspired immense curiosity and wonder. But Darwin's descriptions were almost unremittingly negative: 'hideous...filthy...greasy...discordant...violent and without dignity'.[10] Living for the most part without dwellings, uncovered, clinging to rocks, they appeared not even to desire what the Brazilian slaves most longed for, the comforts of home, the domestic affections. Janet Browne claims that Darwin viewed the 'Fuegians' in the same light as the slaves, with an 'unflinching egalitarian gaze'.[11] But it was only the civilised natives who aroused some compassion, the three who were 'passengers' on the *Beagle* and were now being returned to their birthplace. Darwin felt for 'poor Jemmy', who hoped to rejoin his mother and other relations, having been absent for some years.[12] Later Darwin witnessed Jemmy's terrible descent to a savage state: 'I never saw so complete and grievous a change'.[13] The closeness to animality that Darwin witnessed in the peoples of Tierra del Fuego, and the reversion of the Yhagan natives who had been transported to England, civilised, and then returned, was felt with revulsion.

Darwin's encounters with slaves and natives on the *Beagle* voyage have often been interpreted as seminal experiences for his later species theory, on the one hand deepening his passionate conviction in the unity of humankind, and on the other, awakening his sense of the close proximity of human to animal nature.[14] Yet it seems that these two equations, between the races of man, and between the human and the animal, could not be mutually sustained. The question, 'are slaves or savages men, are they brothers?' was not just one of taxonomy. Darwin moved between sympathy and the extremes of antipathy according to whether the people he met conformed to the standards of European civility. However much he might identify with them as members of the same species, his fellow-feeling vanished when they displayed characteristics that he regarded as animal. An almost unbridgeable divide thus seemed to separate the human and the animal when (and only when) the animal was man: 'viewing such men, one can hardly make oneself believe they are fellow-creatures'.[15]

The antipathy for the animal in man in Darwin's *Beagle* diary may be an artifact of the discourse of civilisation and savagery that structured so many voyage accounts in this period, as well as the more recent imperialist framework of British natural history, in which exotic animals were often aligned with native peoples or territories as objects of possession and mastery.[16] The several decades before and after the *Beagle* voyage saw an enormous growth in zoological collections, public displays, and print. New insects, birds, fish, and mammals were added to Britain's growing empire of fauna. On returning to England, Darwin prepared his *Beagle* diary for publication, and began to establish connections with leading experts and institutions in British natural history. He divided his many specimens among the specialists in different fields, taking charge of the animals himself by editing the voyage's official *Zoology*. Darwin obtained public money for the work, emphasising its value for the nation. The lavish volumes contained a menagerie of exotic animals, individually described and hand-colored, new creatures that Darwin and his navy colleagues had captured for the British Admiralty and reading public.[17] Darwin's name was occasionally attached by the taxonomist to the discovery – a cat (*Felis darwinii*), a mouse (*Mus darwinii*) and the aforementioned ostrich (*Rhea darwinii*).

In the same years, immediately following the voyage, Darwin's work on animals entered a more private space: a series of notebooks in which he began to speculate on the nature of species and development. The conventional abstractions and juxtapositions of human and animal were largely absent here, replaced by new combinations and relations as Darwin investigated the particularity of species on islands, or the genealogy of extinct forms. Animals became problems of classification, spatial distribution, and position in time: Galapagos finches and iguanas, megatherium, armadillos, and sloths. Animals and humans intertwined in questions about sterility ('Eunuchs...cut stallions...nuns') and the production of hybrids: domestic and wild rabbits may interbreed, 'White men & Hottentots or Negros cross'.[18] If species were mutable, then new questions were raised about directionality and hierarchy. What set humans apart? 'Each species changes. Does it progress? Man gains ideas. the simplest cannot help'.[19] As many scholars have noted, Darwin came to disregard the 'human' as a standard of comparison or fixed vantage point as he speculated about species, though he admitted privately: 'difficult for man to be unprejudiced about self'.[20]

In bridging the emotional divide between humans and animals, another notebook of observations was perhaps more important. In 1839, Darwin's first child (William) was born, and he recorded the infant's first movements and expressions, continuing through early childhood. As more children were born (about one per year for the next ten), more observations were added and similarities or contrasts noted. Darwin observed his children as he might an animal, this in striking contrast to the more personal approach of his wife, Emma, who also made observations in the same notebook. For Darwin, the children were de-familiarised, described sometimes as 'it' or as sentient bodies. He wrote of William in his first weeks: 'cries, squalls, jerks its foot when touched'.[21] The children were animalised in another way, through a series of comparisons that Darwin drew between human and animal behaviours. William playing with a prickly hair brush, 'pawed at it, like a dog at a hot coal'. The boy cried for food like nestling birds. He tried to slap his sister's face when she took his cake, 'like just-born crocodile from egg, learns to snap'.[22] Darwin was already making visits to the Zoological Gardens in Regent's Park, where wild and exotic animals were available to the British public for amusement and education. But unlike many of his countrymen, he did not go to the zoo to see animals as surrogate savages or trophies of captured territories; nor to see primates dressed like English ladies, sipping cups of tea. As a member of the Zoological Society, Darwin was given special privileges and allowed to experiment with some of the animals. He placed a looking glass on the floor of the orang-utan cage and watched as the creatures approached to gaze at their image, their lips protruding as if to kiss it.[23] Back at home in the children's nursery, Darwin observed William, now one year of age, kiss himself in the glass and press his face against his image 'very like Ouran Outang'.[24]

Darwin travelled from the nursery to the zoo and back, experimenting in each sphere, trying to trace the origins of the human kiss in an animal grimace, the same expression monkeys made when angry, sulky, or pleased. In one of his species notebooks, he wrote: 'ones tendency to kiss, & almost bite, that which one sexually loves is probably connected with flow of saliva'. Later he would describe, more tenderly, female dogs licking their puppies, 'the dearest object of their love'.[25] Love itself was rooted in animal affections, expressed through bodily contact, touching, rubbing, and stroking, in the act of procreation, the nursing and tending of children, and a lover's caresses. Unlike the animality that Darwin observed in the South American natives and slaves, the behavioural resemblance of his own children to monkeys or dogs, birds or reptiles, brought no revulsion. To see the child as an animal, or the animal in the child, was to see progressive stages of development, not a loss of civility. Their habitat was the home, full of familial affection. Darwin's children were now evidence in support of a theory of transmutation. In such conditions, the children would thrive, gain in strength and assertiveness, as well as tenderness. Darwin continued his experiments with the kiss. What happened when affection was extended and then withdrawn? Could the little fellow sense the needs of another and fulfil them? William, he discovered, did not like his father saying 'Wont kiss Doddy' and so Darwin repeated the reprimand in a reproving voice, adding: 'Doddy wont give poor Papa a kiss, – naughty Doddy'.[26] The boy then protruded his lips, shook his hands angrily, and in so doing, drew his father toward him to receive affection. At two years and three months, Darwin witnessed the dawn of sympathy.

Instinctive Sympathy

Darwin's programme of animal studies, embracing apes, infants, savages, and mammals of all sorts, extended also to moral philosophy. In a set of notes on the work of James Mackintosh, made about the same time as the birth of his first son, Darwin briefly sketched the origins of the moral sense from the instinctive sympathy of animals.

> Looking at Man, as a Naturalist would at any other Mammiferous animal, it may be concluded that he has parental, conjugal and social instincts, and perhaps others.... These instincts consist of a feeling of love & sympathy or benevolence to the object in question. Without regarding their origin, we see in other animals they consist in such active sympathy that the individual forgets itself, & aids & defends & acts for others at its own expense. –[27]

Robert Richards has discussed in detail how Darwin's moral theory developed together with his study of instincts in different animal species, especially social insects, eventually leading to the concept of group selection, the process whereby characteristics of no direct benefit to the individual may nonetheless favour the community as a whole and thus be preserved.[28] Richards, like many who have examined Darwin's moral theory, focuses on the origins of altruism: did Darwin succeed in giving a naturalistic foundation for altruistic behavior? is such altruism only selfishness in disguise? As Thomas Dixon has recently noted, however, Darwin's moral theory was not strictly about altruism, but rather about sympathy and love.[29] The discourse of sympathy on which Darwin drew, especially in Scottish moral philosophy, was quite different from the philosophy of altruism as it had been formulated by August Comte and some of his English commentators. Altruism was defined as self-abnegation and the dissolution of the individual into the social, whereas sympathy for Smith, Hume, and other eighteenth-century writers was a process of extending or enlarging the self through bonds of obligation and affection.[30]

For Smith, one of Darwin's chief reference points for moral theory, sympathy was largely an imaginative projection: 'as we have no immediate experience of what other men feel, we can form no idea of the manner in which they are affected, but by conceiving what we ourselves should feel in the like situation'.[31] But there were other models of sympathy in the eighteenth century that emphasised more immediate, affective bonds. Darwin found Smith's rarified theory inadequate because it did not explain the emotional qualities of sympathy. Why, for example, was sympathy stronger between loved ones? Why was the discharge of sympathy a pleasurable release? In his reading notebook he remarked of Smith: 'very unsatisfactory because does not like Burke explain pleasure'.[32] Darwin read Edmund Burke's *Essay on the Sublime and Beautiful* and highlighted passages that described the delight people took in the misfortunes of others: theatrical tragedies, natural disasters, public executions. Unless accompanied by some degree of pleasure, Burke argued, sympathy would not be efficacious, for otherwise people would be so overwhelmed by pain on witnessing a calamity that they would simply turn away or flee rather than offer assistance.[33] Burke's attention to the delights of sympathy agreed with what Darwin had observed in his own family, 'the extreme pleasure children show in the naughtiness of brothers'.[34]

Burke's view was also consistent with Darwin's approach to sympathy as an animal instinct. For Burke, sympathy was a mechanism of fellow-feeling that had been designed by a benevolent God to secure the bonds of human society. But the same providential approach to behaviour was a feature of much natural historical writing on animals: the sagacity of the beaver in building its dam, the cell-making powers of hive bees, the navigational abilities of migrating birds. In William Paley's *Natural Theology*, such instincts were described as wonderful or beautiful 'contrivances', the behavioral equivalent of anatomical structures such as the hand or the eye.[35] As Darwin developed his species theory, he struggled to explain these marvellous instincts, devoting a chapter in the *Origin of Species* to the problem. He argued that most instincts could be explained by the operation of natural selection because, like other organismal structures, they had a tendency to vary, they conferred advantages or disadvantages, and they were heritable. Darwin did not discuss sympathy or social instincts, per se, in *Origin*, but he did describe certain behaviours in social insects that were beneficial to the group, rather than to the individual, thus implying a notion of the community good.

When Darwin eventually published his theory of the moral sense in *Descent of Man*, a great number of animals were found to have instinctive sympathy. 'It is curious to speculate on the feelings of a dog', Darwin wrote, 'who will rest peacefully for hours in a room with his master or any of the family, without the least notice being taken of him; but if left for a short time by himself, barks or howls dismally'.[36] Horses, sheep, and cows were similarly miserable when separated from their 'companions', affectionate on their reunion, or mournful upon their death. Animals living in groups performed a variety of services for each other, grooming, removing parasites, thorns or burs. 'Many animals...certainly sympathise with each other's distress or danger', Darwin remarked.[37] They hunted together and defended one another, positioning themselves as sentinels, giving warning signs at the approach of a predator. Some animals looked after the weak and helpless. Blind pelicans and crows were fed by their fellow birds, elder baboons tried to save infants of the group from attack. Such social feelings, Darwin suggested, probably evolved from the parental and filial affections, and conferred great advantages to the animal group.

Darwin's discussion of animal sympathy in *Descent* built upon his extended argument about the kinship of humans and animals. Darwin gradually established a 'community of descent' through structural homologies, shared diseases, bodily responses to various substances (monkeys who enjoy rum and tobacco), embryology, and common behaviors such as courtship and nurturing. Examples of emotional affinity abounded: the pleasure of puppies, kittens, and lambs at play, ants that chased each other and pretended to bite, the intense grief of female monkeys on the loss of their young, their keenness to adopt orphans, even of other species. The full range of 'human' emotions – suspicion, courage, sulkiness, jealousy, pride, wonder, curiosity – were exhibited in turn, some by example, others by simple assertion: 'the fact … is so well established' Darwin remarked, 'it is almost superfluous to state …'.[38] For Darwin's wide readership, the emotional lives of animals would have been familiar from a variety of sources, travel writing, natural histories of animal character, zoology and animal husbandry, children's stories and fables, the growing culture of pet keeping. The sweetness, gentleness, and affection of household animals, the bonds of friendship and companionship that they could form with their owners, were described in loving detail in letters, handbooks, treatises on various breeds, even occasionally in works authored by the creatures themselves. In such literature and culture, the correspondence between human and animal sensibility was often remarkable: animals who recalled and rejoiced in the company of their masters' friends, animals who sniffed out sexual impropriety and punished the perpetrator, animals who displayed such religious devotion that they attended church, or tried to convert their masters to Methodism by disrupting communion.[39]

Reviews of *Descent* show that many readers were charmed by Darwin's examples, which drew on this common currency of stories, anecdotes, and experiences in rural life and the home. Yet most rejected his theory that the moral sense was rooted in the instinctive sympathy of animals. Darwin had anticipated some hostility, writing to his friend Joseph Hooker, 'I shall be well abused, for as my son Frank says, 'you treat man in such a bare-faced manner'.[40] An anonymous review in *The Times* condemned the book for undermining the 'elementary principles of duty'; the affinity between humans and animals extended only to the 'flesh', and thus not properly to 'man' at all, who was a 'spiritual inhabitant' in an 'animal abode'.[41] Darwin condemned the author of the review as 'a windbag full of metaphysics and classics'.[42] But the windbag had a point. The superior status attributed to 'higher' mental powers and productions (metaphysics, classical literature, and moral philosophy, for example), was potentially undermined by Darwin's account of human virtue as an animal emotion.

Moral Mastery

At a late stage in his work on *Descent,* Darwin read a bit of Kant. He remarked to a correspondent on the contrast between the 'great philosopher looking exclusively into his own mind', and himself 'a degraded wretch looking from the outside thro' apes & savages at the moral sense of mankind'.[43] Darwin referred to the philosopher briefly in *Descent*, acknowledging Kant's view of the imperious 'ought', a 'law in the soul' that resides in every man, and before which 'all appetites are dumb'.[44] But Darwin elided the distinction between Kant's rational and deliberative structure of ethics and one based on immediate emotional impulse: the 'deep feeling of right or duty' that impelled a creature to act 'without a moment's hesitation'. Surely the categorical imperative resided, for example, in dogs? 'We hardly use the word ought in a metaphorical sense', Darwin suggested, 'when we say hounds ought to hunt...if they fail thus to act, they fail in their duty and act wrongly'.[45] Kant had been recommended to Darwin by Frances Power Cobbe, a great lover of dogs, an author of works extolling their virtues, and an activist on their behalf against human cruelty. But Cobbe's enormous regard for dogs rested on their being lower creatures. She was willing to grant them an abundance of 'natural sympathy'. But the possession of sympathy did not make dogs moral, for they had no choice in the matter. In a critical review of *Descent*, she reaffirmed the essential distinctness of man, rooted in his fallenness and fallibility, his need for absolute moral principles that had no equivalent in the animal world.[46]

Cobbe's criticism rested on a widespread distinction between animal instinct and human volition. Leading works in human and comparative physiology described a chain of being extending from lower animals who were ruled exclusively by instinct, to man who alone possessed the power of will.[47] William Carpenter, Thomas Laycock, Henry Maudsley and other physiologists, also wrote in detail about human and animal emotional response as an elaborate reflex mechanism. Emotion, regarded as a 'lower' function of the body, was thus accorded a subordinate place in relation to the will, analogous to the ranking of animals in relation to man.[48] The analogy was in fact explicitly drawn by Carpenter: man had the ability to 'keep his appetites and passions under subordination to his higher nature', just as the 'trained horse' responded to 'the will of its master'.[49] Darwin's research on emotional response was heavily indebted to the physiology of Carpenter and others. This debt was most explicit in Darwin's detailed study of expression, which presented sympathy, together with the whole emotional fabric of man and animals, as a series of reflex mechanisms.[50] But if, according to Darwin's theory, human moral behaviour was essentially like that of animals, an instinctive emotional response, what was there to take the place of the master in man? Or was it enough to be sympathetic and caring without hesitation, and whenever the circumstances demanded it?

The question of moral mastery was taken up in a lengthy correspondence between Darwin and his cousin, Hensleigh Wedgwood, who disputed Darwin's account of the origin of conscience. Once again, the moral subject was a dog. 'A pointer', Darwin had written in *Descent*, 'if able to reflect on his past conduct would say to himself, I *ought* (as indeed we say for him) to have pointed to that hare and not have yielded to the passing temptation of hunting it'.[51] Darwin argued that dogs possessed sufficient memory, probably in the form of 'images of past actions and motives', to recall wrongful behaviour to mind, and such recollection would give rise to feelings of dissatisfaction owing to the persistence of social instincts. Conscience thus arose through a conflict between enduring social feelings and more fleeting desires such as hunger or the urge to give chase. Wedgwood, however, denied that a simple comparison between instincts, one enduring and one fleeting, could produce regret or remorse, for neither of the behaviours (pointing or hunting) carried any moral stigma in itself. The true essence of conscience, according to Wedgwood, was shame, and this indeed was manifest when a dog disobeyed its master and was rebuked. 'He knows that he has not done what he had to do...to gain his masters approbation', Wedgwood insisted, 'the sense of the disapprobation of a mind to which we look with reverence or respect is shame'.[52] Wedgwood went so far as to attribute 'a distinct sense of sin' to the naughty dog.[53] In the case of humans, the role of the master was played by a community of moral observers, a tribunal who presided over one's actions and whose approval or disapproval was registered in the pangs of conscience.

A debate with Wedgwood unfolded in a series of letters, with Darwin reaffirming his belief that moral impulses were fundamentally automatic. 'It seems to me', he replied, 'that there is an instinct to aid our fellows as blind as when a beaver sees a rill of water in a court yard where he is confined & tries to dam it'. Darwin had already considered the role of praise and blame in *Descent*, and given it a subordinate function to that of instinctive sympathy. Wedgwood's community of approval was in effect a version of Adam Smith's impartial spectator, which Darwin had long ago rejected as the basis of conscience. The desire for approval, Darwin wrote to his cousin, often determined 'what line of conduct shall be followed', but it was not the chief motive for such conduct. Humans sought the approbation of others because they were social animals. The value attached to public opinion, Darwin insisted, was itself based on 'enduring social instincts which include sympathy'.[54] Darwin thus inverted the hierarchy in which will, abstract principles, and the judgment of superiors served as moral masters, insisting instead on a government from 'below'.

In the course of defending himself against Wedgwood, Darwin frequently took up the position of his cousin, placing himself in question. He admitted to being 'muddled' about the word 'ought'. He wished he had said more about shame. In a postscript, he remarked parenthetically on how Wedgwood's comments have shaken him: '(I feel much doubt, but yet cannot quite give in)'. Was Darwin just being polite? In one particularly long letter he alludes to the pain of disagreement and the pleasure of consensus, trying to establish common ground with his cousin, often through considerable labour: 'we seem to agree about...' he writes, then 'I think we also agree about...' then again, 'am I right in supposing that we agree about...' and finally, 'every point of agreement is a satisfaction'.[55] Such letters, while rendering sympathy an object of philosophy and science, also allow a glimpse of sympathy as a social practice. They suggest how sympathy might operate within a community of readers and critics, some of whom were also members of a family circle. Darwin's correspondence with Wedgwood was circulated within the family and commented upon by his wife and children. Darwin's eldest daughter Henrietta, who had worked extensively on drafts of the chapters of *Descent* on mental and moral development, remarked that the two men seemed to pay too little heed to each other's arguments, while his wife could not abide his moral theory: 'I am something of a traitor in the camp'.[56] Because his work was situated in the home, because he relied extensively upon friends and family for scientific assistance, Darwin was perhaps unusually mindful of this immediate circle of readers to whom he was emotionally bound. In letters he revealed his acute desire to be in accord with those close to him, and he struggled to sustain this community of sympathy amidst marked differences.

In the early modern period, when affects were conceived as substances that could inhere in physical objects (letters, books), there was a material model for the commerce of feeling mediated by 'text': the 'ink of sympathy'. The transmission of feeling between author and reader was a crucial feature of the epistolary novel, embodied in characters whose bodies and souls could cross through correspondence as through conversation or touch. A further parallel could be drawn with the Victorian novel, often regarded as an organ of sympathy, seeking to activate and extend the affections of readers across increasingly pronounced divides of class, education, and circumstance.[57] We might not expect a 'scientific' text to engage the sympathies of readers, yet Darwin's manner of correspondence was not unlike his style of authorship. *Descent of Man* followed the pattern of *Origin of Species*, with its frequent shifts from a highly personalised narrator to the position of a critical reader, whose points were then addressed, deflected, or sometimes conceded. It is tempting to regard Darwin's conciliatory style as a strategy for commanding the reader, his professed humility as a pose, displaying the caution and equivocation appropriate to sound scientific method, diplomatically avoiding controversy, retaining respectability as a gentleman, and so forth. Darwin occasionally remarked to friends that his works had 'shaken' readers, 'converted' them to his cause.[58] Yet his writing enacted sympathy by surrendering ground to the reader, entertaining assumptions, introducing doubt and uncertainty that was then filled by a more open, speculative discourse. Sometimes he created fictitious characters, the bee who reflects on its conduct, the dog who utters the word 'ought', inviting readers to bridge the difference between species, but not always in ways that were comforting and familiar. In his long discussion of language, Darwin worked from both sides to draw the animal and human together, enabling readers to sense the material connections between species by tracing the origins of human speech in animal sounds, the animal cries that humans still made at moments of intense feeling, the ability of animals to learn new sounds, to mimic human voices, to communicate even across species boundaries.

Some friends and supporters urged Darwin to adopt a more assertive style of writing, claiming that in giving such consideration to readers, he lent too much strength to his 'opponents'.[59] But Darwin rarely opposed his critics directly, choosing instead to incorporate them, opening his text to other voices, quoting from letters, conversations, and other authors, enlarging and revising his argument in turn. The form helped to bind readers to him through the promise of participation, readers who would later become devoted correspondents (rather than 'converts'), offering more contributions to be included in a future work or edition, despite substantial theoretical differences. The promise was extended to Wedgwood, 'In a 2^d^. Edit. I might say from several discussions & correspondence with H.W. I have modified my views –'.[60] The kind of fellow-feeling that informed Darwin's respectful acknowledgement of his cousin could help sustain a dynamic process of exchange and incorporation, transformative of both author and reader.

Metamorphosis

In his chapter on the moral sense in *Descent*, Darwin remarked briefly on sympathy across the species divide, and especially of human sympathy for animals. It involved a widening of the circle of affections that began with parent and child and moved outward from the immediate family to other kinship relations, the local community, the nation, all of mankind, and only then to other species.[61] This form of sympathy for animals, designated 'humanity', was not an immediate emotional impulse, but rather a product of abstract thought, and of comparatively recent acquisition. Darwin followed several contemporary philosophers in regarding it a mark of human progress.[62] Darwin also aligned it with the kindness and generosity extended to 'lesser' members of society, to imbeciles and the sickly, noting that animal and savage societies often banished the weak or left them to die. Here then was a distinctly human and philanthropic form of sympathy, very familiar to Victorians, and deeply entrenched in Victorian attitudes toward the poor and unfortunate, the victims of cruelty and abuse. Darwin seems to return to that framework of fellow-feeling in which he had first regarded slaves and Jemmy Button, a form of sympathy that extended to social lessers who were nonetheless worthy of kindness and generosity if they conformed to the standards of English civility. This form of sympathy, based on the one hand on likeness and conformity, and on the other hand on human superiority and English condescension, seems in marked contrast to the instinctive sympathy that had passed from animals to man in the course of evolutionary history.

But if Darwin's moral theory was correct, then the social feelings that were at the heart of moral behaviour in animals had been to a considerable degree eroded in man. Humans relied upon language, abstract thought, and their by-products (laws, religions, philosophies), because in the course of evolution they had lost many of the special instincts that directed most animal activity.[63] Their social feelings were more circuitous and flawed. Selfish impulses often prevailed, and moral character had to be rebuilt through education, repetition, and other devices such as praise and blame. Eventually habits were formed that enabled humans to act morally without hesitation. These habits could in turn be inherited, and thus revert to the condition of instincts. Indeed, the main role of the 'higher faculties' in shaping human moral behaviour was to return man to an instinctive condition. Darwin thus seems to offer two versions of sympathy, one typified by the highly civilised European (ideally English) gentleman and lady, creatures of surpassing will, charitable to the poor and kind to (lower) animals. Another embodied in various animals themselves, for whom the discharge of sympathy and love was purely instinctive. Perhaps at some more advanced period of history, man would regain what he had lost in his ascent, surpass that stage of evolutionary ethics known as 'humanity', and come again to resemble his animal ancestors in moral perfection.

In a recent work on the history of sympathy, Jonathan Lamb looks to Darwin's theory of evolution as providing a new foundation for an expansive and transformative kind of emotion that could bind humans with other forms of life precisely by rendering the human (together with all species) more provisional and mutable.[64] The point is suggestive but it needs clarifying. Darwin's evolutionary theory, which emphasised the artificiality of species distinctions, the graduation of structures, and the conversion of one form into another, required great reaches of time to work its transformations. But what sort of metamorphosis was effected by sympathy? When extended to ethics, Darwin's evolutionary theory implied a model of identity and community that was radically open. The Darwinian 'individual' was not a Cartesian self, a mind enclosed and autonomous even from its own body; but rather a material compound, an accretion of structures, impulses, and behaviours evolved from and shared with other species.[65] What then are Darwinian 'communities'? Darwin's great and only moral principle – the community good – nowhere prescribes what counts as a community. Is it a family, a species group, an animal pack? Yes, but not only these. The boundaries of community are not ultimately determined by biological kinship. Darwin gives a number of examples of bonds across the species divide: dogs and cats, monkeys and their keepers. What makes communities is sympathy, and sympathetic communities are dynamic. Sympathy is strong where identity is weak. Sympathy provides an opening, a vulnerability to the influences of others, to change, and so can help build communities that are not based on likeness and conformity alone.[66]

In his famous essay, *The Animal That Therefore I Am*, Derrida encounters his cat in the kitchen and is disarmed by its indifferent gaze, prompting some very long reflections on how animals have been disregarded by modern philosophy. In the tradition usually traced from Descartes to Agamben by way of Heidegger, man is held apart from and opposed to animals who have no soul, no reason, no history, no progress (or in Heideggerian terms, no negation, no profound boredom).[67] Man only becomes human by splitting himself and opposing the animal in him, transcending it, destroying it, or as Victorians would say, mastering it through the will. Derrida links these denigrating presumptions about the animal to the erasure of human sympathy for those myriad creatures who suffer on a massive scale on our behalf, for our food, health, beauty, and knowledge: 'men do everything they can to dissimulate this cruelty'.[68] Darwin remarked on the capacity of animals to love in the face of such cruelty, and of humans to silence their sympathies in the course of animal study: 'everyone has heard of the dog suffering under vivisection, who licked the hand of the operator; this man, unless he had a heart of stone, must have felt remorse to the last hour of his life'.[69]

Yet Derrida, like a number of other writers on 'the question of the animal', seems to continue the instrumental approach in which nonhuman creatures are considered only in the abstract (as wild, non-individuated, beyond good and evil and so forth), and so remain objects of pity, or tools for deconstructing the human.[70] It is ironic that in their aversion for biology and its alleged reductivism, many posthumanist authors on the 'animal' fail to engage with animals or animal–human relations at all. Poetry and the novel may offer alternatives to this metaphysics and morality of abstraction. In J.M. Coetzee's *Elizabeth Costello*, the heroine (a novelist on an interminable lecture tour), has recourse to poems by Ted Hughes, and to her own feelings for animals as they suffer and die, in order to render the experience of animal bodies, and the bodily connection between humans and animals, in emotional terms that the academic philosophers who interrogate her fail to register.[71] A different approach, in some ways closer to Darwin's own method of animal studies, is offered by Donna Haraway in her recent work on 'companion species'. Haraway examines a variety of social settings in which human–animal relations are configured (breeding, farming, the laboratory, sport, pets), but in ways that are surprisingly 'open'. She offers a highly personalised narrative of agility training that she undertook with her dog, a process she characterises as 'transforming partnership', for the training she insists is mutual, and involves not just learning about and feeling with the animal, but evolving as a person. The evolutionary model for Haraway is co-adaptation, cooperation, and cross- or multi-species organisms, a model increasingly prevalent in ecology and post-genomic biology.[72]

It is doubtful whether Darwin, at least in his older and notoriously sickly years, could have survived an agility course. But he was, from an early age and to the end of his life, almost always in the company of animals. He had long imagined the human in animal form, what it would be like to inhabit their bodies, their experiences of love and beauty, their sublime pleasures of sense, and more sober feelings of conscience. What kind of sympathy was this, that could allow a man to become an animal without degrading him? An animal with whom he might work or play; an animal to think and grow old with, like Darwin in his study, or on his daily walk round the 'thinking path' with his beloved terrier, Polly.

Notes

1 Letter from D. Thomas, [after 11 March 1871], in Frederick Burkhardt et al (eds), *The Correspondence of Charles Darwin*, 20 vols (Cambridge: Cambridge University Press, 1985–2011), xix. 169.

2 *Illustrated London News*, 22 March 1871. For Darwin portraits and cartoons, see Janet Browne, 'Looking at Darwin: Portraits and the Making of an Icon', *Isis*, 100 (2009), 542–70; and Constance Areson Clark,

'"You are here": Missing Links, Chains of Being, and the Language of Cartoons', *Isis*, 100 (2009), 571–89.

3 Darwin, *Life*, 1838, in James A. Secord (ed.), *Charles Darwin: Evolutionary Writings* (Oxford: Oxford University Press, 2008), 351–4; and Darwin, *Recollections of the Development of my Mind and Character*, 1876, in Secord, 2008, 356.

4 Darwin, 1876, 378.

5 Darwin, *Journal of Researches into the Geology and Natural History of the Various Countries Visited During the Voyage of H.M.S. Beagle* (London: John Murray, 1839), 108–9, 465.

6 R.D. Keynes (ed.), *Charles Darwin's Beagle Diary* (Cambridge: Cambridge University Press, 1988), 42. On Darwin's emotional engagement with nature, see David Kohn, 'The Aesthetic Construction of Darwin's Theory', in Alfred I. Tauber (ed.), *The Elusive Synthesis: Aesthetics and Science* (Dordrecht: Kluwer, 1996), 13–48; and Paul White, 'Darwin, Concepción, and the Geological sublime', *Science in Context,* 25 (2012), 49-71. DOI: http://dx.doi.org/10.1017/S0269889711000299.

7 Darwin, 1839, 323–4, 26–8.

8 Burkhardt *et al,* 1985–2011, i. 312–13, 320. On Darwin's extreme antipathy for slavery, see Adrian Desmond and James Moore, *Darwin's Sacred Cause: Race, Slavery and the Quest for Human Origins* (London: Allen Lane, 2009).

9 Darwin, 1839, 27–8; see also Janet Browne, *Charles Darwin: Voyaging* (London: Jonathan Cape, 1995), 214.

10 Darwin, 1839, 235.

11 Browne, 1995, 245.

12 Darwin, 1839, 204, 325.

13 Darwin, *Journal of Researches into the Natural History and Geology of the Countries Visited During the Voyage of H.M.S. Beagle Round the World*, 2nd edn (London: John Murray, 1845), 228.

14 Gillian Beer, 'Travelling the Other Way', in N. Jardine, J.A. Secord, and E.C. Spary (eds), *Cultures of Natural History* (Cambridge: Cambridge University Press, 1996), 332–3; Browne, 1995, 249–50; and Desmond and Moore, 2009.

15 Darwin, 1839, 239.

16 On civilisation and savagery in travel writing, see Mary Louise Pratt, *Imperial Eyes: Travel Writing and Transculturation* (London: Routledge, 1992). On animals as emblems of empire, see John M. MacKenzie, *The*

Empire of Nature: Hunting, Conservation and British Imperialism (Manchester: Manchester University Press, 1998); and Harriet Ritvo, *The Animal Estate: The English and Other Creatures in the Victorian Age* (Cambridge, MA: Harvard University Press, 1987).

17 Darwin, *The Zoology of the Voyage of H.M.S. Beagle*, 5 vols (London: Smith Elder and Co, 1839–43).

18 Paul H. Barrett, Peter J. Gautrey, Sandra Herbert, David Kohn and Sydney Smith (eds), *Charles Darwin's Notebooks 1836–1844* (London: British Museum (Natural History), 1987), 171, 178, Notebook B.

19 *Ibid.*, 175.

20 *Ibid.*, 182.

21 Burkhardt *et al*, 1985–2011, iv. 418. On Darwin's emotional management, see George Levine, *Darwin Loves You: Natural Selection and the Re-enchantment of the World* (Princeton: Princeton University Press, 2006), 230–51; and Paul White, 'Darwin's Emotions: the Scientific Self and the Sentiment of Objectivity', *Isis*, 100 (2009), 811–26.

22 Burkhardt *et al*, 1985–2011, iv. 420.

23 Darwin, *The Expression of the Emotions in Man and Animals* (London: John Murray, 1872), 140.

24 Burkhardt *et al*, 1985–2011, iv. 419.

25 Barrett *et al*, 1987, 574, N Notebook; Darwin, 1872, 120.

26 Burkhardt *et al*, 1985–2011, iv. 419.

27 Barrett *et al*, 1987, 619.

28 Robert J. Richards, *Darwin and the Emergence of Evolutionary Theories of Mind and Behavior* (Chicago: University of Chicago Press, 1987), 110–22, 142–52.

29 Thomas Dixon, *The Invention of Altruism: Making Moral Meaning in Victorian Britain* (Oxford: Oxford University Press, 2008), 131–3.

30 On models of sympathy in the eighteenth century, see Jonathan Lamb, *The Evolution of Sympathy in the Long Nineteenth Century* (London: Pickering and Chatto, 2009).

31 Adam Smith, *The Theory of Moral Sentiments*, 6th edn (1790; Oxford: Clarendon Press, 1976), 9.

32 Barrett *et al*, 1987, 546, Notebook M.

33 Edmund Burke, *A Philosophical Enquiry into the Origin of Our Ideas of the Sublime and the Beautiful*, 2nd edn (1759; London: Routledge & Kegan Paul, 1958), 44–8.

34 Barrett *et al*, 1987, 531, Notebook M.

35 William Paley, *Natural Theology or, Evidence of the Existence and Attributes of the Deity, Collected from the Appearances of Nature*, 15th edn (London: F.C. and J. Rivington; 1815), 299–319.

36 Darwin, *The Descent of Man, and Selection in Relation to Sex*, 2 vols (London: John Murray; 1871), i. 74.

37 *Ibid.*, i. 77.

38 *Ibid.*, i. 39, 45.

39 On animal autobiography, see Frances Power Cobbe, *The Confessions of a Lost Dog* (London: Griffith and Farran, 1867); on animal piety, see George R. Jesse, *Researches in the History of the British Dog* (London: Hardwicke, 1866), 1, 132–3; on animals as friends and companions, see W. Elwin, 'Blaze's History of the Dog', *Quarterly Review*, 72 (1843), 488–516; Mary Russell Mitford, *The Life of Mary Russell Mitford: Related in a Selection from Her Letters to Her Friends* (London: R. Bentley; 1870), 1: 207; and William Youatt, *The Dog* (London: Charles Knight, 1845).

40 Letter to Joseph D. Hooker, 21 January [1871], in Burkhardt *et al,* 1985–2011, xix. 21.

41 *The Times*, 8 April 1871, 5.

42 Letter to John Murray, 13 April [1871], in Burkhardt *et al,* 1985–2011, xix. 281

43 Letter to Frances Power Cobbe, 23 March [1870?], in Burkhardt *et al,* 1985–2011, xviii. 81–2. The book Darwin read was Immanuel Kant, *Metaphysics of Ethics*, trans. by J.W. Semple (Edinburgh: T. Clark, 1836).

44 Darwin, 1871, i. 70–1.

45 *Ibid.*, i. 92

46 Frances Power Cobbe, 'Darwinism in Morals', *Theological Review*, 8 (187), 167–92.

47 William B. Carpenter, *Principles of Human Physiology*, 5th edn (London: Churchill, 1855); William B. Carpenter, 'The Physiology of the Will', *Contemporary Review*, 17 (1871), 192–217; and George Rolleston, *Forms of Animal Life* (Oxford: Clarendon Press, 1870).

48 William B. Carpenter, *Principles of Mental Physiology* (London: Henry S. King & Co, 1874); Thomas Laycock, *Mind and Brain: or, the Correlations of Consciousness and Organisation*, 2 vols (Edinburgh: Sutherland and Knox, 1860); Henry Maudsley, *The Physiology and Pathology of Mind* (London: Macmillan and Co, 1867). On the bodily and moral hierarchies imposed by reflex physiology, see especially, Roger Smith, *Inhibition: History and Meaning in the Sciences of Mind and Brain* (Berkeley: University

of California Press, 1992).

49 Carpenter, 1874, 106.

50 Darwin, 1872. Darwin intended to include this material in *Descent*, but for reasons of size it had to be published as a separate volume the following year. On Darwin's use of reflex physiology, see Paul White, 'Darwin Wept: Science and the Sentimental Subject', *Journal of Victorian Culture*, 16 (2011), 195–213.

51 Darwin, 1871, ii. 392. On Victorian dog training, see *The Dog, in Health and Disease* (London: Longmans, Green, Reader & Dyer, 1867).

52 Hensleigh Wedgwood to Darwin, [before 3 March 1871], in Burkhardt *et al*, 1985–2011, xix. 124.

53 Wedgwood to Darwin, [3-9 March 1871], in Burkhardt *et al*, 1985–2011, xix. 133.

54 Darwin to Wedgwood, 3 March [1871], in Burkhardt *et al*, 1985–2011, xix. 133.

55 Darwin to Wedgwood, 9 March [1871], in Burkhardt *et al*, 1985–2011, xix. 149.

56 On Henrietta Darwin's remark, see Wedgwood to Darwin [after 9 March 1871]; and Emma Darwin to Frances Power Cobbe, [25 February 1871], in Burkhardt *et al*, 1985–2011, xix. 106.

57 On the ink of sympathy, see Lamb, 2009, 50–1, 98–100; on sympathetic reading practices in the eighteenth century, see Ildiko Csengei, '"I will not weep": Reading through the Tears of Henry MacKenzie's Man of Feeling', *Modern Language Review*, 103 (2008), 952–68; on sympathy in the Victorian novel, see Rae Greiner, 'Sympathy Time: Adam Smith, George Eliot, and the Realist Novel', *Narrative*, 17 (2009), 291–311; and Marianne Noble, 'The Courage to Speak and Hear the Truth: Sympathy and Genuine Human Contact in Harriet Beecher Stowe's The Minister's Wooing', *New England Quarterly*, 81 (2008), 676–702.

58 For example, letters to Jean Louis Armand de Quatrefages, 25 April [1861] and Joseph D. Hooker, 19 January [1865], in Burkhardt *et al*, 1985–2011, ix. 102 and 13: 29.

59 For example, letter from Ernst Haeckel, 12 May 1867, in Burkhardt *et al*, 1985–2011, xv. 254–7.

60 Draft of letter to Hensleigh Wedgwood, Darwin Archive, Cambridge University Library, DAR 88:67.

61 Darwin, 1871, i. 101–4.

62 William Edward Hartpole Lecky, *History of European Morals*, 2 vols, 3rd

edn (London: Longmans, Green and Co, 1877), ii. 161-–77.

63 Darwin, 1871, i. 37–8.

64 Lamb, 2009, 11.

65 On the composite nature of subjectivity that emerges in Darwin's work on emotional expression, see Tiffany Watt-Smith, 'Darwin's Flinch: Sensation Theatre and Scientific Looking in 1872', *Journal of Victorian Culture*, 15 (2010), 101–17.

66 See the discussion of different political models underpinned by sympathy, in Lamb, 2009.

67 Jacques Derrida, *The Animal that Therefore I Am* (New York: Fordham University Press, 2008) comments on Heidegger, Lacan, and Levinas. See also Giorgio Agamben, *The Open: Man and Animal* (Stanford: Stanford University Press, 2004); and John Berger, *Why Look at Animals?* (London: Penguin, 2009).

68 Derrida, 2008, 25. On the erasure of sympathy for animal suffering, see also Cary Wolfe, *Animal Rites: American Culture, the Discourse of Species, and Posthumanist Theory* (Chicago: University of Chicago Press, 2003).

69 Darwin, 1871, i. 40.

70 For example, Matthew Calarco, *Zoographies: The Question of the Animal from Heidegger to Derrida* (New York: Columbia University Press, 2008); and Gilles Deleuze and Félix Guattari, *A Thousand Plateaus: Capitalism and Schizophrenia* (London: Althone, 1987), 232–43.

71 J.M. Coetzee, *Elizabeth Costello* (London: Vintage, 2004); and the commentaries in Cora Diamond, 'The Difficulty of Reality and the Difficulty of Philosophy', in Stanley Cavell, Cora Diamond, Ian Hacking, John McDowell and Cary Wolfe, *Philosophy and Animal Life* (New York: Columbia University Press, 2008), 43–90.

72 Donna J. Haraway, *When Species Meet* (Minneapolis: University of Minnestota Press, 2008). On Darwin as a resource for non-reductive biology, see also Elizabeth A. Grosz, *The Nick of Time: Politics, Evolution, and the Untimely* (Durham, NC: Duke University Press, 2004); Elizabeth A. Wilson, *Psychosomatic: Feminism and the Neurological Body* (Durham, N.C.: Duke University Press, 2004); and Angelique Richardson, 'Darwin and Reductionisms: Victorian, Neo-Darwinian and Post-Genomic Biologies', 19: *Interdisciplinary Studies in the Long Nineteenth Century*, 11 (2010), http://19.bbk.ac.uk/index.php/19.

5

George Eliot, G.H. Lewes, and Darwin: Animals, Emotions, and Morals

Angelique Richardson

> 'a truth-seeker at the present day will do ill to turn a deaf ear to the teachings of science'[1]

George Eliot the novelist, G.H. Lewes the writer, man of science and drama critic, and Charles Darwin, naturalist, geologist, and Fellow of the Royal Society, read each other's work and became personally acquainted. Before the *Descent* was out they were calling at each other's houses. Following Lewes's warm and lengthy exposition of Darwin's development theory in the *Fortnightly Review*,[2] Darwin visited them in Regent's Park. He told Lewes that he had '*thoroughly* enjoyed my hour's talk with you & M[rs] Lewes'.[3] Later, Emma Darwin called on them and they accepted an invitation to lunch at Down.[4]

Eliot, Lewes and Darwin shared a deep interest in the material nature of mind, and the relations between humans and animals. This chapter draws on Eliot and Lewes's engagement with these subjects in their shared reading and writing over more than twenty years. Following close on the *Descent*, *The Expression of the Emotions in Man and Animals* had an even wider imaginative appeal and was acclaimed as more interesting than the latest novel (see chapter two). It was especially valuable to novelists, providing them with new ways of conceiving the moral sense and the emotional life, and of rethinking human–animal relations, and I suggest ways in which, in foregrounding instincts and unconsciousness, Darwin opened up new possibilities for fiction, undermining the epistemological certainty of the realist novel. For Darwin, the moral sense had its origins in social instincts and feelings.

Animals possessed qualities that could be seen as moral and were, Darwin argued, different not in kind but degree. Likewise, animals were crucial to Darwin's study of the emotions, demonstrating evidence of the physical basis of emotions. While Eliot and Lewes both maintained human distinction in their writing, Eliot saw a new kinship with animals, expressing this through anthropomorphism. Registering Darwin's growing emphasis on the physical basis of the emotions, and on the centrality of community to the origins of the moral sense, she increasingly gave a more physiological treatment to the emotions in her fiction, and her focus on unconscious, physically felt emotion integrated her characters into the natural world in a way that was not only novel but seen to be such by her readers.[5] This chapter also offers an alternative to the dominant reading of Gwendolen Harleth in *Daniel Deronda* (1876) as hysteric, arguing, instead, that Eliot provides in Gwendolen an embodiment of Darwinian expression in a hazardous environment.[6] It concludes that Eliot's revisions to Lewes's final work provided her with another space to reflect on animal kinship.

Pug Dog and Pig

In 1859, Eliot's publisher John Blackwood sent her a pug dog.[7] Eliot seems to have had a higher opinion than Lewes of its intellectual capacity. She wrote to Blackwood: 'I hope the photograph does justice to his physiognomy. It is expressive, full of gentleness and affection, and radiant with intelligence when there is a savoury morsel in question – a hopeful indication of his mental capacity'.[8] In the carte-de-visite photograph of Lewes and Pug (see Figure 5.1) Pug sits, anthropomorphically, on his own chair, next to Lewes. Eliot would write later to Blackwood: 'Since his introduction into literary society he has developed an aristocratic audacity and refinement, not to mention originality of genius, of which he before had no gleam', and, again, 'I am proud of the advance he has made in three months of our intellectual society'.[9]

Do these playful musings come from, and in turn bolster, a sense of a secure boundary between humans and animals – surely a dog would be impervious to intellectual society? Eliot is drawing here at least in part on Victorian conventions of talking about pets. But is she also, with a seriousness belied by the irony, questioning the boundar-

Figure 5.1: G.H. Lewes and Pug. Photograph, John & Charles Watkins (carte-de-visite), ca. 1864. Reproduced courtesy of National Portrait Gallery.

ies that Darwin would challenge in the *Origin* the same year? Darwin was similarly openly effusive about his dogs, declaring that he loved them 'with all my heart', and remarking to the mathematician William Pengelly 'I can believe almost anything about them'.[10] He drew on them in his work; his fox terrier Polly was the 'small dog watching a cat on a table' (see Figure 2.2 of this volume), its raised paw an illustration of the principle of serviceable habits: 'dogs of all kinds when intently watching and slowly approaching their prey, frequently keep one of their fore-legs doubled up for a long time, ready for the next cautious step'; his retriever Bob was the dog with the hot-house face, a neat illustration of the principle of antithesis, showing disproportionate dejection when Darwin took a detour to visit his experimental plants, in case this should mean the end of a walk.[11] In 1872 he told Frances Power Cobbe that since writing the *Descent* he had come to believe dogs have a conscience.[12]

Returning to Pug, Lewes was more circumspect about his intelligence, observing to one of his three surviving sons (with Agnes Jervis): 'Pug grows fatter and more fascinating every day; but his intellect does not develope [sic]. He is decidedly no genius and will never set the Thames on fire nor even the Wimbledon ponds'.[13] In his journal, he noted 'he turns out to be a great beauty, though not wise'.[14] A few years earlier, thanking Bessie Rayner Parkes for *The History of our Cat Aspasia* (1856), Eliot had remarked of Lewes 'he dotes on animals – I mean animals with four legs',[15] the humour again seeming to mark an ease with human–animal kinship. She told Parkes that she had been up to London only twice that year: 'once to call on Mr Chapman, and once to call on the Molluscs in the Zool Gardens. I have no other visits to pay and I shudder at shopping'.[16] At about this time Lewes remarked in the periodical press: 'I have no doubt that the mollusc is a moral individual, but you cannot consider him one greatly impassioned'.[17]

How to relate to other species raised questions of how to read them. When in Ilfracombe for *Sea-side Studies* Eliot and Lewes encountered a pig by the Braunton Road. Eliot wrote up the incident in charmingly anthropomorphic language:

> we were descried by a black pig, probably of an amiable and sociable disposition. But as unfortunately our initiation in porcine physiognomy was not deep enough to allow any decisive inferences, we felt it an equivocal pleasure to perceive that piggy had made up his mind to join us in our walk without the formality of an introduction.[18]

She recorded Lewes resorting 'in vain to hishes, and, at last, instigated as he says by me, threw a stone and hit piggy on the chop. This was final. He trotted away, squealing, as fast as his legs would carry him'. While Lewes 'was remorseful that he had bruised the cheek of a probably affectionate beast, and the sense of this crime hung about him for several days', Eliot recalled satisfying her conscience 'by thinking of the addition to the pig's *savoir vivre* that might be expected from the blow; he would in future wait to be introduced'.[19] She moves in her account between a language of manners and one of the inner, affective life. While the idea that a pig might introduce itself is appealingly absurd, the pig is nonetheless not without affection, sociability and codes of conduct, as the passage suggests. Darwin would write at length about these aspects of animal life in the *Descent*, emphasising 'Animals of many kinds are social'.[20] There is a connection between humans and pigs at this deeper level. Seeing the moral sense as developing from social instincts, Darwin drew parallels between human morality and aspects of social behaviour in animals. He asserted 'Besides love and sympathy, animals exhibit other qualities which in us would be called moral',[21] and ascribed to them reason:

> Of all the faculties of the human mind, it will, I presume, be admitted that *Reason* stands at the summit. Only a few persons now dispute that animals possess some power of reasoning. Animals may constantly be seen to pause, deliberate, and resolve. It is a significant fact, that the more the habits of any particular animal are studied by a naturalist, the more he attributes to reason and the less to unlearnt instincts.[22]

Darwin had set out to show that 'there is no fundamental difference between man and the higher mammals in their mental faculties',[23] and throughout the *Descent* he emphasised kinship and commonality: 'We have seen that the senses and intuitions, the various emotions and faculties, such as love, memory, attention, curiosity, imitation, reason, &c., of which man boasts, may be found in an incipient, or even sometimes in a well-developed condition, in the lower animals'.[24]

Foreshadowing Darwin, Eliot and Lewes also drew on physiognomy in their observations on the Braunton Road pig. This practice, which had been declining in scientific circles from the early nineteenth century, had materialist implications. Allowing both pre-evolutionary hierarchies and teleology, it was essentially dualistic, emphasising human distinction and conceiving of the body as an envelope to the soul. Eliot's phrase 'porcine physiognomy' does little to lend it credence, and she signals ambivalence here about the possibility of reading character and intention from appearance as she would in her novels. The following decade, in *Middlemarch* (1871–2), Celia may appear innocent but, compared with her older sister Dorothea, she is in fact 'knowing and worldly-wise'; 'so much subtler is a human mind than the outside tissues which make a sort of blazonry or clock-face for it'.[25] These are telling images: blazonry and clock-faces give limited, uni-dimensional information.

Physiognomy was succeeded in the second quarter of the nineteenth century by a more materialist phrenology which shifted the focus from character to mental processes.[26] In 1854, Eliot had a phrenological cast made of her head by the renowned phrenologist George Combe.[27] She also developed a close friendship with the phrenologists Charles and Caroline Bray, with whom she shared phrenological observations. On encountering Dickens, for example, she reported: 'his appearance is certainly disappointing – no benevolence in the face and I think little in the head – the anterior lobe not by any means remarkable'.[28] But she was sceptical.[29] While her first novel *Adam Bede* (1859) is steeped in the language of phrenology, it is absent from her later work, resurfacing only in *Felix Holt* (1866), displaced with more Darwinian ways of relating mind and body.[30] Darwin's *Descent* and *Expression* would provide her with a more dynamic way of reading, grounding emotions in physiological processes which were intimately connected to social relations. Even phrenology, stressing the physicality of mind, still worked with established hierarchies, separating what were termed 'animal' propensities (such as acquisitiveness, combativeness, inhabitiveness) from the moral and intellectual. For Darwin these distinctions did not hold. Eliot engaged early in her fiction with ideas of human-animal kinship, and they infused *Adam Bede*. 'In truth, it is a beautiful most *human* Book! Every *Dog* in it, not to say every man woman and child in it, is brought home to one's "business and bosom", an individual fellow-creature', wrote Jane Welsh Carlyle.[31] Reading it as soon as it was out, Darwin thought it 'excellent'.[32] Undergoing the water treatment at Moor Park, before embarking on proofs of the *Origin,* he wrote to his botanist friend J.D. Hooker 'entire rest & the douche & Adam Bede have together done me a world of good'.[33] When Hooker later told Darwin he liked *The Mill on the Floss* (1860) even better than *Adam Bede* and remarked 'how evidently the authoress belongs to the class of life of her heroines, with whom first love is an animal passion with nothing to elevate it', Darwin replied 'we will read it again'.[34]

But Eliot also showed an ambivalence about kinship. The idea of humans behaving like animals was often used to indicate a lack of self-control, a failure in the exercise of reason, which was, especially before Darwin, considered the preserve of humans. In *Adam Bede*, communication between the shepherds Alick and Kestner is 'confined to an occasional snarl. Alick, indeed, was not by any means a honeyed man. His speech had usually something of a snarl in it, and his broad-shouldered aspect something of the bull-dog expression'.[35] The heroic Adam Bede shows a masterful will, faced with his rival in love, Arthur Donnithorne: 'If he had moved a muscle, he must inevitably have sprung upon Arthur like a tiger; and in the conflicting emotions that filled those long moments, he had told himself that he would not give loose to passion, he would only speak the right thing'.[36] But when Arthur deals him a blow, 'the two men, forgetting the emotions that had gone before, fought with the instinctive fierceness of panthers in the deepening twilight darkened by the trees'.[37] Instincts, assumed here to be animal, and portrayed, negatively, as uncontrollable, are distinguished from, and seen as inferior to, emotions, which are assumed to be human. Darwin would collapse this hierarchy, referring to emotions as social instincts.

In *The Mill on the Floss* animality is something perhaps to be left behind. The narrator remarks:

> We learn to restrain ourselves as we get older. We keep apart when we have quarrelled, express ourselves in well-bred phrases, and in this way preserve a dignified alienation, showing much firmness on one side, and swallowing much grief on the other. We no longer approximate in our behaviour to the mere impulsiveness of the lower animals, but conduct ourselves in every respect like members of a highly civilized society. Maggie and Tom were still very much like young animals. (39)[38]

But, there is an ambivalence here. The freely expressive children show a greater authenticity, and, Eliot suggests, civilisation is not of unquestionable benefit. When she describes mourning practices at St Ogg's, she comments: 'It is a pathetic sight and a striking example of the complexity introduced into the emotions by a high state of civilisation', but her remarks have a satirical edge, as she homes in on buckram sleeves and bonnets (56). In both instances the idea of civilisation as 'higher' is questionable. Nonetheless, human distinction is also noted: young Maggie is already 'gifted with that superior power of misery which distinguishes the human being, and places him at a proud distance from the most melancholy chimpanzee' (46–7).

The early anthropomorphism of Eliot and Lewes seems both to have pleased and troubled them. It served at one level to highlight human superiority, to distance the human from other species. In his *Sea-side Studies*, published serially in 1856 in *Blackwood's Magazine* and subsequently as a book, Lewes warned against 'all anthropomorphic interpretations'; 'we are incessantly at fault in our tendency to anthropomorphise, a tendency which causes us to interpret the actions of animals according to the analogies of human nature' – how, for instance 'could an anemone be as liable to emotions of alarm as a fine lady?' – and argued that 'before we can truly understand the Physiology and Psychology of animals, we must acquire the most minute knowledge of their structure'.[39] But while anthropomorphism has the potential to mislead, its more positive aspects have recently been embraced by evolutionary biologists such as Marc Bekoff, philosophers such as Jessica Pierce, the neuroscientist Robert Sapolsky and the psychologist Gordon Burghardt as a way of understanding human–animal continuity rather than making reductive assumptions. In its more considered forms, it enables closer relating between humans and animals.[40] As Bekoff argues in the afterword to this volume, 'denying animals emotional and moral intelligence is bad biology'; he points out that 'being anthropomorphic is a tool to make the thoughts and feelings of other animals accessible to humans'.[41] This does not seem inconsistent with how Eliot and Lewes deployed anthropomorphism in their early work and private recollections. Certainly, the reality of animals permeated their domestic lives. Lewes remarked in *Sea-side Studies* that the observation of pets counted as natural history.[42] They would continue to think through the relation of humans and animals and, as discussed in the final section of this chapter, Darwin's research on the moral sense in humans and animals informs Lewes's *The Study of Psychology. Its Object, Scope, and Method* which Eliot completed in 1879.

Eliot and Lewes's reading

Through their often shared reading, Eliot and Lewes engaged extensively with contemporary debates on the physical basis of mind. Encountering Johann Kaspar Lavater's voluminous *Physiognomic Fragments for the Advancement of the Knowledge and Love of Man* (1775–8; first English translation, *Essays on Physiognomy*, 1789-98) in her late twenties

Eliot was sceptical, referring to 'queer sketches of physiognomies and still queerer judgements on them',[43] though she uses the language of physiognomy in her letters: Mr Forbes's 'fine moral and intelligent head' was 'spoiled by immense pride and vanity, which betray themselves in his thin affected-looking lips'.[44] Phrenology offered a more materialist approach and the Eliot-Lewes library housed a number of key works on the subject, including Franz Joseph Gall and Johann Spurzheim's *Anatomie et physiologie du systême nerveux en géneral, et du cerveau en particulier: avec des observations sur la possibilité de reconnoitre plusiers dispositions intellectualles et morales de l'homme et des animaux, par la configuration de leurs têtes* (1810–12).[45] This foundational text studied the relationship between abilities or attributes (faculties) and the brain in both humans and animals. They also owned Gall's *Sur les fonctions du cerveau et sur celles de chacune de ses parties* (1822–6), and *Sur l'origine des qualités morales et des frailtes intellectuelles de l'homme: et sur les conditions de leur manifestation* (1822), Gall and Johann Caspar Spurzheim's *Phrenology, or, The Doctrine of the Mind; and of the Relations Between its Manifestations and the Body* (4th edition, 1833), and several works by George Combe, including his *System of Phrenology* (signed and dated 1836 by Lewes), and *Phrenology Applied to Painting and Sculpture* (1855). The Eliot and Lewes library held numerous works by their friend Herbert Spencer, the philosopher, biologist and sociologist.[46] These included two copies of *Principles of Psychology* (1855), which argued for entirely physical beginnings of life, with mind developing in response to external pressures. Eliot wrote to her friend Sara Sophia Hennell (Caroline Bray's sister): 'Mr Lewes is *nailed* to the book by his interest in it. I have not had time to do more than glance at it'.[47] (Spencer had recently declined to marry Eliot, finding her too 'morbidly intellectual' and insufficiently attractive, and was relieved by her relationship with Lewes, to whom he had introduced her.[48]) In 1856 she and Lewes would, as Eliot put it, 'wind up the evening' with Thomas Rymer Jones's eight-volume *General Outline of the Animal Kingdom, and Manual of Comparative Anatomy* (1841).[49] The following year Eliot was reading John William Draper's *Human Physiology, Statical and Dynamical* (1856) aloud to Lewes during evenings in Jersey.[50] They also owned Maudsley's *Body and Mind: An Enquiry into their Connection and Mutual Influence, Specially in Reference to Mental Disorders* (1870) and his *Physiology of Mind* (1876), which Eliot marked in the margins.[51] In 1875 Lewes wrote: 'I am hard at work on the 'Physical Basis of Mind' with ever increasing clearness, and consequent enjoyment'.[52] This work, which formed the second series of his five-volume *Problems of Life and Mind*, drew on his reading with Eliot from phrenology through Spencer and Darwin to Maudsley. Lewes assumed the evolutionary origins of mind and was a leading exponent of monism, the idea that mind and matter are made of the same substance. He challenged the idea of mind as 'an entity inhabiting the organism, using it as an instrument';[53] instead the two were continuous.

On Darwin and Comte

Eliot and Lewes read the *Origin* as soon as it appeared. Lewes referred favourably to it in *John Bull*,[54] subsequently writing in the *Fortnightly Review*: 'No work of our time has been so general in its influence'.[55] He declared: 'The Development Hypothesis is an inevitable deduction from the Monistic conception of the world; and will continue to be the battle-ground of contending schools until the opposition between Monism and Dualism ceases'; 'It is possible that the hypothesis of Natural Selection, which Mr. Darwin opposes to that of creative fiat and fixity of Plan, may be an imperfect explanation, but at any rate it has the immense merit of bringing the question within the region of Research'.[56] He also set out his differences from Darwin, for whom kinship was extensive, offering far-reaching explanation:

> Mr. Darwin assumes a community of kinship as the explanation of *all* organic resemblances, whereas I assume it only as the explanation of *many*, the others being due to similarities in the causal nexus. He assumes Adaptation and Natural Selection as the explanation of *all* organic diversities, whereas I assume in many cases an initial diversity in substance, and consequent diversity in epigenesis; so that we may say that many plants and animals were originally 'created' on distinct foundations, and that their superstructures retain the marks of this distinction. It is here that an approach can be made towards reconciling Natural Selection with the facts which have been held to imply distinct 'creations' and fixity of Species.[57]

In the same year Lewes praised Darwin anonymously in the *Pall Mall*

Gazette for his equanimity: 'We must call attention to the rare and noble calmness with which he expounds his own views, undisturbed by the heats of polemical agitation which those views have excited, and persistently refusing to retort on his antagonists by ridicule, by indignation, or by contempt'.[58] Lewes also noted that Darwin was 'lavish in acknowledging the smallest debts he may owe; and his book will make many men happy'.[59] The following year, in the fifth edition of the *Origin*, Darwin quoted Lewes on salamander and newt tadpoles and, in the sixth edition, on timescales for changes in environmental conditions.[60]

On the *Origin* Eliot was more hesitant than Lewes. She wrote to Charles Bray: 'we are reading Darwin's Book on Species, just come out, after long expectation. It is an elaborate exposition of the evidence in favour of the Development Theory, and so, makes an epoch'.[61] Shortly afterwards, she wrote to Barbara Bodichon noting, again, that it made an epoch but remarking:

> the book is ill-written and sadly wanting in illustrative facts – of which he has collected a vast number, but reserves them for a future book of which this smaller one is the avant-courier. This will prevent the work from becoming popular, as the 'Vestiges' did, but it will have a great effect in the scientific world, causing a thorough and open discussion of a question about which people have hitherto felt timid. So the world gets on step by step towards brave clearness and honesty!

She added: 'But to me the Development theory and all other explanations of processes by which things came to be, produce a feeble impression compared with the mystery that lies under the process'.[62]

Elements of the *Origin* are taken up ambivalently in *The Mill on the Floss* on which Eliot was working at the time of its appearance. The selfishness of Tom Tulliver is questioned as he shoots peas 'with the distinct aim of hitting a superannuated blue-bottle which was exposing its imbecility in the spring sunshine, clearly against the views of Nature, who had provided Tom and the peas for the speedy destruction of this weak individual'.[63] Commenting wryly on ways in which Darwinian ideas were being misapplied to social situations, Eliot later wrote to her American admirer Elma Stuart, glad that her advice had hindered Stuart from marrying a problematic man, and impatient with widowers who are:

> always expecting women to take compassion on them, and never themselves take compassion on women really forlorn. They must always have the best, forsooth – are always good enough for the best to accept. This is the fine principle of Natural Selection, they will say. I admit it, but it is also the selection of conceited gentlemen.[64]

But the ethical implications of evolution were also emerging at this time, made clear in Darwin's account of the development of the moral sense in the *Descent* and explored in detail in *The Data of Ethics* (1879) in which Spencer examined cooperation and the growth of conscious altruism. The Eliot-Lewes library held both. Eliot's copy of *The Data of Ethics* was inscribed: 'Mrs Lewes with the kindest regards of the author', and contains her pencil marginalia and linings.[65]

Eliot and Lewes also owned several books by Auguste Comte, whose positivism, emphasising the ascendancy of emotion over intellect, of feeling over mind, would be a major influence on their views on morality and emotion. 'Knowledge', Lewes wrote in his *Problems of Life and Mind*, was 'a presentation of feeling'.[66] Comte declared in his chapter on the social aspect of positivism in *A General View of Positivism*, that the primary principle of positive morality was 'the preponderance of Social Sympathy'; it was brought into 'a coherent and systemic form by its principle of universal love'.[67] Although he rejected Lamarckian transformism, arguing for limits to variation, he saw social progress as an extension of Lamarck's animal series, also praising Lamarck's contribution to understanding of the environment.[68] Comte insisted that animals were capable of progress and interspecies altruism, writing in his *Theory of the Great Being* (1854):

> our conception of the constitution of the Great Being remains defective unless we associate with man all the animal races which are capable of adopting the common motto of all the higher natures: *live for Others*. Without the animals, the Positive Synthesis could but imperfectly form the permanent alliance of all voluntary agents to modify the external conditions of our life so far as they are modifiable.[69]

In his *System of Positive Philosophy, or Treatise on Sociology, Instituting the*

Religion of Humanity (1851–4; begun in 1830), he wrote 'the happiness of living for others is not entirely monopolised by Man. Many animals possess it likewise, and indeed give evidence of a higher degree of sympathy than our own'.[70] Elements of Comte would combine with Darwin in Eliot and Lewes's understanding of the moral sense.

Physical Emotions

Eliot insisted on the need for art to 'work on the emotions'.[71] Writing to the Comtean positivist Frederic Harrison she describes this as a physical process: 'get breathing, individual forms, and group them in the needful relations'; the presentation needs to 'lay hold on the emotions as human experience'; to '"flash" conviction on the world by means of aroused sympathy'.[72] As Darwin emphasised the role of the social instincts, including sympathy (see chapter four), in the development of morality, Eliot stressed the role of the emotions in developing desire for social justice. Through working with the emotions the aesthetic was able to reshape the moral sense. She wrote to the social reformer Clementia Taylor: 'My function is that of the *aesthetic*, not the doctrinal teacher – the rousing of the nobler emotions, which makes mankind desire the social right, not the prescribing of special measures, concerning which the artistic mind, however strongly moved by social sympathy, is often not the best judge'.[73]

Emotions in Eliot's fiction burst and quiver, reminding us of their physical nature, of the unconsciousness of expression that Darwin would make the subject of his research. In *The Mill on the Floss* Book Third (The Downfall) the reader learns: 'Maggie's cheek flushed and her lip quivered with conflicting resentment and affection. Struck with the unusual emotion in Philip's tone, she had turned quickly to look at him; and as he went on speaking, a great change came over her face, – a flush and slight spasm of the features' (334). Her feelings register bodily. Later, 'At Philip's name she had blushed, and the blush deepened every instant from consciousness, until the mention of the Red Deeps made her feel as if the whole secret were betrayed, and she dared not even hold her tea-spoon lest she should show how she trembled' (339). In turn, Philip's whole frame 'is shaken by violent emotions' (345). It is unclear whether Maggie consciously wills a union with her cousin's betrothed, Stephen Guest; instead, she quivers and trembles, 'a fierce battle of emotions' under a 'torpor of manner', lapsing in duty and causing mayhem (458). Even Stephen, singing to Maggie, was perhaps 'not distinctly conscious that he was impelled to it by a secret longing – running counter to all his self-confessed resolves – to deepen the hold he had on her' (459). 'Watch your own speech', the narrator remarks, and notice how it is guided by your less conscious purposes, and you will understand that contradiction in Stephen' (459). Stephen, we are told in a chapter called 'The Laws of Attraction', 'meant to choose Lucy; she was a little darling, and exactly the sort of woman he had always admired' (371). But 'A mad impulse seized on Stephen; he darted toward the arm, and showered kisses on it, clasping the wrist' (442). Later, Maggie allows herself to drift, actually and metaphorically, on the river with Guest, 'dimly conscious of the banks, as they passed them...At all times she was so liable to fits of absence, that she was likely enough to let her waymarks pass unnoticed' (464). Emotion flows through the novel, but by its close Maggie is brought back to the moral strictures of St Ogg's; an order is restored which depends on right choice and reason.

A decade later, in December 1871, the first of eight instalments of *Middlemarch* appeared. The novel charts the growth of sympathy, exploring the possibility of emerging from 'moral stupidity'. Such a process mirrors the moral development of the species – ontogeny recapitulating phylogeny (198). For Darwin sympathy was informed by a nexus of bodily responses, and the moral sense emerged in humans and animals from the social medium, from instincts and emotions which developed from life within the community. Eliot's characters are profoundly affected by community, responding, reacting and relating to each other, growing in sympathy and understanding, or failing to do so: 'there is no creature whose inward being is so strong that it is not greatly determined by what lies outside it' (784–5). *Middlemarch* continues Eliot's emphasis on the physical nature of emotions. Bulstrode goes through 'a crisis of feeling almost too violent for his delicate frame to support' (683). Rosamond, 'with an overmastering pang, as if a wound within her had been probed, burst into hysterical crying' (747). Casaubon fails in part because of his lack of emotion: he 'determined to abandon himself to the stream of feeling, and perhaps was surprised to find what an exceedingly shallow rill it was' (58). Henry James, brother of the Spencer-averse Darwinian psychologist

William James, published an impassioned imaginary conversation on the novel in the *Atlantic Monthly,* in which Theodora remarked on Casaubon's dryness.[74]

Critics were struck by the physiological language of *Middlemarch.* In the words of R.H. Hutton, editor of the *Spectator,* and anti-vivisectionist, 'we all grumble at *Middlemarch*; we all say that the action is slow, that there is too much parade of scientific and especially physiological knowledge in it, that there are turns of phrase which are even pedantic' and yet 'we feel that we cannot do without it'.[75] It was singled out as the most scientific of Eliot's novels, having, as it did, a doctor at its centre. The jurist A.V. Dicey referred to the 'flavour of scientific allusion',[76] and Henry James declared that the novel was 'too often an echo of Messrs Darwin and Huxley'.[77] Eliot's engagement with Comtean positivism, no less than with natural history, also contributed to the assessment of her work as scientific. Sidney Colvin, a regular visitor of Eliot and Lewes, observed that her writing was 'so full of her time... saturated with modern ideas' and that 'if scientific thought and the positive synthesis are indubitably to rule the world, then any one, it should seem, might speak boldly enough to George Eliot's place'; she had taken the lead in 'expressing and discussing the lives and ways of common folks – *votum, timor, ira, voluptas* – in terms of scientific thought and the positive synthesis'.[78] He argued that her 'studies in science and physiology will constantly come in to suggest for the spiritual processes of her personages an explanation here or an illustration there'.[79]

Strikingly, following the publication of Darwin's *Expression,* readers brought a new awareness and expectation to fiction. This is apparent in late reviews of *Middlemarch.* Colvin saw a new emphasis on the body acting independently of consciousness – an idea central to the *Expression.* He quoted in full the description of Bullstrode: 'many of these misdeeds were like the subtle muscular movements which are not taken account of in the consciousness; though they bring about the end that we fix in our minds and desire',[80] pointing up Eliot's new treatment of expression: 'she seizes on critical points of physical look and gesture in her personages, in a language which is scientific and her own...'.[81] His review brings together the fading art of physiognomy and the latest of Darwin's ideas, and in doing so also suggests Darwin's own residual use of physiognomical ideas:

> look at Caleb Garth; he is a complete physiognomical study in the sense of Mr. Darwin, with the 'deepened depression in the outer angle of his bushy eyebrows, which gave his face a peculiar mildness'; with his trick of 'broadening himself by putting his thumbs into his arm-holes', and the rest. Such are Rosamond's ways of turning her neck aside and patting her hair when she is going to be obstinate. So, we are not allowed to forget 'a certain massiveness in Lydgate's manner and tone, corresponding with his physique'; nor indeed, any point of figure and physiognomy which strike the author's imagination as symptomatic. Symptomatic is the best word. There is a medical strain in the tissue of the story. There is a profound sense of the importance of physiological conditions in human life. But further still, I think, there is something like a medical habit of the author, of examining her own creations for their symptoms, which runs through her descriptive and narrative art and gives it some of its peculiar manner.[82]

With *Middlemarch* complete, Eliot embarked on Darwin's *Expression*, taking notes for *Daniel Deronda*. Lewes noted that she began reading it on 4 December 1872 and continued until 23 December.[83] It is no surprise she read it with such fascination. It was immensely popular, even among those who resisted its implications. Darwin's theory of descent through modification plunged humans into a world of impulses, instincts and unconscious habits, and the *Descent* and the *Expression* observed this close up. Eliot took notes on the three principles of expression in her *Daniel Deronda* notebook:

> 1 Serviceable associated habits
> 2 Antithesis
> 3 Actions due to the constitution of the Nervous system, independently from the first of the Will, & independently to a certain extent of Habit.[84]

This third chimed with Lewes's argument that sentience was dependent not simply on the brain but the whole nervous system, lending further significance to the centrality of the body in generating emotion.

Daniel Deronda

One of the questions Darwin addressed in the *Expression* was why does emotion manifest itself in involuntary bodily changes? Reviewers of the *Expression* were especially drawn to these changes. For one reviewer in the *Literary World*: 'Blushing is an emotional expression of this kind, as also trembling – due to a certain state of the nervous system and not under the direct control of the will'.[85] Darwin speculated on the likely origins of certain bodily expressions. One was the sneer:

> He who rejects with scorn the belief that the shape of his own canines, and their occasional great development in other men, are due to our early progenitors having been provided with these formidable weapons, will probably reveal by sneering the line of his descent. For though he no longer intends, nor has the power, to use these teeth as weapons, he will unconsciously retract his 'snarling muscles' (thus named by Sir C. Bell) so as to expose them ready for action, like a dog prepared to fight.[86]

In *Daniel Deronda* the narrator informs us of Lush, right-hand man to the villainous Henleigh Mallinger Grandcourt: 'This sneer slipped off his tongue without strict intention'.[87] Deronda's mother reports: 'women and men seemed to look at me with a sneering smile' (213). The novel contains numerous examples of such involuntary, unconscious, expressions. Lush is taken in by Gwendolen's seeming self-composure early in the novel – 'I like a mouth that trembles more' – but for much of the novel she trembles, palpitates and blushes in spite of her attempts to remain outwardly composed (8). A biology comes to the fore that is not deterministic but responsive, immediate and at times predominant, as the novel challenges easy dichotomies between feeling and intellect. Deronda is able to combine emotion and thought, but he places greater value on emotion: 'He was ceasing to care for knowledge – he had no ambition for practice – unless they could both be gathered up into one current with his emotions' (365).

The bodied and impulsive nature of feeling is emphasised through the novel: 'Some feelings are like our hearing: they come as sounds do, before we know their reason' (218).[88] At various points Gwendolen is overcome by emotion to a greater extent than any of Eliot's earlier heroines: 'her pallid lips were parted; her eyes, usually narrowed under their long lashes, were dilated and fixed' (61). While she screams or cries hysterically at various points, this is usually in response to

Grandcourt or his deeds, and hysteria is not a state unique to her, or to women, in the novel; Lapidoth, we are told, has an 'hysterical excitability' (773).[89] The physical nature of Gwendolen's experiences is brought out through a language of instinct and sensation: 'The ideas it raised had the force of sensations'. 'The defeated clutch of struggling hope gave her in these first moments a horrible sensation' (142, 346). Likewise, 'she had hardly had more cruel sensations than now; for emotion was at the acute point, where it is not distinguishable from sensation', and later, 'It came as an instinct rather than a thought' (602, 589).

The critics noted a difference, picking up on the new treatment of emotion. For Dicey, the reader learned too much about 'the expressiveness' of Deronda's countenance.[90] Robert Francillon, writing in the *Gentleman's Magazine*, saw the predominance of emotion as a backward step. The characters of Eliot's earlier novels, he observed, 'have to do with the broad passions and emotions common to us all: this with exceptional moods and passions, brought out by exceptional circumstances, special to individuals. They develop the study of healthy anatomy: this of pathology'.[91] From the outset *Daniel Deronda* pulls at the boundaries between reason and emotion. The opening epigram compares the two, finding similarities where difference is expected: 'Men can do nothing without the make-believe of a beginning. Even science, the strict measurer, is obliged to start with a make-believe unit...His less accurate grandmother Poetry has always been understood to start in the middle; but on reflection it appears that her proceeding is not very different from his' (1).[92]

Daniel Deronda begins with expression. It emphasises uncertainty, providing its readers with an inconclusive attempt to read Gwendolen: 'Was she beautiful or not beautiful? and what was the secret of form or expression which gave the dynamic quality to her glance?' (1).[93] In his *Atlantic Monthly* piece on the novel, an imagined conversation between a hostile Pulcheria, an enthusiastic Theodora and a more restrained Constatius, Henry James encapsulated the complex and ambivalent response of a readership that didn't quite know how to respond to Eliot's experimental novel. Theodora is so affected she declares 'I couldn't have read them aloud myself; I should have broken down'. [94] But Pulcheria is unimpressed by the new emphasis: 'there is plenty about his emotions. Everything about [Deronda] is 'emotive'.

That bad word occurs on every fifth page' (as it happens it occurs on more like each hundredth page).[95] Theodora also approves of the scientific edge of the novel: 'how can she be too scientific? She is merely permeated with the highest culture of her age'.[96]

Eliot's reading of *Expression* seems to inform a new self-consciousness about expression in *Daniel Deronda*. The Princess Alcharisi, the eponymous hero's mother, abandons motherhood for acting. Her studied self-consciousness enables her both to convey and represent emotion at the same time:

> The speech was in fact a piece of what may be called sincere acting; this woman's nature was one in which all feeling – and all the more when it was tragic as well as real – immediately became matter of conscious representation: experience immediately passed into drama, and she acted her own emotions. (629)

This process, of becoming conscious of the expression of the emotions, and thus acting them out, is apparent not only in Alcharisi but also the child Jacob Cohen, son of Ezra Cohen: 'Jacob, feeling the danger well-nigh over, howled at ease, beginning to imitate his own performance and improve upon it – a sort of transition from impulse into art often observable' (478). Some responses to the *Expression* underlined relations between the science of expression and the art of drama. Drama offers a self-conscious, embodied and immediate representation of emotion. *The Times* remarked:

> Any one who has seen a great play well acted will know that even the most powerful words convey only half their meaning without a visible embodiment. To artists and actors this work must become a sort of text-book on the facts of facial and bodily expression, and the details are full of interest to ordinary readers.[97]

A number of the illustrations in the *Expression* had been acted out by one of the photographers, Oscar Rejlander, and, in turn, the study was taken up in drama criticism and seen as likely to benefit the stage. The leading drama critic William Archer drew closely on the study in his *Masks or Faces? A Study in the Psychology of Acting* (1888), first published serially in *Longman's Magazine*.[98]

For the poet and critic Edward Dowden, in a review welcomed by Lewes, *Daniel Deronda* provided evidence that 'the manner of few great artists – if any – becomes simpler as they advance in their career'; 'their emotions receive more numerous affluents from the other parts of their being'.[99] Dowden acknowledged an increase in Eliot's work of 'scientific habits of thoughts and expression', noting that other reviewers had objected to such terms as 'dynamic' and 'natural selection' in a work of literature.[100] He observed that *Daniel Deronda* showed a 'passionate selection of facts of life' and he justified Eliot's use of science: 'reason is to her imagination both law and impulse. And therefore her art is not a mere luxury for the senses, not a mere aesthetic delicacy or dainty. It was chosen for its part to be founded in truth, to nourish the affections, to quicken the conscience'.[101] He stressed that it was essential to the highest characteristics of her art that she should not isolate herself from the chief intellectual movement of her time'; 'a truth-seeker at the present day will do ill to turn a deaf ear to the teachings of science'.[102] The review challenges clear-cut divisions between law, measured and methodical, and impulse, spontaneous and experienced bodily. Dowden also saw Eliot's emphasis on emotion as a way of defending her against critics who accused her writing of being too scientific, remarking:

> But what renders singular the warning addressed to George Eliot that her work is undergoing a 'scientific depravation' is that the whole of her last book is a homage to the emotions rather than to the intellect of man. Her feeling finds expression not only in occasional gnomic utterances in which sentiments are declared to be the best part of the world's wealth, and love is spoken of as deeper than reason, and the intellect is pronounced incapable of ascertaining the validity of claims which rest upon loving instincts of the heart, or else are baseless. The entire work possesses an impassioned aspect, an air of spiritual prescience, far more than the exactitude of science. The main forces which operate in it are sympathies, aspirations, ardours, and ideas chiefly as associated with these'.[103]

Eliot's evangelical beginnings, transformed here as 'spiritual prescience', were finding expression in a Darwinian morality in her final writing. The emphasis on feelings, on the social instincts and on sympathy, to which the critics were drawing attention, was precisely the subject of the new Darwinian science.

Eliot seems in *Daniel Deronda* to be seeking to challenge the dichotomy between intellect and emotion, going against more conventional beliefs such as Dowden's insistence that the novel was a homage to the emotions rather than to the intellect.[104] She explores the possibility of an emotional intellect, remarking:

> Men may dream in demonstrations, and cut out an illusory world in the shape of axioms, definitions, and propositions, with a final exclusion of fact signed Q.E.D. No formulas for thinking will save us mortals from mistake in our imperfect apprehension of the matter to be thought about. And since the unemotional intellect may carry us into a mathematical dreamland where nothing is but what is not, perhaps an emotional intellect may have absorbed into its passionate vision of possibilities some truth of what will be—the more comprehensive massive life feeding theory with new material, as the sensibility of the artist seizes combinations which science explains and justifies. (514)

As part of this challenging of boundaries, the relation between humans and animals in *Daniel Deronda* is less one of analogy than continuity. This marks a shift from Eliot's earlier fiction. Gwendolen, frequently described in animal terms, also shows kinship with animals, though some distinction is also noted:

> Thus Gwendolen felt no check on the animal stimulus that came from the stir and tongue of the hounds, the pawing of the horses, the varying voices of men, the movement hither and thither of vivid colour on the background of green and gray stillness: – that utmost excitement of the coming chase which consists in feeling something like a combination of dog and horse, with the superadded thrill of social vanities and consciousness of centaur-power which belongs to humankind. (71)

In the *Descent* Darwin had observed:

> Most of the more complex emotions are common to the higher animals and ourselves. Every one has seen how jealous a dog is of

> his master's affection, if lavished on any other creature; and I have observed the same fact with monkeys. This shows that animals not only love, but have the desire to be loved.[105]

The expression of dogs is accentuated in *Daniel Deronda.* One of Grandcourt's many dogs, Fetch, 'the beautiful liver-coloured water-spaniel, which sat with its forepaws firmly planted and its expressive brown face turned upward', watches Grandcourt 'with unshaken constancy' (125). Fetch, like Pug, has an 'expressive' face which registers the feelings that Darwin saw conjoining humans and animals. Grandcourt, noticing its jealously, lifts the tiny Maltese dog he has on his lap 'and gave it caressing pats, all the while gravely watching Fetch, who, poor thing, whimpered interruptedly, as if trying to repress that sign of discontent' (126). Spaniels were known for their submissiveness – Mary Wollstonecraft had referred in her *Vindication of the Rights of Woman* (1792) to the 'spaniel-like affection' of submissive women.[106] Lush himself is compared to a dog; Grandcourt has a confirmed sense 'that he might kick Lush if he chose – only he never did choose to kick any animal, because the act of kicking is a compromising attitude, and a gentleman's dogs should be kicked for him' (125). Fetch shows some capacity for self-restraint; but at last she 'rested her head beside the appealing paw, looking up with piteous beseeching. So, at least, a lover of dogs must have interpreted Fetch, and Grandcourt kept so many dogs that he was reputed to love them; at any rate, his impulse to act just in that way started from such an interpretation' (126). Grandcourt's failing lies in his lack of both empathy and sympathy.

The Moral Sense

Steeped in Comte and Darwin, Eliot and Lewes were both thinking about the moral sense when Lewes died in 1878, leaving *The Study of Psychology: Its Object, Scope, and Method* incomplete. Revising the manuscript for publication, Eliot removed some of the direct references to Darwin, but the Darwinian ideas remain. She retained Lewes's statement that 'if we admit the Intelligence of animals to be a rudimentary Intellect, we may admit the Emotions of animals to furnish a rudimentary Moral Sense', though she changed 'furnish' to 'be', thus making the possession of the moral sense in animals more explicit, albeit rudimentary. She then expanded on this. Considering the self-repressing effort induced by 'sexual and parental' instincts, and the capability of attachment of birds and animals, she emphasised 'we may recognize the same germs as those which in man the social life has developed into devoted affection, passionate sympathy, and self-denying forethought'.[107] Eliot removed Lewes's reference to Darwin's argument that dogs refraining from stealing food was an example of self-command 'not wholly the result of fear', but she added a passage on the origins of morality which emphasised continuity between humans and animals: 'the moral dispositions of men have manifold roots, and a great deal of 'right' action is sure to be done by very simple beings from healthy impulse, under the guidance of fact, without the idea of an uplifted rod'.[108] The emphasis on nature as a source of morality is retained, but a caution as to the limits of the study of animals is added: 'how short a way the consideration of animal life alone will take us in the appreciation of the moral life of mankind, which is wrought out of innumerable closely interwoven threats of feeling and knowing'. She went on, however, to stress the importance of recognising the link with animals, and the *origins* of morality: 'such reduction of the sympathetic or moral life to its primary manifestations is not merely useful, it is indispensible to a true analysis and natural history of morals'.[109] She then added a reference to a pointer: 'in the pointer dog we observe the effect of a trained impulse; a native tendency restrained and fashioned to a specific variety of action by external influence or the presentation of motive', and underlined a necessary cooperation 'of the impulsive and the perceptive, the emotional and the intellectual in the development of morality'; this was the same combination she had emphasised in *Daniel Deronda*. Here, again, she brings the two together. In doing so she suggests that animals, like humans, possess both. Darwin had used the pointer as an example of conscience, of the resolve 'to act differently for the future': a pointer dog, if able to reflect on his past conduct, would say to himself, 'I ought (as indeed we say of him) to have pointed at that hare and not have yielded to the passing temptation of hunting it'.[110]

Eliot retained Darwin's thought experiment of humans living as hive-bees, where unmarried females saw it a duty to kill their brothers, and mothers their fertile daughters. Darwin's argument was that conscience is developed from context. Eliot expanded on this section, emphasising the importance of the social environment, and thus underlining Darwin's argument: 'Social Sanction', she wrote, has dominated 'all other influences'.[111] She introduced other notable differences. Lewes had referred to the discussion in the *Descent* of the development of conscience, in which Darwin had remarked, in a section otherwise emphasising continuity, that 'a moral being is one who is capable of comparing his past and future actions or motives, and of approving or disapproving of them. We have no reason to suppose that any of the lower animals have this capacity' and that man alone 'can with certainty be ranked as a moral being'.[112] Lewes followed this with the statement: 'Now this looking back upon actions & judging of their social instincts is impossible to an animal, whose life is therefore regulated by desires & intelligence, never by moral conceptions & organised ethical tendencies', and referred to Darwin's remark, which he also quoted, that 'conscience looks backwards and judges past actions'.[113] In her revision, Eliot retained Lewes's reference to the 'desires & intelligence' that regulate the life of an animal, but removed the idea that retrospection was 'impossible to an animal', emphasising the original connection between humans and animals and introducing the concept of kinship: 'man, in his moral beginnings, has a marked kinship with the animals, whose life, like his own, is regulated by desires and intelligence' before writing 'he stands apart at the attainment of moral conceptions and organised ethical tendencies'.[114] Eliot, then, describing the *development* of humans, chose, more than Lewes, to emphasise kinship with animals.

In her revisions to 'The Affective States' in the second volume of Lewes's *The Study of Psychology*, Eliot took out references to animals in general and tigers in particular (Lewes had written that tigers would be 'gentle & sociable were animal food as abundantly accessible as vegetable food').[115] Thomas Dixon points out that she makes the account of emotions more Comtean by emphasising that both the selfish and sympathetic emotions have their roots in forms of egoism.[116] Lewes saw sympathy coming from the reproductive instincts while for Eliot sympathy (like selfishness) was rooted in egoism.[117] As she put it, appetite 'is also the ancestor of Love'.[118] Depopulated of animals, the account nonetheless remains consistent with Darwin. Darwin had stated that even with those animals 'which live permanently in a body, the social instincts are ever present and persistent'; such animals 'feel at all times...some degree of love and sympathy' for 'their fellows'. The fellow-feeling that is constantly stressed in Eliot's novels exists in Darwin's animal world, and her 'healthy impulse' is resonant with the moral impulses he finds in animals.[119]

For Henry James the influence of Lewes on Eliot had been regrettable: 'the contagion of his studies pushed her further than she would otherwise have gone in the direction of *scientific* observation'.[120] But the assumption that Lewes was responsible for Eliot's engagement with science undervalues the independence of that engagement. And, on the question of continuity between humans and animals, Eliot would at times go further than Lewes. When it came to the origins of human morality, Eliot insisted on its 'primary manifestations' in animals and would draw on anthropomorphism to signal continuity. As noted earlier, Lewes, avid in his study of animals, saw anthropomorphism as having little value. Quoting Darwin, he clarified his own, distinct, position: 'it has, I think, now been shown that man and the higher animals have some few instincts in common. All have the same senses, intuitions, and sensations', adding 'for same read similar' before continuing the quotation:

> similar passions, affections, emotions, even the more complex ones: they feel wonder and curiosity; they possess the same faculties of imitation, attention, memory, imagination, and reason, though in very different degrees. Nevertheless, many authors have insisted that man is separated through his mental faculties by an impassable barrier from all the lower animals.[121]

Lewes declared himself to be in this camp, arguing that language, unique to humans, would educate animal passions into human sentiment:

> With these authors I agree. I hold, indeed, that the mental faculties of man are developed out of mental functions which animals share with man; but these faculties, when developed, constitute as broad a line of demarcation, a barrier as impassable, as that between the vertebrate and invertebrate structure. The moral and higher intellectual faculties of man can no more be explained by reference to the animal functions alone than the flight of birds can be explained by the creeping

> of reptiles, though both are reducible to mechanical and physiological principles. Just as birds have wings, man has Language. The wings give the bird its peculiar aptitude for aerial locomotion. Language enables man's intelligence and passions to acquire their peculiar characters of Intellect and Sentiment.[122]

Lewes remained reluctant to relinquish the idea of human separateness. Writing to Elma Stuart from Godalming about their new home he remarked:

> it is a small paradise. … I turn into my woods to meditate, and the squirrels contemplate me with astonished contempt: 'What, little Sir, are *you* there, straddling among the fir cones and unable to run up among these branches – pondering problems and unable to enjoy nuts – a two-fisted, two-legged, troublous and troubled animal!' The thrushes and blackbirds have their musical remarks to make – and the cuckoo occasionally throws in a distant observation – the pigeons coo and the moles dart into their burrows. The sweet sunshine irradiates the scene – and the blessed sunshine of affection irradiates my interior.[123]

He presents himself as emotionally enraptured but intellectually distant from his animal environment. Darwin, by contrast, was more ready to throw his lot in with the squirrels, writing to Emma shortly before he wrote the *Origin*:

> At last I fell asleep on the grass, & awoke with a chorus of birds singing around me, & squirrels running up the trees, & some woodpeckers laughing, & it was as pleasant & rural a scene as ever I saw, & I did not care one penny how any of the beasts & birds had been formed.[124]

Eliot was increasingly drawn to this Darwin, who emphasised mental and emotional affinities with animals and integrated humans into a world of impulses and physical sensation. The predominance of emotion in *Daniel Deronda* helps to explain the extent to which it troubled its early readers. Darwin's emphasis on instinct, unconsciousness, and human-animal continuity led, in Eliot's fiction, to a new questioning of boundaries between body and mind, emotion and intellect, passion and reason, and of the ability of an individual to offer a full account of his or her actions. Challenging the possibility of a steady observation of objective realities, the *Expression* pulled at the neat divisions on which the realist fiction dominant at the mid-century had depended.

Notes

1 Edward Dowden, '*Middlemarch* and *Daniel Deronda*', *Contemporary Review*, 39 (1877), 348–69, in David Carroll (ed.), *George Eliot: The Critical Heritage* (1971; London: Routledge, 2009), 444.

2 G.H. Lewes, 'Mr Darwin's Hypothesis', *Fortnightly Review*, 3 (1868), 353–73, 611–28; *Fortnightly Review*, 4 (1868), 61–80, 492–509.

3 Lewes, 'Mr Darwin's Hypothesis', *Fortnightly Review*, 3 (1868), 353–73, 611–28; *Fortnightly Review*, 4 (1868), 61–80, 492–509. Darwin to Lewes, 18 November 1868, Darwin Correspondence Database, http://www.darwinproject.ac.uk/entry-6465 (accessed 7 October 2011).

4 Emma Darwin visited on 11 October 1873 and 23 October 1873. See Gordon S. Haight (ed.), *The George Eliot Letters*, 7 vols (New Haven and London: Yale University Press, 1954–5), v. 449. On Eliot and Lewes's engagement with contemporary science see, especially, Gillian Beer, *Darwin's Plots: Evolutionary Narrative in Darwin, George Eliot and Nineteenth-Century Fiction*, 3rd edn (Cambridge: Cambridge University Press, 2009); George Levine, *Darwin and the Novelists: Patterns of Science in Victorian Fiction* (Cambridge, MA: Harvard University Press, 1988); Rick Rylance, *Victorian Psychology and British Culture 1850–1888* (Oxford: Oxford University Press, 2000); Sally Shuttleworth, *George Eliot and Nineteenth- Century Science: The Make-Believe of a Beginning* (Cambridge: Cambridge University Press, 1984), and Shuttleworth, *The Mind of the Child: Child Development in Literature, Science, and Medicine, 1840–1900* (Oxford: Oxford University Press, 2010).

5 My reading contrasts with K.M. Newton, *Modernizing George Eliot: The Writer as Artist, Intellectual, Proto-Modernist, Cultural Critic* (London: Bloomsbury Academic, 2011); Newton sees Eliot seeking to highlight and oppose Darwinian individualism, and argues that she emphasises difference between humans and animals.

6 On Gwendolen as 'the original literary hysteric' see Jacqueline Rose, *Sexuality in the Field of Vision* (London: Verso, 1986), 115–16. Evelyne

Ender reflects on a gender-marked consciousness in *Daniel Deronda* and proposes that Gwendolen's hysteria is the symptom of the inability of language to express a particularly feminine history; see Ender, *Sexing the Mind: Nineteenth-Century Fictions of Hysteria* (Ithaca: Cornell University Press, 1995), 271. More recently, Kaara L. Peterson has argued that Gwendolen evidences a Charcotian repressed hysteria: Peterson, *Popular Medicine, Hysterical Disease, and Social Controversy in Shakespeare's England* (Aldershot: Ashgate, 2010), 184. See also Athena Vrettos, 'From Neurosis to Narrative: The Private Life of the Nerves in *Villette* and *Daniel Deronda*', *Victorian Studies*, 33 (1990), 551–79.

7 GHL Journal, 29 July 1859, in Haight, 1954–5, iii. 119.

8 Eliot to John Blackwood, 30 July 1859, in Haight, 1954–5, iii. 124–5.

9 Eliot to John Blackwood, 6 September 1859 and 16 October 1859, in Haight, 1954–5 iii. 145, 185.

10 Darwin to unidentified correspondent 19 May [1871–4], Darwin Correspondence Database, http://www.darwinproject.ac.uk/entry-13889 (accessed 17 October 2011); 19 March [1873], to William Pengelly, http://www.darwinproject.ac.uk/entry-8817 (accessed 17 October 2011). On Darwin and dogs see Paul White in this volume and Emma Townshend, *Darwin's Dogs: How Darwin's Pets Helped Form a World-Changing Theory of Evolution* (London: Frances Lincoln, 2009).

11 Darwin, *The Expression of the Emotions in Man and* Animals (London: John Murray, 1872), 42, 57–8.

12 Darwin to Frances Power Cobbe, 28 November 1872, Darwin Correspondence Database, http://www.darwinproject.ac.uk/entry-8652 (accessed 17 October 2011).

13 Lewes to his sons, 10 November 1859, in Haight, 1954–5, iii. 196.

14 GHL Journal, 29 July 1859, cited in Haight, 1954–5, iii. 124.

15 Eliot to Bessie Rayner Parkes, 1 April 1856, cited in Haight 1954–5, iii. 235.

16 Eliot to Bessie Rayner Parkes, 22 March 1856, cited in Haight, 1954–5, ii. 232.

17 Lewes, *Sea-side Studies at Ilfracombe, Tenby, The Scilly Isles & Jersey* (Edinburgh and London: William Blackwood and Sons, 1858), 51.

18 George Eliot Journal, Ilfracombe, 8 May–26 June 1856, in John Walter Cross (ed.), *George Eliot's Life, as Related in her Letters and Journals*, 3 vols (1885; Cambridge: Cambridge University Press, 2010), iii. 396–406; Haight, 1954–5, ii. 250.

19 *Ibid.*, ii. 250.

20 Darwin, *The Descent of Man, and Selection in Relation to Sex*, 2 vols (London: John Murray; 1871), i. 74.

21 Darwin, 1871, i. 78.

22 *Ibid.*, i. 46, emphasis in original.

23 *Ibid.*, i. 35.

24 *Ibid.*, i. 105.

25 Eliot, *Middlemarch*, ed. by David Carroll (1871–2; Oxford: Oxford University Press, 2008), 9. Subsequent page references to *Middlemarch* are given parenthetically in the text.

26 See Roger Cooter, *The Cultural Meaning of Popular Science: Phrenology and the Organization of Consent in Nineteenth-Century Britain* (Cambridge: Cambridge University Press, 1984); and Robert M. Young, *Mind, Brain and Adaptation in the Nineteenth Century: Cerebral Localization and Its Biological Context from Gall to Ferrier* (Oxford: Clarendon Press, 1970). See also The History of Physiognomy: A Leverhulme Trust International Network, http://physiognomy.history.qmul.ac.uk/about.html.

27 See Haight, *George Eliot: A Biography* (Oxford: Clarendon Press, 1968), 51.

28 Eliot to Mr and Mrs Charles Bray, 27 April 1852, Haight, 1954–5, ii. 21; 5 May 1852, ii. 23, 24.

29 See Rosemary Ashton, *George Eliot: A Life* (London: Penguin, 1996), 221; N.N. Feltes, 'Phrenology: from Lewes to George Eliot', *Studies in Literary Imagination*, 1 (1968), 13–22; see also Beryl Gray, 'Pseudoscience and George Eliot's "The Lifted Veil"', *Nineteenth-Century Fiction*, 36 (1992), 407–23, and 'Afterword', in Eliot, *The Lifted Veil* (1859; London: Virago, 1985), 69-91.

30 See Hugh Witemeyer, *George Eliot and the Visual Arts* (New Haven: Yale University Press, 1979), chapter 5; Feltes, 1968; and Karen Chase, *Eros and Psyche: The Representation of Personality in Charlotte Brontë, Charles Dickens and George Eliot* (New York: Methuen, 1984) on Eliot and phrenology. While Eliot engaged to varying degrees with both physiognomy and phrenology she rejected the idea of mesmerism and of hypnotism as dualistic and unempirical, and though these discourses had found expression in the characters of sensation literature they do not feature in her novels. Of mesmerism she wrote to George Combe, the educationalist and populariser of phrenology in Britain, that its 'laws are so latent as to forbid even the formation of a hypothesis

concerning them', Eliot to Combe, in Haight (ed.), *The George Eliot Letters: 1840–1870. Supplementary Letters* (New Haven and London: Yale University Press, 1978), 45. On Eliot's rejection of mesmerism and spiritualism see Vanessa D. Dickerson, *Ghosts in the Noontide: Women Writers and the Supernatural* (Columbia and London: University of Missouri Press, 1996), 90–1. See also Feltes, 1968 and Chase, 1984, 105–6.

31 Jane Welsh Carlyle to Eliot, 20 February 1859, in Haight, 1854–5, iii. 18, emphasis in original.

32 Darwin's reading notebook, July 1859; see Frederick Burkhardt et al (eds), *The Correspondence of Charles Darwin*, 20 vols (Cambridge: Cambridge University Press, 1985–2011), iv. Appendix IV, 496.

33 Darwin to J.D. Hooker, 26 May 1859, Darwin Correspondence Database, http://www.darwinproject.ac.uk/entry-2464 (accessed 25/2/2011).

34 J.D. Hooker to Darwin, 26 September 1865, Darwin Correspondence Database, http://www.darwinproject.ac.uk/entry-4899 (accessed 25 February 2011); Darwin to J.D. Hooker, 27 or 28 September 1865, Darwin Correspondence Database, http://www.darwinproject.ac.uk/entry-4901 (accessed 25 February 2011).

35 Eliot, *Adam Bede*, ed. by Carol A. Martin (1859; Oxford: Oxford University Press, 2008), 463.

36 *Ibid.*, 268.

37 *Ibid.*, 272.

38 See also Shuttleworth, 2010, 97, on this passage. Stéphanie Dupouy argues that in *Expression* Darwin characterises human expression as less pronounced in 'normal civilized adults' ('The Naturalist and The Nuances: Sentimentalism, Moral Values, and Emotional Expression in Darwin and The Anatomists', *Journal for the History of the Behavioral Sciences*, 47 (2011), 352), though Darwin's notebooks suggest a less hierarchical ordering. This reading suggests further scope for agreement between Eliot and Darwin.

39 Lewes, 1858, 255, 365, 368. On Lewes's later resistance to anthropomorphism see Lewes, *Problems of Life and Mind*, 5 vols (London: Trübner & Co, 1875–79), First Series, *The Foundations of a Creed*, 1875, i. 124. *Problems of Life and Mind* was published in three series: First Series: (1875), *The Foundations of a Creed*, 2 vols; Second Series: (1877), *The Physical Basis of Mind*; Third Series: (1879), 2 vols: *Problem the First: The Study of Psychology. Its Object, Scope, and Method*); *Problem the Second: Mind as a Function of the Organism.*

40 For a broad historical and contemporary consideration of anthropomorphism see Lorraine Daston and Greg Mitman (eds), *Thinking with Animals: New Perspectives on Anthropomorphism* (New York: Columbia University Press, 2005); on the more negative aspects, see James A. Serpell, 'People in Disguise: Anthropomorphism and the Human–Pet Relationship', in Daston and Mitman, 121–36.

41 See Marc Bekoff in this volume; see also Bekoff, *Animal Passions and Beastly Virtues: Reflections on Redecorating Nature* (Philadelphia: Temple University Press, 2006); and Bekoff, 'Animal Emotions and Animal Sentience and Why They Matter: Blending "Science Sense" with Common Sense, Compassion and Heart', in Jacky Turner and Joyce D'Silva (eds), *Animals, Ethics, and Trade* (London: Earthscan Publishing, 2006), 27–40.

42 Lewes, 1858, 396.

43 Eliot to John Sibree, February 1848, Haight, 1954–5, i. 250.

44 See Graeme Tytler, *Physiognomy in the European Novel: Faces and Fortunes* (Princeton: Princeton University Press, 1982), 113.

45 All references to this library are from William Baker, *The George Eliot–George Henry Lewes Library: An Annotated Catalogue of their Books at Dr William's Library, London* (New York and London: Garland Publishing, 1977).

46 On Eliot in relation to Spencer see Shuttleworth, 1984; Nancy L. Paxton, *George Eliot and Herbert Spencer* (Princeton: Princeton University Press; 1991); and Thomas Dixon, *The Invention of Altruism: Making Moral Meanings in Victorian Britain* (Oxford: Oxford University Press for the British Academy, 2008), 110–11.

47 Eliot to Sara Sophia Hennell, 25 August 1855, in Haight, 1954–5, ii. 213, emphasis in original. Lewes's 'Review of Spencer, Principles of Psychology' appeared in the *Leader*, 20 October–3 November 1855, 1012–13, 1036–1037, 1062–3.

48 Herbert Spencer, *An Autobiography*, 2 vols (London: Williams and Norgate, 1904), i. 395.

49 Eliot to Hennell, 19 February 1856, in Haight, 1954–5, ii. 228.

50 19 February 1856, Haight, 1954– 5, ii. 228; George Eliot Journal, [26?] July 1857, in Haight, 1954–5, 369. See also E.M. Tansey, 'George Eliot's Support for Physiology: The George Henry Lewes Trust, 1879–1939', *Notes and Records of the Royal Society of London*, 44 (1990), 221– 40; and Tansey, '"... the science least adequately studied in England": Physiol-

ogy and the George Henry Lewes Studentship, 1879–1939', *Journal of the History of Medicine and Allied Sciences*, 47 (1992), 163–85, on Eliot and Lewes's interest in physiology.

51 See Baker, 1977, 132.

52 Lewes to Alexander Main, 22 July 1875, in Haight, 1954–5, vi. 157.

53 Lewes, *Mind as a Function of the Organism*, 1879, 7.

54 Lewes, 'Studies in Animal Life', *John Bull*, 24 December 1859; reprinted in *Cornhill Magazine*, 1 (1860), 438–47, http://darwin-online.org.uk/supplementary_works.html (accessed 21 February 2011); and Lewes, 'Mr Darwin's Hypothesis', *Fortnightly Review*, 3 (1868), 353–73, 611–28; *Fortnightly Review*, 4 (1868), 61–80, 492–509.

55 Lewes, *Fortnightly Review*, 3, 1868, 353.

56 *Ibid.*, 364.

57 Lewes, *Fortnightly Review*, 4, 1868, 75.

58 Lewes, 'The Variation of Animals and Plants under Domestication', unsigned review, *Pall Mall Gazette*, 10, 15, and 17 February 1868, in Francis Darwin (ed.), *The Life and Letters of Charles Darwin, including an Autobiographical Chapter*, 3 vols (London: John Murray, 1887), ii. 259. See also Ernest A. Parkyn, *Darwin, His Work and Influence: A Lecture Delivered in the Hall of Christ's College Cambridge* (London: Methuen, 1894), 30– 1.

59 Lewes, 'Variation of Animals and Plants', 1868, in Francis Darwin, 1887, ii. 259.

60 Darwin, *On the Origin of Species by Means of Natural selection, or the Preservation of Favoured Races in the Struggle for Life*, 5th edn (London: John Murray,1869), 356; Darwin, *The Origin of Species by Means of Natural selection, or the Preservation of Favoured Races in the Struggle for Life*, 6th edn (London: John Murray, 1872), 425.

61 Eliot to Charles Bray, 25 November 1859, in Haight, 1954–5, iii. 214.

62 Eliot to Barbara Bodichon, 5 December 1859, in Haight, 1954–5, iii. 277.

63 Eliot, *The Mill on the Floss*, ed. by Haight (1860; Oxford: Oxford University Press, 2008), 87. Subsequent page references to *The Mill on the Floss* are given parenthetically in the text.

64 Eliot to Elma Stuart, 22 September 1874, in Haight, 1954–5, vi. 81.

65 See Baker, 1977, 192. For an excellent study of Victorian ethics and evolution see Dixon, 2008, chapter 4. On the ethical implications of Darwin in contemporary biology see Bekoff, *The Emotional Lives of Animals. A Leading Scientist Explores Animal Joy, Sorrow, and Empathy and Why They Matter* (Novato, California: New world Library, 2007); Bekoff,

Animal Passions, 2006; 'Animal Emotions', 2006; and the afterword to this volume; Bekoff and Jessica Pierce, *Wild Justice: The Moral Lives of* Animals (Chicago: University of Chicago Press, 2009); Frans de Waal, *The Age of Empathy: Nature's Lessons for a Kinder Society* (New York: Harmony Books, 2009); 'Putting the Altruism Back into Altruism: The Evolution of Empathy', *Annual Review of Psychology*, 59 (2008), 279–300); *Primates and Philosophers: How Morality Evolved* (Princeton, NJ: Princeton University Press, 2006); and *Good Natured: The Origins of Right and Wrong in Humans and Other Animals* (Cambridge: Harvard University Press, 1996); Pierce, 'Mice in the Sink. On the Expression of Empathy in Animals', *Journal of Environmental* Philosophy, 5 (2008), 75-96; and 'Morality in Humans and Animals', in Bekoff, ed., *Encyclopedia of Human-Animal Relationships* (Westport, Connecticut: Greenwood Press; 2007); and Elliott Sober and David Sloan Wilson, *Unto Others: The Evolution and Psychology of Unselfish Behavior* (Cambridge, MA: Harvard University. Press; 1998). On new research on the evolution of empathy and cooperation see Sarah Blaffer Hrdy, *Mothers and Others: The Evolutionary Origins of Mutual Understanding* (Cambridge: Harvard University Press, 2009).

66 Lewes, *The Study of Psychology*, 1879, chapter 1.

67 Auguste Comte, *A General View of Positivism*, trans. by J.H. Bridges (1848; London: Trubner and Co., 1865), 103, 104. Eliot and Lewes owned an 1880 edition of the translation by Bridges.

68 See Comte, *The Positive Philosophy of Auguste Comte, Freely Translated and Condensed by Harriet Martineau*, 2 vols (London, Chapman, 1853), i. 414. On ways in which Lamarck informed Comte's sociology, see Michael Bourdeau, 'Comte's Lamarckian Heritage', in Jorge Martinez-Contreras (ed.), *Darwin's Evolving Legacy* (Mexico: Siglo XXI, 2011), 232–42. On Comte, Lamarck and the environment, see J.F. Braunstein, 'Le concept de milieu, de Lamarck à Comte et aux positivistes', in G. Laurent (ed.), *Jean-Baptiste Lamarck* (Paris: Editions du CTHS, 1997), 557–71.

69 Cited in Rod Preece, *Awe for the Tiger, Love for the Lamb: A Chronicle of Sensibility to Animals* (Vancouver: University of British Columbia Press, 2002), 254.

70 Comte, *System of Positive Polity, or Treatise on Sociology, Instituting the Religion of Humanity*, 4 vols (Paris: Carillian–Goeury and Vor Dalmont, 1851–4; English translation, London: Longmans, 1877), i. 495. Dixon has recently argued persuasively that Eliot was more drawn to the idea of

sympathy and empathy than to Comtean altruism; see Dixon, 2008, 112; and, on Eliot and Comte more generally, 101–4; 107–13. Suzy Anger suggests that Eliot was more drawn to the later Comte of the *System of Positive Polity*, in which Comte engaged more directly with moral questions and began to dilute his empiricism, foregrounding the affective, while Lewes was more interested in the *Cours de philosophie positive*, which was committed to establishing a realist and objectivist account of knowledge (Suzy Anger, 'George Eliot and Philosophy', *The Cambridge Companion to George Eliot,* ed. by George Levine (Cambridge: Cambridge University Press, 2001), 76–97: 85). Anger notes that Eliot kept two notebooks in which she selected passages from Comte's *System of Positive Polity* which showed Comte's attempts to combine objective and subjective methods (Anger, 2001, 86). See also Lewes, *The History of Philosophy from Thales to Comte*, 4th edn (London: Longman, Green & Co., 1871); and T.R. Wright, *The Religion of Humanity: The Impact of Comtean Positivism on Victorian England* (Cambridge: Cambridge University Press, 1986).

71 Eliot to Frederic Harrison, 15 August 1866, Haight, 1954–5, iv. 300.

72 *Ibid.*, 301.

73 Eliot to Clementia Taylor, 18 July 1878, in Haight, 1954–5, vii. 44, emphasis in original.

74 Henry James, '*Daniel Deronda*: A Conversation', *Atlantic Monthly*, 38 (1876), 684–94, in Carroll, 2009, 432.

75 R.H. Hutton, 'The Melancholy of *Middlemarch*' (Review of *Middlemarch*, Book 4) *Spectator* (1872), in Carroll, 2009, 297.

76 A.V. Dicey, unsigned review of *Middlemarch*, *Nation*, 16 (1873), 60–2, 76–7, in Carroll, 2009, 350.

77 Henry James, unsigned review of *Middlemarch*, *Galaxy*, 15 (1873), 424–8, in Carroll, 2009, 359.

78 Sidney Colvin, review [of *Middlemarch*], *Fortnightly Review*, 13 (1873) 142–7; in Carroll, 2009, 331, 332. Colvin quotes Juvenal, *Satire*, I, 85, 'their wishes, fears, angers, pleasures'.

79 Colvin, 1873, in Carroll, 2009, 333.

80 *Ibid.*, 334.

81 *Ibid.*

82 *Ibid.* On Darwin's own use of physiognomy in the *Descent*, noted in the introduction to this volume, see Lucy Hartley, *Physiognomy and the Meaning of Expression in Nineteenth-Century Culture* (Cambridge: Cambridge

University Press, 2001) chapter 5; and Richard Twine, 'Physiognomy, Phrenology and the Temporality of the Body', *Body and Society*, 8 (2002), 67– 88. Jonathan Smith argues instead for the absence of physciognomy in Darwin's work in *Charles Darwin and Victorian Visual Culture* (Cambridge: Cambridge University Press, 2009), 182, linking Darwin instead to the visual conventions of portrait photography.

83 GHL Diary, ms diaries for 1869-1876, Beinecke Rare Book and Manuscript Library, Yale University, cited in Eliot, *George Eliot's Daniel Deronda Notebooks*, ed. by Jane Irwin (Cambridge: Cambridge University Press, 1996), 242, note 2.

84 Eliot, 1996, 242.

85 Unsigned review, 'Mr Darwin's New Book', *Literary World*, 29 November 1872, 337 –8: 337.

86 Darwin, 1872, 127.

87 Eliot, *Daniel Deronda*, 1876, ed. by Terence Cave (London: Penguin, 1995), 284. Subsequent page references to *Daniel Deronda* are given parenthetically in the text.

88 The remark recalls Spencer's *Principles of Psychology*, 2 vols (1855; London: Williams and Norgate, 1872), ii. 138.

89 See note 6 for readings of Gwendolen as hysteric.

90 Dicey, unsigned review of *Daniel Deronda*, *Nation*, 23 (1876), 245-6, in Carroll, ed., 2009, 402.

91 R.E. Francillon, review of *Daniel Deronda*, *Gentleman's Magazine*, 17 (1876), 411–27, in Carroll, 2009, 396.

92 For a recent evaluation of science and emotion see Paul White, 'Darwin's Emotions. The Scientific Field and the Sentiment of Objectivity', *Isis*, 100 (2009), 811–26.

93 See Beer, 2009, 139, for discussion of this passage.

94 James, 1876, in Carroll, 2009, 425.

95 *Ibid.*, 425.

96 *Ibid.*, 427.

97 *The Times*, 13 December 1872, 4. On acting and expression see Smith, 2006, 229–33. See also chapter two, 65 and 82 n. 73 in this volume.

98 On Rejlander see Phillip Prodger, *Darwin's Camera: Art and Photography in the Theory of Evolution* (Oxford: Oxford University Press, 2009), 308–9.

99 Dowden, 1877, in Carroll, 2009, 444.

100 *Ibid.*, 443.

101 *Ibid.*

102 *Ibid.*, 444

103 *Ibid.*, 445.

104 *Ibid.*,

105 Darwin, 1871, i. 41– 2.

106 Mary A. Wollstonecraft, *Vindication of the Rights of Men; A Vindication of the Rights of Woman; An Historical and Moral View of the Origin and Progress of the French Revolution* (1792; Oxford: Oxford University Press, 2009), 100; see also 155 and 222.

107 K.K. Collins, 'G.H. Lewes Revised: George Eliot and the Moral Sense', *Victorian Studies*, 21 (1978), 463–92: 484, contains all revisions to the manuscript of Lewes's *The Study of Psychology*, 1879.

108 Darwin, 1871, i. 78; Collins, 1978, 485.

109 Collins, 1978, 486.

110 Darwin, 1871, ii. 392.

111 Collins, 1978, 487.

112 Darwin, 1871, i. 89, 88–9.

113 Collins, 1978, 490; Darwin, 1871, i. 91.

114 *Ibid.*, 490.

115 *Ibid.*, 491.

116 See Dixon, 2008, 110.

117 Collins, 1978, 491.

118 *Ibid.*, 492.

119 Darwin, 1871, i. 89.

120 Henry James, review of Cross's Life, *Atlantic Monthly* 55 (1885), 668-78, in Cox, ed., 2009, 490-504 in Carroll, 2009, 498–9, emphasis in original.

121 Lewes, *The Study of Psychology. Its Object, Scope, and Method*, 142. Quotation from Darwin, 1871, i. 48–9.

122 Lewes, *Ibid.,* 143.

123 Eliot to Mrs Elma Stuart, 12 July 1877, in Carroll, 2009, 393.

124 Darwin to Emma Darwin, April 1858, in Francis Darwin, 1887, ii. 114.

6

Between Medicine and Evolutionary Theory: Sympathy and Other Emotional Investments in Life Writings by and about Charles Darwin

David Amigoni

In his *Autobiography*, first published posthumously in 1887 in *The Life and Letters of Charles Darwin*, edited by Francis Darwin (his own son), Darwin famously recalls his father as a patriarchal colossus, physically imposing (6 ft 2", twenty-four stone), and intimidating ('many people were much afraid of him'). Darwin recalls his father complaining, bitterly, to his 'ordinary' schoolboy son, who seemed to operate 'rather below the common standard of intellect', that he appeared to care 'for nothing but shooting, dogs, and rat catching, and you will be a disgrace to yourself and all your family'.[1]

Bringing disgrace upon the Darwin family was to take in vain the familial inheritance that bore down upon young Charles. Firstly, there was the legacy of medicine; Robert Waring Darwin had a thriving and prestigious medical practice in Shrewsbury and its environs. He himself was the son of Dr Erasmus Darwin of Lichfield, the most famous physician in later eighteenth-century England, the man who had declined the offer to become court physician to George III. Erasmus also handed down a powerful legacy of scientific invention, and speculation about the origins and workings of life in publications such as *Zoonomia* (1794), and *The Temple of Nature* (1802). Erasmus's intellectual affiliations, cultivated and sustained through the Lunar Society of midlands inventors, intellectuals and industrialists, were consolidated through marital alliances: in 1796, Robert Waring Darwin married Susannah Wedgwood, daughter of Josiah, the pottery manufacturer of Etruria, Staffordshire. There was, thus, also a legacy of business acumen, so that when Robert Waring Darwin berated his ordinary little son for threatening to bring disgrace upon 'all your family', he would have had in mind the dynasty comprising the industrial aristocracy of the north midlands, with its connections to fashionable London, the regional aristocracy and gentry – Robert Waring Darwin was to all intents and purposes their source of finance – as well as the wealthy and prestigious line of doctors.[2] When Charles Darwin decided not to join the line of doctors, abandoning medicine after spending two years studying the subject at Edinburgh, he notes that his father 'perceived' his sense of unhappiness at the prospect of entering the profession.[3] Given that the doctor 'had the art of making every one obey him to the letter' it is remarkable that he sympathised with his son's feelings sufficiently to permit him to abandon a medical career.[4]

To credit Robert Waring Darwin's perceptiveness is to invoke one of the chief character traits that Darwin recalled of his father: what he remembered above all was the patriarch's capacity for observation, reading character, and the emotional attribute of 'sympathy': 'His sympathy', the son recalled, 'was not only with the distresses of others, but in a greater degree with the pleasures of all around him'.[5] In recognition of the intersubjective dynamics of sympathy, the placing of the self in the place of another to experience, vicariously, the pleasures and pains subjecting another's body, Darwin insists that his father was always 'scheming to give pleasure to others' through generous actions. The pleasure generated by sympathy becomes linked first and foremost in Darwin's autobiographical narrative to the sharing of accumulated wealth. The example that Darwin gives of the 'scheming pleasure' motivated by sympathy that his father could bestow is of an unsecured loan of £10,000 to a local Shrewsbury manufacturer threatened with bankruptcy. The doctor's 'intuitive perception of character' leads him to believe that the manufacturer could repay the money, and, although then only a young man for whom such a sum of money is enormous, he is proved right.[6] This is a neat anecdote; however, like the story about Erasmus Darwin depositing Robert Waring Darwin in Shrewsbury with only £20 in his pocket before abandoning him, sink-or-swim, to his professional fate, it is possibly one embellished by family tradition. Robert Waring Darwin's account books indicate that, among the vast array of cautious but lucrative investment transactions that he conducted between the 1790s and 1830s, unsecured loans motivated solely by 'sympathy' and the spread of pleasure were probably uncommon.[7]

Nonetheless, this anecdote is valuable for beginning to illuminate the arguments that this chapter will pursue concerning sympathy and its relations to ideas about wealth, privilege, and heredity. To this end the concept of sympathy accumulated a varied range of significations in Darwin's life writings about himself, his grandfather and father. If the anecdote about the unsecured loan signifies the scheming pleasures that Robert Waring Darwin's wealth enabled, Darwin also invokes the idea of a sympathetic bearing as the peculiar advantage that his father, and grandfather, brought to medical practice. Furthermore, Darwin also conceived sympathy as an emotional state that bonded men of science together into a class of intellectual practitioners. A version of this multi-faceted construction of sympathy also figured in Darwin's science, for Darwin theorised sympathy as a naturally selected emotional foundation to human morality in *The Descent of Man* (1871). Thus a complex idea of sympathy was doubly suspended: between medicine, science and the evolution of social intercourse on the one hand; and biological and cultural stories about heredity on the other. In broader context, the chapter examines evolutionary accounts of the emotions as a platform for the Darwin family's sense of its members as elite, self-conscious, socially networked practitioners of medicine and scientific speculation, endowed with entry qualifications into a Victorian intellectual class.

The idea of sympathy bonds three generations of the Darwin family across three acts of life writing: Charles Darwin's biography of his grandfather Erasmus, published in 1879; his own autobiographical recollections written around 1876; and, the text in which those recollections first came to public attention, Francis Darwin's posthumous life and letters of his own father. These acts of life writing 'guided' reading and understanding of the different strains comprising the Darwinian theory of evolution (including the theoretical legacy of Erasmus Darwin). Three out of four generations of Darwin males were thus suspended between medicine and evolutionary science. The career paths of Erasmus and Charles are well known, but Francis Darwin also undertook medical training before going on to develop a career as a botanist at Cambridge University, while simultaneously shaping and defending his father's legacy through the publication of the 'life and letters', and the vigilant monitoring of his estate's copyright.[8]

My chapter will explore the importance of the concept of 'sympathy' in the biographical representations of the Darwins' intersecting family and professional lives as set out in Charles Darwin's biography of his grandfather, the great Lichfield physician Erasmus Darwin; and, as we've seen, Darwin's own autobiographical sketch; and the three-volume life and letters work compiled by Francis Darwin about his father, Charles. It is important to remember that sympathy was both an important concept in Charles Darwin's own account of human evolution, and an active, subtly evolved emotional trait that Darwin had to exercise in eliciting information, building a reputation in a field other than medicine; and, not least, observing and writing.

George Levine, in *Darwin Loves You* (2006), his heroic work about the ethical and emotional basis of Darwin's 're-enchantment' of a disenchanted world, acknowledges in his chapter on 'Darwin and Pain' that 'spending time with the many biographies and biographical narratives about him, I have found it hard not to feel a sentimental fondness for the man'.[9] In other words, Levine comes to feel sympathy for this 'nice guy', who belongs 'almost unself-consciously' to the elite. Most would agree with Levine's assessment that Darwin, though by no means a saint, was a fundamentally sympathetic and feeling man. Levine's chapter is organised around the emotionally devastating effect that the death of Darwin's eldest daughter, Annie, had on him in 1851. The long-term resonance of this loss to Darwin cannot be underestimated, but its effects have been most movingly acknowledged and brought to public attention by his most recent biographers, notably Adrian Desmond and James Moore (1991), Janet Browne (1995); and, especially, Randal Keynes (2001), in a work that did so much to explain the relations between this loss and Darwin's work on human evolution. Indeed, the recent focus on an emotional Darwin actually emphasises the fact that the biographical images that we have of Darwin's emotional and intellectual disposition have been products of history and selection, as the most recent research of Browne (2010) and Bernard Lightman (2010) has demonstrated.[10] In a letter to John Murray the publisher about the production of Charles Darwin's life and letters, Francis Darwin agreed with the publisher 'as to the necessity of showing especially what my father was as a <u>man</u>'.[11]

However, this was no straightforward task: Francis Darwin had to solicit and arrange letters; edit Darwin's autobiographical recollections; and defend the image created against incursions of copyright, and items of 'rogue' correspondence that resonated with discordant notes.[12] My point is that Levine's 'unself-conscious ... nice guy' who generates sympathy was self-consciously fashioned from a much more fraught and conflicted emotional landscape. If Darwin felt the loss of Annie deeply, he avoided talking about it to his family; and his children knew not to talk to him about her. This produced its own repressive effects: Randal Keynes writes about Maud du Puy, his American great-grandmother who married Darwin's second son George in 1883. On meeting the family, she noted their kindness and cordiality, but a fundamental reticence in manners and a sense of outward-facing, public mission: 'there is no personal talk, it is more about the world at large and everything in general'.[13] Such a complex emotional family topography can translate, unconsciously, into biographical construction. Thus, the most recent *Oxford Dictionary of National Biography* entry on Darwin, by Desmond, Moore and Browne, which replaces Francis Darwin's original, records the death of Annie and its wrenching emotional effect; before switching immediately to the state of the family finances and investments in 1851.[14]

A Line of Doctors: Life Writing and the Inheritance of Sympathy

According to Charles Darwin, sympathy motivated Robert Waring Darwin in his issuing of financial loans. It also propelled his success in medicine: 'I suppose', Darwin reflected, 'that it was his sympathy which gave him unbounded power of winning confidence, and as a consequence made him highly successful as a physician'. This sympathy paid dividends with 'ladies', who treated the doctor as 'a sort of Father-Confessor':

> He told me that they always began by complaining in a vague manner about their health, and by practice he soon guessed what was really the matter. He then suggested that they had been suffering in their minds, and now they would pour out their troubles, and he heard nothing more about the body. … Owing to my father's skill in winning confi-

> dence he received many strange confessions of misery and guilt. He often remarked how many miserable wives he had known. In several instances husbands and wives had gone on pretty well together for between twenty and thirty years, and then hated each other bitterly; this he attributed to their having lost a common bond in their young children having grown up.[15]

This was a practised sympathy on which Robert Waring Darwin built one plank of the family's wealth and influence.[16] Dr Darwin's sympathetic ruse enabled him to understand the breakdown of love between partners, as the bonds created to sustain selfless parental investment are prone to fall apart when no longer needed by once-needy young offspring. If sympathy was an inherited emotional instinct that brought special utility in child-rearing, then it could also bring with it cruel reversals.

Indeed, for the medical practitioner, the capacity to visualise and enter into the pain inflicted on another brought its own forms of horror, and professional inhibition:

> His great success as a doctor was the more remarkable, as he told me that he at first hated his profession so much that if he had been sure of the smallest pittance, or if his father had given him any choice, nothing should have induced him to follow it. To the end of his life, the thought of an operation almost sickened him, and he could scarcely endure to see a person bled – a horror which he has transmitted to me – and I remember the horror which I felt as a schoolboy in reading about Pliny (I think) bleeding to death in a warm bath.[17]

Life writing can be constructed as a record of inheritances. So far as Charles Darwin is concerned, the 'horror' of blood that prevented him from becoming a physician has been 'transmitted' directly to him from his father. But the status of life writing as a record of inheritance brings both powers of revelation and problems. The biography by Desmond and Moore has become a key source for the study of Darwin since its publication in 1991, and it is important to remember that while it narrates the life of its subject with generous amplitude, it also weaves a story of relations across generations and between the Darwins and the Wedgwoods. In fact, the opening pages of Desmond and Moore's magnum opus allude to a key, mediating biographical text:[18] Charles Darwin's 1879 sketch of the life of his grandfather, Erasmus. Charles's biographical sketch began thus:

> As the character of a man depends in some degree on the circumstances under which he has been brought up, it will be advisable to give a very short account of the family to which Erasmus Darwin belonged. It is more important to show to what extent a man inherits and transmits his characteristic qualities; for every addition, however small, to our knowledge on this head is a public benefit, as well as spreading a belief in inheritance.[19]

Inheritance happens through biological descent between generations – but it is also something that one 'believes' in, and it is a 'public benefit' which must be disseminated, through biography: writing a family memoir, Darwin has a strong sense of himself as a public moralist. Except that, in this instance at least, the public benefit went publicly unuttered. For this was one of the passages that was cut by Henrietta Darwin, Charles's daughter, restored only recently by Desmond King-Hele's edition. As Desmond and Moore point out, Henrietta was keen to remove material from the biography that was detrimental to the reputation of the family. In this instance, she placed the whole question of biography as a source of evidence about inheritance off limits. If inheritance was biological, then perhaps respectable Charles Darwin FRS of Down House had inherited old Erasmus's militant atheism, his sexual excesses, and his tendency to 'sarcasm'.

Life writing inscribes silent struggles around identities and reputations. In the case of the Darwin family the struggle was conducted around the moral status of both doctors and evolutionary theory: 'sympathy' emerged as a contested feature of the character of Erasmus Darwin: the doctor had perhaps been wanting in the more refined virtues. Darwin's biography quotes in order forthrightly to rebut Anna Seward's biography of 1804 which held that Erasmus was a 'large mass of genius and sarcasm' that repulsed 'the confidence and sympathy of friendship'.[20] Seward, his early and controversial biographer, thought Erasmus to be emotionally barren, bereft of sympathy. Darwin's biography of his grandfather insisted on its re-insertion into his stock of virtues. Quoting a supporter, Darwin's biography insisted that for his grandfather, 'sympathy and benevolence were the most striking features. He felt very sensibly for others, and, from his knowledge of human nature, he entered into their feelings and sufferings into the different circumstances of their constitution, character, health, sickness, and prejudice' and 'he extended his humanity to every sentient being'.[21] This was precisely the identity that Darwin had earlier ascribed to his father, in the autobiographical recollections that would go on to frame Francis Darwin's life and letters of his own father. This was the moral inheritance that Charles Darwin wished to inscribe and extend. However, in practice there was a stranger, more difficult story to narrate in which versions of sympathy still played a role.

Francis Galton, in his account of the Darwin family in *English Men of Science* (1874), an argument about the role of nature over nurture in the emergence of the Victorian scientific elite, identified another perspective on the capacity for 'love' passed down the generations by the Darwin family. 'Speaking from private sources of knowledge', Galton observed from his position as second cousin, it was 'love for natural history and theory, and of an aptitude for collecting facts in business-like but peculiar ways' that marked the Darwin males as distinctive. Asserting a stronger role for nature rather than nurture, he was 'sure that these characteristics are hereditary rather than traditional', though the evidence he cites could be said equally to support the case for nurture, in so far as it was amenable to *resistance*. For he adds, 'there is also a strong element of individuality in the race which is averse to traditional influence'.[22] Robert Waring Darwin may have turned on the sympathy in order to collect 'business-like' facts about the mental condition of lady patients which he sought to pass onto his son, destined as he seemed for medicine: though what the son took from the father owes, perhaps, more to the 'individuality' and aversion to 'traditional influence' discerned by Galton.

Galton's formulation poses a general problem: is identity evidence of transmission of biological material between the generations, or instead what we might call, after the sociologist Pierre Bourdieu, evidence of the transmission of 'cultural capital'? Life writings are a source for exploring this question, and thus have wider implications for the way in which we examine the relationship between medicine, science and literature. A rather strange picture of inheritance emerges from the selection of anecdotes that Darwin relays about his father's capacity for sympathy in the autobiography that frames Francis Darwin's edited life and letters. Taken together, these anecdotes about Robert Waring Darwin negotiate a peculiar emotional and intellectual path away from medicine, and towards a displaced, slightly mis-recognised version of scientific speculation.

Cheese out of Place: A Place in the Athenæum, a Place in Science

When Charles Darwin reminisced of his father's uncanny powers of sympathy and perception in his autobiographical recollections, the most crisply re-told anecdote had little to do with the sympathetic moments between confessing patient and physician at the bedside, though it had as its context a medical consultation with one of Robert Waring Darwin's elite aristocratic patients. Instead it was more concerned with a 'queer concatenation' between the unexpected discovery of English cheese in Holland (cheese out of place, one might say), cognitive leaps, and the sympathies exchanged between eminent families:

> The following story shows what good guesses my father could make. Lord Shelburne, afterwards the first Marquis of Lansdowne, was famous (as Macaulay somewhere remarks) for his knowledge of the affairs of Europe, on which he greatly prided himself. He consulted my father medically, and afterwards harangued him on the state of Holland. My father had studied medicine at Leyden, and one day [while there] went a long walk into the country with a friend who took him to the house of a clergyman (we will say the Rev. Mr. A——, for I have forgotten his name), who had married an Englishwoman. My father was very hungry, and there was little for luncheon except cheese, which he could never eat. The old lady was surprised and grieved at this, and assured my father that it was an excellent cheese, and had been sent her from Bowood, the seat of Lord Shelburne. My father wondered why a cheese should be sent her from Bowood, but thought nothing more about it until it flashed across his mind many years afterwards, whilst Lord Shelburne was talking about Holland. So he answered, 'I should think from what I saw of the Rev. Mr. A——, that he was a very able man, and well acquainted with the state

> of Holland'. My father saw that the Earl, who immediately changed the conversation, was much startled. On the next morning my father received a note from the Earl, saying that he had delayed starting on his journey, and wished particularly to see my father. When he called, the Earl said, 'Dr. Darwin, it is of the utmost importance to me and to the Rev. Mr. A—— to learn how you have discovered that he is the source of my information about Holland.' So my father had to explain the state of the case, and he supposed that Lord Shelburne was much struck with his diplomatic skill in guessing, for during many years afterwards he received many kind messages from him through various friends. I think that he must have told the story to his children; for Sir C. Lyell asked me many years ago why the Marquis of Lansdowne (the son of grandson of the first marquis) felt so much interest about me, whom he had never seen, and my family. When forty new members (the forty thieves as they were then called) were added to the Athenæum Club, there was much canvassing to be one of them; and without my having asked any one, Lord Lansdowne proposed me and got me elected. If I am right in my supposition, it was a queer concatenation of events that my father not eating cheese half-a-century before in Holland led to my election as a member of the Athenæum.[23]

What can one say about this anecdote, reflecting on a capacity for diplomacy, making 'good guesses', and, in effect, being able to extend one's powers of perception into the 'life' of an object (cheese) and the forms of agency and motive power that transport it from one location to another? In Charles Darwin's skilful re-telling of the story of imported cheese, an inspired guess about the origins of a statesman's knowledge ends in Charles's election to the Athenæum in 1838 as one of the so-called 'forty thieves': arriviste writers who had started to make a mark in science and literature in the early Victorian period, and who were elected to the leading intellectual club in London life. Charles Dickens was another.

Francis Darwin's publication of his father's subsequent life and letters adds, without consciously meaning to, another layer of context and complexity to this anecdote. A long letter to Charles Lyell, dated 9 August 1838 contains Darwin's only significant, contemporary reference to his election to the Athenæum (the present edition of the correspondence seems not to have uncovered another). In the course of a letter describing the present state of his geological work following his return from the *Beagle* voyage, Darwin effectively thanks Lyell for his recent election to the club: 'Your helping me into the Athenæum has not been thrown away, and I enjoy it the more because I fully expected to detest it'.[24] Thus, while Lord Lansdowne, grandson of Lord Shelburne, may have played a formal role in the election of Darwin, it seems to have been Lyell who did the moving and shaking. Darwin's attitude to belonging to the club is interesting. Recalling Levine's remarks concerning Darwin's unself-consciousness about belonging to the elite, his anecdote about his father is careful to observe that he never canvassed to join the club. However, in his letter he informs Lyell that, at the end of new working day (modelled on Lyell's) 'I go and dine at the Athenæum like a gentleman, or rather like a lord, for I am sure the first evening I sat in that great drawing-room, all on a sofa by myself, I felt just like a duke'.[25] It is an interesting expression, as though Darwin is feeling his way by emotional increments into the intellectual aristocracy.

The letter names numerous members of the scientific elite that Darwin has become acquainted with at the Athenæum, and Darwin shares with Lyell his own recent geological speculations, formulated during field work in Scotland on the Glen Roy problem, on which he was shortly to publish a paper.[26] The Glen Roy formation, and the sea-action thesis that Darwin formulated against prevailing theories (wrongly as it turned out), famously required Darwin to speculate on agencies and motive powers that seemed out of place. Glen Roy posed Darwin the problem of erratic rocks, located 2,200 feet above sea level. Darwin's identity as a geologist took him to landlocked gravel pits in an effort to tell stories about sea shells and fossils that, due to the present position of the sea, were seemingly out of place, like the 'Bell-Stone', the ancient and seemingly inexplicable rock feature in Shrewsbury.[27] There is a sense, then, in which Darwin's much later anecdote about his father's perceptive powers in working out the true significance of English cheese out of place in Holland, and the networks of aristocratic sympathy into which it placed him, is a sort of displaced allegory of the kind of intellectual powers that enabled the younger Darwin to become a scientific theorist, and more precisely of what would qualify him to join the ranks of the intellectual aristocracy in the Athenæum.

There is a final point to make about Darwin's letter to Lyell in August 1838: its tone of sympathetic identification, and exclusive belonging. Darwin playfully refers to this long epistle as 'unmerciful' in its delivery of a profusion of news, comment and observation. In one sense, Darwin's letter is unmerciful, or at least unsympathetic, in its recounting of sights observed during his passage to Edinburgh on the packet steamer: 'I enjoyed the spectacle, wretch that I am, of two ladies, and some small children quite sea-sick, I being well'.[28] This is the Darwin who survived and mastered sea-sickness on the *Beagle*, but it is also the Darwin who can humorously, and rather more exclusively, share comic reading tips with the scientifically powerful, and their wives: 'Tell Mrs. Lyell to read the second series of 'Mr. Slick of Slick-ville's Sayings'. ... He almost beats 'Samivel', that prince of heroes'.[29] If Darwin's recommendation has become obscure, the source of the comparison, responsible for Sam Weller the 'prince of heroes', has not. Darwin is confident that Mrs Lyell will sympathise with his powers of discrimination regarding Dickens – another, after all, of the 'forty thieves' elected to the Athenæum in 1838.

As Francis Darwin's life and letters indicates, by 1838, Darwin had started to secure his place in the scientific elite, and to belong by sharing what he refers to, in a letter of September 13 1838, as Lyell's 'geological sympathy'.[30] In fact, the letters selected by Francis Darwin reinforce a construction of sympathy as an approach to scientific enquiry that appears in Darwin's framing autobiographical recollections: of Lyell, Darwin observes that 'He would advance all possible objections to my suggestion, and even after these were exhausted would long remain dubious. A second characteristic was his hearty sympathy with the work of other scientific men'.[31] 'Sympathy' as a bond holding together the scientific elite also figures much later in the life when Francis Darwin deals with Samuel Butler's dispute with Charles Darwin. Butler's accusation of 'falsehood' was an attack on the integrity of natural selection, and was in part triggered by Darwin's publication of the life of his grandfather in 1879, accompanying as it did an interpretive essay on the significance of Erasmus's science by Errnst Krause.[32] Francis Darwin concludes that the 'affair gave my father much pain, but the warm sympathy of those whose opinion he respected soon helped him to let it pass into a well-merited oblivion'.[33]

In the letter that immediately follows this commentary on Darwin's pain and the sympathy of intellectual like-mindedness that consoles it, Darwin writes to the French entomologist J.H. Fabre. Darwin's letter seeks politely and gently to correct Fabre's misunderstanding of one of Erasmus Darwin's entomological observations in *Zoonomia*: thus, there was (in parallel to the Butler–Darwin dispute) an attempt to defend the family's scientific legacy, which had become much more powerful as a result of Darwin's own evolutionary labours: thus, Darwin feels 'sorry that you are so strongly opposed to the Descent theory; I have found the searching for the history of each structure or instinct an excellent aid to observation; and wonderful observer as you are, it would suggest new points to you'. Thus, Darwin gestures to a future in which Fabre might come to sympathise with the best evolutionary minds in science. In the meantime, Darwin concludes by adding that he 'sympathized deeply' with the last sentence in Fabre's book. As Francis Darwin's commentary indicates, the book was a memorial to Fabre's son, who had been his scientific assistant, and had died prematurely.[34] Of course, this poignantly mirrored the situation of biographer and subject: Francis Darwin had worked as his father's assistant during Darwin's botanical researches of the 1870s. Deeper still was a sympathetic identification based on his own loss of a child, his beloved Annie. However, poignancy was always tempered by science, and sympathy remained an emotional instinct with a hard-edged evolutionary history. Darwin attempted to narrate it in the *Descent* and in doing so articulated further justifications for a certain kind of intellectual class.

The Descent of Sympathy: An Evolutionary Rationale for an Intellectual Class and the Role of Biography

It is important to locate the complex construction of an idea of sympathy in Darwin family life writings in the context of Charles Darwin's extended discussion of the concept in the *Descent*. The evolution of the emotion of sympathy as a moral faculty is discussed in volume I.[35] In chapter 3 it is established as further evidence of a continuum between humans and animals and, following Adam Smith, conceived in its highly developed human form as the ability to sense, memorise and recall sensations of pain/pleasure, and to enter imaginatively, vicariously and mimetically into the sensory worlds of others who are undergoing them. Darwin extends the discussion of sympathy in chapter V of the *Descent* (vol. I) where it becomes part of a discussion of biological and cultural modes of inheritance, sympathy being a variable, heritable trait proving that 'these faculties are variable; and we have every reason to believe that the variations tend to be inherited', perfected and advanced by natural selection (159). Darwin places the emotional and moral 'instinct' of sympathy in the context of the human and animal capacity for imitation (161): both sympathy and imitation propel the invention of 'superior arts' that bestow emotional and intellectual advantages on a tribe. 'The social and moral faculties of 'sympathy, fidelity, courage' develop social coherence, and they are capable of advancement and diffusion throughout the world (163). Darwin will return to the diffusion of these 'arts' – what we perhaps think of as 'culture' – in a later phase of the discussion.

Darwin's discussion of the advances brought about by sympathy nonetheless goes on to rehearse problems that he saw arising from the refinement of the same emotion: at this point, a Galtonian, eugenic imperative intervenes. 'There is reason to believe', he observes, that

> vaccination has preserved thousands, who from a weak constitution, would formerly have succumbed to the small-pox. Thus the weak members of civilised societies propagate their kind… excepting in the case of man himself, hardly any one is so ignorant as to allow his worst animals to breed.

And yet, Darwin contends that the instinctual basis of sympathy makes it an impossible drive to resist for

> the aid which we feel impelled to give to the helpless is mainly an incidental result of the instinct of sympathy … [now] more tender and widely diffused. Nor could we check our sympathy, if so urged by hard reason, without deterioration in the noblest part of our nature. The surgeon may harden himself whilst performing an operation, for he knows that he is acting for the good of his patient; but if he were to neglect the weak and the helpless, it could only be for the contingent benefit, with a certain and great present evil. Hence we must bear without complaining the undoubtedly bad effects of the weak surviving. (168–9)

Sympathy produces paradoxes; an instinct that promotes social cohesion and strength must perforce protect the weak. It is striking that Darwin's most paradoxical understanding of sympathy – that certain professionals have to inflict terrible pain precisely to alleviate the greater pains that their powers of sympathy can foresee and imagine – is advanced from the example of medicine and surgery: the very profession which, as his autobiography attests, his father both practised and recoiled from in horror.

Darwin's Galtonian and Malthusian perspective will not, however, let it rest there, with the salvation of 'poor stock' by medicine and a kind, if tortured, heart. He concludes this phase of the discussion by reminding readers of the stringent checks of sexual and natural selection: the weak in body and mind are most likely to be restrained from marriage and reproduction. Darwin follows this with a contrasting observation that builds from the 'natural' logic of successful sexual selection, reproduction, and the inheritance of property down the generations because 'in all civilised countries man accumulates property and bequeaths it to his children', and Darwin admits that this creates inequalities: yet, 'without the accumulation of capital the arts could not progress' (169). Darwin thus allies the discussion of sexual selection as a positive natural check to a statement about the accumulation of wealth in successful families. Darwin's argument about sympathy also turns into a discussion of inheritance – material inheritance rather than biological inheritance, though Darwin's point is that the two are linked. Of course, Robert Waring Darwin left his children, including Charles, a very considerable inheritance.[36] We have seen how Robert Waring Darwin's financial transactions and investments were the source of this inheritance, and Darwin specifically remarks on its scale in the autobiography. Furthermore, Darwin's first example in the autobiography of his father's capacity for sympathy is specifically linked to his ability to award monetary loans.

The move that follows draws the reader back into his discussion of sympathy and imitation being at the basis of the arts of 'invention' – or what came to be seen as the development of culture in a modern civilisation. For in discussing the social ethics of the inheritance of substantial material wealth – which classes should, and should not, be expected to labour from scratch for their livelihood, the means of reproduction – Darwin makes a distinctive contribution to a common nineteenth-century debate about the importance of culture as a good.[37] Darwin distinctively conceives culture as an heritable acquisition and its links to material wealth. In a sense, Darwin elaborates an evolutionary understanding of, and rationale for, a clerisy, an endowed intellectual class, when he argues that 'The presence of a body of well-instructed men, who have not to labour for their daily bread, is important to a degree which cannot be over-estimated; as all high intellectual work is carried on by them, and on such work material progress of all kinds mainly depends' (169). Darwin's argument reminds us, firstly, of where Darwin had moved to, professionally and emotionally, as he entered the last full decade of his life in the 1870s: he had become a producer of high intellectual work, and the work that he produced was supported by inherited material wealth that granted him life-long independence. Secondly, it reminds us of our very constitution as readers of high intellectual work about evolution by natural and sexual selection and its public benefits, including its capacity for spreading a scientific *understanding* of the emotion of sympathy, elaborated from within a class of *sympathetic*, like-minded intellectuals.

There is a key role, of course, for biography and life writing in this conception of intellectual work. Charles Darwin himself conceived of it when he wrote the biography of his grandfather as a 'public benefit' that would spread belief in the importance of inheritance. Darwin's anecdotal material on his doctor father, included in his autobiography, was perhaps both more deliberate and curiously allegorical than readers have so far been inclined to recognise. In turn, Francis Darwin became more than his father's assistant: he also became Charles Darwin's biographer, providing the autobiography with pride of place in a work that defended the family's professional and intellectual legacy. Though Francis Darwin's communications with the house of John Murray are unfailingly courteous, sympathetic ('I am obliged for your kindness'), often self-deprecating (in ways learned, presumably, from his father), there is still little doubt that the biography was an act of territorial defence. Francis Darwin and his siblings were acutely aware of the possibility of rival biographers, such as Darwin's devoted younger disciple and correspondent, George J. Romanes, staking a claim;[38] and rival versions of the life, carrying un-policed correspondence, drifting in from the continent.[39]

The preparation of the life and letters became something of a pan-European cultural campaign, waged on behalf of the English scientific elite. The translation of the work into a variety of European languages was expected.[40] As early as 1885, two years before the actual publication of the work, Francis Darwin wrote to the publisher requesting that, following his attendance at a dinner in Paris in honour of the anticipated publication of the life, he make a note to send two copies of the work to two French academicians because 'The Academy is slowly coming round to Evol^{n}'.[41]

If the state-sponsored French Academy was coming around to the idea of evolution, it was coming around to a high intellectual theory of life that that had been cultured, endowed and disseminated by means of a very specific set of family virtues and models of acuity, inherited from the habits of capital accumulation exemplified by Robert Waring Darwin. Upon publication of the life and letters, Francis Darwin could, consequently, drive a hard bargain in ensuring that the family retained its interest in Charles Darwin's 'life', in every sense:

> where you supply the trade with 25 copies at the price of 24, you shall only account to me for 24. In all other cases you agree to pay [insert] /the / 11/20 on <u>all</u> copies sold, <u>and that no other deduction from the sum agreed to be paid to me</u>… Will you kindly let me have a line to the above effect? The financing of the book is a family matter and I am responsible. I hope you will not think me unreasonable in asking you to let me have such a note.[42]

Conclusion

While driving a shrewd bargain, Francis Darwin simultaneously conceived his act of life writing, his own contribution to the high intellectual labour supporting the idea of evolution, in a specifically emotional register. In November 1887, he wrote to Murray indicating that he found sales of the three-volume life to be 'very satisfactory', and, on the back of this success, proposed a single-volume selection of letters, based on the autobiography. In proposing this, Francis Darwin reflected on the pleasures and feelings associated with the production of the brief life: 'I wrote the article in Leslie Stephens [sic] Dict of nat Biogr on my father (soon to come out) and I found I could do the kind of thing I am thinking of with pleasure + 'go' as far as my own feelings went...'[43] If Robert Waring Darwin's sympathetic bearing was exemplified by his capacity to spread 'pleasure' through monetary loans, then Francis Darwin could satisfy his own feelings of usefulness and value to others through the production of a life. The published life would, presumably, further disseminate these pleasures publicly. A complex economy of 'sympathy' was thus in operation, and it was one facet of it that George Levine glimpsed when he perused the ever-burgeoning pages of Darwin biography to encounter again 'the nice guy', the man who invites sympathy. As I have suggested, there were, however, the Darwins, inter-generationally, who evoked many complex ways in which a culture of sympathy could evolve and be spread around.

I am grateful to Angelique Richardson for providing me with the opportunity to develop this chapter; Paul White, who gave me valuable leads on Robert Waring Darwin's papers; Anna Barton for collaborative work on Victorian biography and its role in publication practices; and the anonymous reviewers of this essay. I would also like to thank the Trustees of the National Library of Scotland and the Syndicate of Cambridge University Library for permission to quote from unpublished material.

Notes

1 Francis Darwin (ed.), *The Life and Letters of Charles Darwin, including an Autobiographical Chapter*, 3 vols (London: John Murray, 1887), i. 18, 32.

2 For instance, in 1845, Robert Waring Darwin loaned to the Earl of Powis [sic] a mortgage on the Llynstwyn (?) estate of the sum of £12000, at 3.5% interest (£203 returned in interest half yearly), Account Book 1831–1848, 77, DAR 265: 9, Darwin Papers, Cambridge University Library; Janet Browne, *Charles Darwin: Voyaging* (London, Jonathan Cape, 1995), 6–8.

3 Francis Darwin, 1887, i. 45.

4 *Ibid.*, i. 18.

5 *Ibid.*, i. 12.

6 *Ibid.*, i. 12.

7 *Ibid.*, i. 8; see Robert Waring Darwin's account books, 1790–1815, DAR

227.5: 38, and 1815–1830, DAR 227.5: 82, Darwin Papers.

8 For Francis Darwin's life see *Oxford Dictionary of National Biography*, http://www.oxforddnb.com/view/article/32717 (accessed 18 August 2010).

9 George Levine, *Darwin Loves You: Natural Selection and the Re-enchantment of the World* (Princeton and Oxford, Princeton University Press, 2006), 138.

10 See Janet Browne, 'Making Darwin: Biography and the Changing Representations of Charles Darwin', *Journal of Interdisciplinary History*, 40 (2010), 347–73; and Bernard Lightman, 'The Many Lives of Charles Darwin: Early Biographies and the Definitive Evolutionist', *Notes and Records of the Royal Society*, 64 (2010), 339–58.

11 Francis Darwin Correspondence, letter to John Murray (1884, no more precise date), MS. 40310, Folios 21–22, John Murray Archive, National Library of Scotland.

12 For minor but interesting edits to the first page of 'Darwin's Recollections', see MS 40310; cf. Folio 23 with the published version in the *Life*: ~~'I have heard my father say that he believed that persons with powerful minds generally had memories extending back to a very early period of life. This is not my case, for~~ my earliest recollection goes back only to when I was a few months over four years old, when we went to near Abergele for sea bathing'; and see 3 October 1899, MS.4031, Folios 61–2, Murray Archive, for the problem of uncollected letters from Darwin appearing without permission.

13 Randal Keynes, *Annie's Box: Charles Darwin, his Daughter, and Human Evolution* (London, Fourth Estate: 2001), 216–17.

14 See *Oxford Dictionary of National Biography*, http://www.oxforddnb.com/view/article/7176 (accessed 12 April 2012).

15 Francis Darwin, 1887, i.12–13.

16 One of Robert Waring Darwin's extant Prescription Books provides some sense of the scale of his practice and his interest in recording the familial relations and networks that it embraced between 1818 and circa 1823; it consists of 341 pages of closely written records, most pages are full, consisting of a name and the site of consultation, followed by a coded prescription; families are grouped on a page, according to the date of the first consultation; that page is then given over to future consultations with either that member of the family, or spouses and kin. There are examples of simultaneous husband and wife consulta-

tions: 1 January 1818, 'Hewitt, Major; Mrs. H., his wife. Same time. nee Lloyd. Perhaps gall stones'; thus, the doctor could note family affiliations prior to marriage, 1, DAR 227.5.67, Darwin Papers.

17 Francis Darwin, 1887, i. 12.

18 Adrian Desmond and Jim Moore, *Darwin* (Harmondsworth: Penguin, 1991), 6–7.

19 Desmond King-Hele (ed.), *Charles Darwin's The Life of Erasmus Darwin: First Unabridged Edition* (Cambridge: Cambridge University Press, 2003), 7.

20 *Ibid.*, 75.

21 *Ibid.*, 79.

22 Francis Galton, *English Men of Science* (London: Macmillan, 1874), 44.

23 Francis Darwin, 1887, i. 15–16.

24 *Ibid.*, i. 294.

25 *Ibid.*, i. 294.

26 *Ibid.*, i. 293.

27 *Ibid.*, i. 41, in Darwin's recollections; he uses reference to the Bell-Stone as an opportunity to critique the quality of geological teaching that he had been subject to during his time as a student in Edinburgh.

28 F. Darwin, 1887, i. 293.

29 *Ibid.*, i. 295.

30 *Ibid.*, i. 296.

31 *Ibid.*, i. 71–2.

32 For an account of this see David Amigoni, *Colonies, Cults and Evolution: Literature, Science and Culture in Nineteenth-Century Writing* (Cambridge: Cambridge University Press, 2007), 146–7.

33 F. Darwin, 1887, iii. 220.

34 *Ibid.*, iii. 221.

35 Darwin, *The Descent of Man, and Selection in Relation to Sex*, 2 vols (London, John Murray, 1871); all page references given parenthetically in the main text.

36 F. Darwin, 1887, i. 19.

37 See Ben Knights, *The Idea of the Clerisy in the Nineteenth Century* (Cambridge: Cambridge University Press, 1978).

38 See Joel S. Schwartz, *Darwin's Disciple George John Romanes: A Life in Letters* (Philadelphia: American Philosophical Society, 2010), section V, 'Darwin's Passing and its Effect on Romanes' for the way in which Romanes, then acting editor of *Nature*, became, by default, a biographer of Darwin for the journal with the blessings of Huxley

and Hooker; also for the remarkably emotional terms in which Romanes expressed his condolences to Francis Darwin; his sense of himself as a correspondent of Darwin who possessed valuable biographical materials; and Francis Darwin's styling of Romanes as an *opponent* of his father's thought, which perplexed Romanes. This did not prevent Romanes from offering Francis Darwin's his poetry occasioned by the death of the latter's father that Romanes felt might properly belong in the biography.

39 3 October 1882; 'A friend of mine saw a notice in the papers [insert] / to the effect/ that a German professor was preparing a memoir of my father with a series of letters – unfortunately, he took no specific notice of the name and cannot recall it. May I ask if you have heard of any such forthcoming publication?', Murray Archive, MS. 40310, Folio 7. In 1882, a controversy ensued when Ernst Haeckel, Darwin's foremost German disciple, delivered a lecture in Germany on the relationship between Darwinism and progress. It publicly quoted from a letter, written by Darwin to one of Haeckel's young students, which included the line 'science has nothing to do with Christ'. See Robert J. Richards, *The Tragic Sense of Life: Ernst Haeckel and the Struggle over Evolutionary Thought* (Chicago and London: University of Chicago Press, 2008), 250–2.

40 See Francis Darwin's extensive correspondence of August 1885, MS. 40310, Murray Archive.

41 2 July 1885, MS. 40310, Folio 30–31, Murray Archive.

42 Francis Darwin to J. Murray: 12 November 1887, MS. 40310, Folios 87–8, Murray Archive; in January 1889, Darwin records receiving a cheque for £2,500 for the life and royalties on Charles Darwin's other works, Folios 142–3.

43 21 November 1887, MS. 40310, Folios 92–3, Murray Archive.

7

From Entangled Vision to Ethical Engagement: Darwin, Affect, and Contemporary Exhibition Projects

Monika Pietrzak-Franger

Contemporary culture produces an irrevocable tension between the pervasive, paradigmatic image of Charles Darwin as an icon of pure intellect and his contemporary reconsideration as a scientist phenomenologically anchored in the world he studied. This tension is best illustrated by a combination of two images of Darwin that were for some time exhibited side by side in the Natural History Museum in London. The first, iconic, one is Joseph Boehm's 1898 sculpture of Darwin that crowns the stairs to the upper galleries in the main hall of the museum; the other, hidden in the east-wing gallery which hosted the project *After Darwin: Contemporary Expressions* (2009), is part of Mark Haddon's *24 Emotions* (2009). Inspired by Darwin's categorisation of emotional expressions, *24 Emotions* is a series of short prose pieces accompanied by Darwin's descriptions of specific emotional states and images that formed part of *The Expression of the Emotions in Man and Animals* (1872). Unlike other images used in this installation, the picture based on the 1842 daguerreotype of Darwin and his son William never appeared in the first edition of the book. In the photograph, Darwin, with William on his lap, sits propped against the painted landscape of the photographic studio. As his hand rests on the boy's legs, he shuns the camera with a reluctant, barely discernible smile. Darwin's beard is missing and his body is clearly visible in this moment of intimate exchange with his son. This is an image of a feeling, experiencing and enchanted Darwin.

While many 2009 bicentennial museum projects addressed the question of Darwin's influence and legacy, be it in scientific circles, art or popular culture, only a few put emphasis on the question of his affect and its contemporary relevance. It was especially the Warsaw exhibition *All Creatures Great and Small* (19 December 2009–21 February 2010) and the London project *After Darwin: Contemporary Expressions* (26 June–29 November 2009) that tackled the above-mentioned tension while paying due attention to the question of emotional and corporeal enmeshment in the world as crucial to Darwin's theories. Continuing the themes of an earlier exhibition, *Hot/Cold – Summer Loving* (2007), the Warsaw exhibition *All Creatures* took up the question of human–animal relations and explored the multifaceted processes of reciprocity and bonding among species.[1] Curated by Maria Brewińska, it took cues from Darwin's works which reposition the human within the landscape of other species. Integrating contemporary philosophical and artistic responses to the animal world, the exhibition contrasted a number of ways of human and animal coexistence. Composed of over seventy works by Polish and international artists, including Grzegorz Kowalski, Artur Malewski, Natalia LL, Carolee Schneemann, The Wooster Group, Douglas Gordon, Marc Quinn, Rudolf Schwarzkogler, Joseph Beuys, Marina Abramovic and Vito Acconci, the exhibition was the first in a series of events devoted to the question of human–animal bond. Much smaller in scope, the London project – which gathered seven works by Mark Haddon, Ruth Padel, Diana Thater, Bill Viola, Tina Gonsalves, Gautier Deblonde and a collaborative work by Jeremy Deller, Matthew Killip and Richard Wiseman – concentrated on the human–animal kinship, with particular emphasis on Darwin's *Expression of the Emotions in Man and Animals,* and was part of the contemporary arts programme initiated by the Natural History Museum.[2] Despite differences in scope, approach, and the way in which Darwin's works were referenced, both exhibitions recognised the significance of the contemporary impact of his vision on our own reflection on human–animal relationships. Their rediscovery of a feeling, indeed even an 'enchanted', Darwin, allowed us to reconsider our knowledge of him, and to rethink our relation to nature and our animal kin. In this chapter, I will be arguing that these two exhibition projects not only directly address the question of affect as integral to Darwin's scientific enquiry, but also make an attempt to reproduce his 'entangled' vision as a way of promoting our ethical engagement with other species.

Post-Millennial Affective Darwin

Darwin's careful crafting of his public persona clearly influenced his contemporaries' perception of the scientist, and continues to shape the views of the generations that followed. His thoughtful use of media not only facilitated laying out and promoting his theories, but also helped create and control his own public image. The media, including his portraits, along with Darwin's painstakingly staged public performances and absences, created an image of him as an icon of disembodied intellect. Janet Browne regards this development as stemming from Darwin's incessant attempts to surmount the frailty of his corporeal frame:[3] attempts which helped him promote an idea of himself as 'a careworn and modest philosopher'.[4] Importantly, Darwin's iconic presence has for a very long time veiled the intrinsic entwining of affect, corporeality and self-reflection as crucial to his work. Suzanne Gapps, who examines contemporary depictions of Darwin on the internet, points out that he is pervasively portrayed as an 'archetypal, heavily bearded Victorian sage', an icon of unattainable grandeur and intellect.[5] Tracing the complex meanings of the imagery that surrounded Darwin in the late-Victorian era, Gapps argues that the modern digital recycling of this imagery not only masks the problematic issues inherent in Darwin's theories but also promotes the 'ethos of disembodied, exulted intellect, and pure mind' as a way of seeing the scientist.[6]

At the same time, academic and popular spaces become increasingly engaged with the affective Darwin. The relevance of Darwin's private life to his scientific pursuits was already recognised in 1888 in 'Charles Darwin, the Man and the Scientist', a review which appeared on the occasion of the publication of *The Life and Letters of Charles Darwin* (1887).[7] The author highlights the necessity of studying Darwin's life writing for a better understanding of his scientific work. Contemporary scholars have also pointed out the need to reconsider the scientist in the context of his feelings and emotions, and have stressed the significance of his affectionate, but also deeply suffering, self to his perception and categorisation of the world, and to his theoretical pursuits. In *Darwin Loves You* (2006), George Levine contends:

> The first step toward a reengagement with the world that Darwin seems so definitely to have evacuated of spiritual meaning and design is recognition of the degree to which his work *and* his life were infused with a sense of the value of things, with a deep emotional engagement in the material world.[8]

Academic and popular projects slowly begin to attend to the question of affect in Darwin's work on a broader scale. In 'The Changing Face of Darwinism' (2003), Michael Ruse illustrates these developments with reference to the evolving portrayal of the scientist in the three successive Norton editions of his works. While the 1970 'cover and the frontispiece were based on the well-known photograph of Darwin around seventy years old but looking more like ninety', the Darwin of the third edition is young and 'living technicolor, framed in eggshell blue and gold'.[9] The 'Technicolor Darwin' mirrors the recent interest in Darwin's formative years and in the significance of his emotional life to his science, as well as reflecting the latest, broader tendencies of 'enlivening' the scientist.

Darwin's affective self has also been addressed in a range of projects aiming to popularise the naturalist among the mass public.[10] Numerous publications and media events nowadays participate in a rediscovery of Darwin's emotional life, which is itself deeply rooted in his corporeality. Randal Keynes's *Annie's Box: Charles Darwin, His Daughter and Human Evolution* (2001) promises a rereading which could transform the traditional perception of the naturalist. The book was hailed by the *Sunday Times* as 'a biography with a difference', and its major achievements were seen as residing in the revelation of Darwin's 'key emotional moment[s]' as well as in indicating the interweaving of his 'experience of fatherhood' with his theories. It was said to throw into relief the convolution of his 'clinical objectivity and compassion'.[11] Ralph Colp's *Darwin's Illness* (2008) offers a chronicle of Darwin's ailments while ruminating on how his work was influenced and channelled by his illness.[12] Reflections on Darwin and emotions also inspired the BBC Radio Four series 'Darwin, My Ancestor' (2009) in which Darwin's great-great-grand-daughter Ruth Padel, in conversation with renowned Darwin scholars such as Gillian Beer and Janet Browne, offered '[a] uniquely personal touch' in exploring Darwin 'as a man'. 'I want to use nature', claims Padel, 'as I understand it, to find Darwin as a writer and find him emotionally in a sense of relationship and loss'.[13] Her four-part personal journey offers the listeners a 'glimpse' of his family bonds, the loss of his daughter, and the sense of wonder that permeated his work. The programme puts us, as listeners, in a situation in which we get a sense of eavesdropping, and encourages personal responses to the issues discussed. This proliferation of the reminiscences of Darwin's descendants and their treatment as Darwin experts, although a manifestation of contemporary quasi-hagiographic reverence of Darwin, also highlights the impact of emotion on his theories and, more importantly, foregrounds our emotional response as a way of rediscovering the scientist. The first feature biopic on the naturalist – Jon Amiel's *Creation* (2009) – clearly manifests the above-mentioned trends. Based on Keynes's book, *Creation* is structured around two key moments in Darwin's life: the recognition of the laws of transmutation in nature and the subsequent decision to publish the findings, and the death of his eldest daughter Annie. It shows Darwin as a tormented father, a loving husband, and a scientist plagued by the recognition of the consequences that his theory would have not only for his family but also for the certainties in his world.

If, as Browne maintains, 'every generation of historians' has its own Darwin,[14] the post-millennial culture seems to be interested in the excavation of the affective Darwin as much as it is preoccupied with his historical public persona. While contemporary biographers have worked to debunk the myth of Darwin as a solitary scientist who singlehandedly created the theory of evolution, they have also put emphasis on the social production and distribution of knowledge, to which affect as well as interpersonal relationships are crucial.[15] This growing interest in Darwin's emotional life also highlights the question of his perceptual powers and their affective conditioning. The glimpses behind the facade of Darwin's pure intellect and sagacity begin to show ways in which we can engage with nature without simultaneously turning it into an object of inhumane vivisection. They allow us to rediscover Darwin's 'entangled vision': a mode of engagement with the natural world which paves a way to a better understanding of natural phenomena as well as advocating a more empathic and ethical approach to nature. It does so by foregrounding a meta-reflection on our own situatedness within nature, and a reconsideration of our embodiment as a way of ethically-engaged learning about the natural world.

Darwin's entangled vision

Darwin's 'entangled vision' is a type of perception that integrates an individual, corporeal situatedness in the world with a high degree of careful (self-)observation. Darwin's mode of seeing represents an alternative to the nineteenth-century 'theological eye', the 'reverent eye' that perceived holiness in nature. The proponents of natural theology believed the human eye to be the ultimate proof of divine creation. Bernard Lightman argues that '[i]n order to undermine [the] design argument, …Darwin, also had to grapple with the theory of vision underlying natural theology'.[16] He abandoned the omniscient eye for one which was characterised by limits and misperception. However, he also fashioned a number of observatory principles which allowed him to pursue an objectified mode of enquiry and eliminate at least some of the obstacles that such 'limited' vision put in his way.

In *The Expression of the Emotions in Man and Animals*, Darwin emphasised the degree to which the ephemeral character of emotions and human empathy constituted major impediments in the study of the manifestation of emotions. In order to avoid these, he used a number of methodological tools that were to optimise his study and make it more accurate. Darwin decided to analyse children, 'for they exhibit many emotions…with extraordinary force'. Likewise, he put under scrutiny 'the insane…as they are liable to the strongest passions and give uncontrolled vent to them'.[17] He also employed the medium of photography as a way of generating a more truthful presentation of emotional expressions, and devised a questionnaire to ascertain whether 'the same expressions and gestures prevail…with all the races of mankind'.[18] While Darwin designed his analytical tools to eliminate the excess of sentiment that was, according to him, bound to distort his scientific findings, he nonetheless painstakingly noted down his first emotional responses in his diaries, only to dilute them into scientific facts in his published works.

In 'Darwin's Emotions: The Scientific Self and the Sentiment of Objectivity' (2009), Paul White shows the processes by which Darwin extracted facts from emotionally-loaded observations and made their objectivity palatable to the reading public. White urges us to make a distinction between Darwin's scientific and private writing, arguing that while the latter was designed to 'evoke, as well as to express, emotion', the former aimed at general intelligibility, stylistic appropriateness, and scientific authentication.[19] These different objectives also show two different Darwins: the loving husband, father and friend, and an ardent scientific observer. And yet, this divide is not as ultimate as one might imagine, as Darwin's scientific performance is always already connected to the performance of his other, private, roles. All of these constitute different modes of his 'technologies of the self':

> Darwin's research on emotions…bears many of the hallmarks of objectivity, even while it allows us to conceive of objectivity quite differently, as a form of scientific engagement in which a whole range of emotions were brought into play: as part of an *emotional economy of science*.[20]

George Levine further emphasises this combination of Darwin's personalised, emotional, and deeply physical, relation to nature, and careful observatory skills, as a crucial epistemological formula underlying his empirical studies:

> In Darwin, intellect and feeling were one. Rather than splitting the world into feeling and reason, a condition of modern disenchantment, Darwin's writing, like his life, reflects a way of thinking and feeling that is distant from the disenchanting austerity of modern interpretations of his work.[21]

While the rewriting of the experiments for the broader public served the objectification of his work, Darwin's methods of observation entailed a corporeal and empathic enmeshment in the world. In *The Phenomenology of Perception* (1945), Maurice Merleau-Ponty envisions perception as conditioned by our embodied and sensory interaction with the world. His account unifies affect, motility and sensory capacities as the basis of human cognition.[22] In this context, Darwin's entangled vision becomes no less than a phenomenological vision that stems from his corporeal and emotional intertwining with nature.

Yet, it is also a vision that is clearly culturally driven and in no way oblivious to the cruelty that, for Darwin, constitutes an indelible characteristic of the natural order. David Kohn, who traces the aesthetic construction of Darwin's theory, sees in his work a tension, stemming from the Romantic movement, between the beautiful and the sublime; this coloured his vision and offered a tangible framework that would accommodate his theories. While Darwin's early writing on the *Beagle* already integrated the reflection on the grandeur of nature, it also intertwined his 'self-consciously incorporated affect and imagination' as prime features of his 'romantic science'.[23] Darwin's reaction to exotic settings that found expression in his metaphorically rich language implies an interweaving of his affective, physical responses with 'high culture'.[24] The 'distanced observer' who narrates the *Origin of Species* likewise invites the reader to enter the landscape in which 'Darwin's metaphysics, aesthetics, and emotions intersect'.[25] The recognition of the beauty and terror as irretrievably intertwined in nature is simultaneously the recognition of the continuously mutating natural forms synthesised in the 'entangled bank' metaphor. And yet, as James Krasner clearly shows, the entangled bank, despite the dynamism of nature that it implies, 'does not approach the formless mutability of the entangled eye' that is suggested in Darwin's writing.[26] It is this entangled vision that Krasner regards as central to Darwin's revolutionary contribution to nineteenth-century scientific literature:

> his representation of nature as though seen through an empirically entangled eye. ... Dazzled by the dynamism and complexity of Darwin's natural world, the reader struggles unsuccessfully to categorize the multitude of fluxing forms into stable or familiar types. By involving the reader in a perceptual chaos that parallels the organic chaos of the entangled bank, Darwin demonstrates the formlessness of evolutionary nature and the artificiality of a theory of distinct species.[27]

Darwin's entangled vision is a fallible, culturally-specific perception founded in a deeply physical, if impaired by illness, experience of the world.

Driven by this entangled, phenomenological vision, Darwin's study yielded the results that transformed the existing epistemological assumptions, and led to an acknowledgement of 'ecological interdependence and "the inextricable web of affinities"' in nature.[28] It is exactly the combination of affect and perception, as a multi-sensory act, and the insistence on the observation of relations between the species, that can be regarded as the basis of Darwin's ethical approach. In 'The Expression of the Emotions in Man and Laboratory Animals' (2008), Jed Mayer sees Darwin's achievement in similar terms: 'Emphasizing the primacy of our emotional connections to animals as a means of persuading readers of our evolutionary connections, Darwin lays a biological foundation for the era's culture of compassion'.[29] This 'culture of compassion' is then strictly bound to a phenomenological entanglement with the flesh of the world that constitutes the basis of Darwin's scientific work. While it does not erase either the beauty or the terror of nature, the culture of compassion, embedded in a physical, affective reaction to nature and our kin, acknowledges the multiplicity of relationships that are made possible, and allows a conceiving of new bonds built not upon any pre-existing ethical grammars but upon individual, physical intersubjectivity.

Phenomenological Entanglement

Entangled vision – vision stemming from a sensory engagement of the viewer in the world they observe – lies at the core of the Warsaw exhibition *All Creatures Great and Small* and the London project *After Darwin: Contemporary Expressions.* The primacy of emotional relations between man and animals is their major concern as they address Darwin's work on human and animal expressions. Despite their diverging missions, both exhibitions point to the embodied experience of the world as the basis for a compassionate and ethical relation between humans and animals. Both highlight the importance of a sensory enmeshment in the animal world as a basis for broader questions of human understanding of nature and a deeper reflection on the human self.

The Warsaw exhibition is concerned with animal subjectivity and with the question of humanity as reflected in the animal world: 'The point of entry for the exhibition is the desire to overcome the dogmatic anthropocentrism that places the human in the centre of the world as a privileged species of the highest ontological status'.[30] Reflecting on Peter Singer's bioethics and his struggle against 'species racism', curator Maria Brewińska points out a growing consciousness of the problem, which this exhibition brings to the fore:

> [It] uncover[s] the mechanisms of 'species racism'. The works...visualize the world of animals in its diverse aspects through the use of more or less engaged observation; they tell of the pleasure of watching, touching or possessing animals, the pleasures experienced by animals themselves.

However, many of these works (for instance Jean-Charles Hue's *Pit-bull Carnival,* 2006, or *Rhyme 'Em to Death,* 1993, by The Wooster Group) also explore the darker side of the human–animal relationship and critically reconsider the pervasive objectification of animals as trophies (for example, Eric Swenson's *Untitled (Ice Head)*, 2004, or Józef Robakowski's *Mr. Deer/Auto-Epitaphy*, 1990–94).[31]

Unlike the Warsaw project, the London exhibition *After Darwin* primarily focuses on contemporary artistic and scientific responses to human–animal relations, and on the commonality of human and animal expressions of emotional states. The curator of the exhibition, Bergit Arends, highlights this preoccupation: 'By asking the same questions as Darwin did, we hoped to gain both a historic and a contemporary view into the research of emotional states, the relation between thoughts and feelings, and the capturing of facial and bodily expressions in motion'.[32] As with the Warsaw exhibition, the question of observation is the major concern of the London project. Already the first exhibit – Jeremy Deller and Matthew Killip's *Aping* (2009) – partly visible from the corridor of the Natural History Museum, draws attention to the question of perception and imitation. Visitors wishing to see the exhibition pass the installation crowned by the headline 'Can you make a face like a chimp' placed over what seems to be a display window. The accompanying description invites spectators to playfully immerge in experimenting with expressions and calls upon them to pay attention to the acquired perceptual habits and the political economies of our observational modes.

Perceptual habits and the supposedly self-evident nature of seeing are also questioned by the thematic engagement of the works displayed in the Warsaw exhibition. Observation, the matrix of power distribution that it invites, and our perspective, are foregrounded by a number of exhibits. The spectator is allowed to watch animal behaviour, but is also confronted by peering, intricate eyes whose expressions will always be, to a certain extent, barred from him. He is also invited to witness mutual scrutiny of men and animals. Thus evoked, human vision loses its neutrality in the exploration of the world. Mircea Cantor's *Deeparture* (2004), a short film depicting mutual alertness and the process of reciprocal reconnaissance between a deer and a wolf locked in a gallery space, emphasises not only the patterns of animal observation but also our own predatory position: the camera/the artist/the spectator unite to observe the animals trapped in a gallery space for the purposes of an art project. Instead of assuming the roles characteristically ascribed to them in nature, the animals stand distraught, caught in intense mutual observation. Their restraint creates tension and unease at the recognition of the position into which the spectator is forced: the role of witness to a looming gory encounter.

Spectators are never at ease as their habits are put under scrutiny. They are called upon to interpret quizzical animal expressions. In a darkened room that belongs to the main gallery space, ardent animal looks become the main theme as they surround the visitors in the form of wall projections. Dušan Skali's film *Cow* (2005) confronts us with a direct, peering animal look that questions our position as spectators. Anna Orzechowska-Barcz draws attention to the uniqueness of this perspective. As the film takes hold of the animal's reciprocal look at the human, it 'marks another perspective, declares a foreign, unfamiliar subjectivity'.[33] Menacing winking of animal eyes undermines the stability of the anthropocentric perspective: 'Flickering looks directed at the spectator …disturb the secure distance which we know from a circus or a zoo, in which the addressee gazes from the position of a privileged colonizer on subjected bodies displayed to be looked at and evaluated'.[34] Spectators are invited to experience the animal world with all their senses. Surrounded by barking noises, they enter the space of the main room only to be confronted with a jungle of artistic works. This intentional crowding, designed to dynamise the space of the gallery to mirror the bustling interlacing of the animal world, also perplexes the visitors. The contiguity of presented artworks makes them exist in a dangerous symbiosis and belies their exclusive interpretation, challenging the spectator instead to attend to their hybrid fabrics. The art critic Marta Lisok points out that such a dizzying combination of works hampers 'the readability of their transmission, in whose ruffled stream the works tread on one another'.[35] This non-traditional display disrupts and complicates spectatorial attention as well as the kinaesthetic patterns appropriate for a museum. Lisok makes clear some of the repercussions of such an arrangement to the spectatorial routine: 'The majority of screens have been placed low, on the level of your knees, so that you have to bend to see particular films. The preponderance of horizontal directions in narrative forces the visitors to abandon the customary rhythm of the visiting route'.[36] This disruption of the assimilated ways of seeing in the gallery throws into relief the question of spectatorial perspective. A further complication of the visiting route by an unorthodox distribution of exhibits beyond the three main rooms, in rarely utilised gallery spaces (for example, the room underneath the main staircase), encourages the reader to observe carefully, paying thoughtful attention to detail.

Similar insistence on embodied experience characterises the London exhibition. The compositional space of *After Darwin,* as well as its specific use of light, emphasise walking and close observation as the bases for an engaged involvement with the world. Carmody Groarke's design throws into relief the spectators' journey of recognition. Divided diagonally, the space of the east-wing gallery of Natural History Museum is darkened so that, after entering, the spectator becomes immerged into Diana Thater's film *gorillagorillagorilla* (2009), which is projected on the right wall of the gallery. The incredibly boastful and captivating natural scenery of the African forest, on whose background playful gorillas make their appearance, enchants with its liveliness. Jointly commissioned by Kunsthaus Graz and the Natural History Museum, *gorillagorillagorilla* uses the existing exhibition spaces as a projection plane. The nineteenth-century terracotta of the Natural History Museum walls highlights the combination of beauty and the sublime present both in Thater's film and in the architectural make-up of the museum itself. Thater insists on bringing to the spectator the bustling, living and dangerous natural world:

> I don't want nature to be contemplative and reassuring, but sublime and terrifying. Not terrifying in that nineteenth-century way, where it is yawning wide, endlessly dangerous and devouring, but in that twenty-first century way – like a dead or dying body because we've killed it and it is gone. And that *is* terrifying.[37]

The sublime of the nineteenth century and the terror of the twenty-first melt together in the beautiful vision of gorillas which, mythologised and vilified in our culture, have become the prime target of deforestation projects. How long will they survive in the natural environment in which they have to withstand the greatly intensified attacks of humans – their archenemy? Walking around in this part of the exhibition, spectators are immersed by the wildlife as the projections dance upon their bodies and their shadows are integrated into the projected images of resplendent greenness.[38] From this space, they are invited to step into geometric spaces which embrace other works of art. These spaces function as recesses of wondrous presences whose beauty can only be experienced in full by a deliberate sensual engagement, be it ocular, tactile or auditory. The labyrinth of these spaces, as well as the multimedia works it encloses, forces us to respond to the theme of animal–human emotion with our whole body.

The narrative insistence of both exhibitions on our affective and motor response to the works assembled there emphasises the importance of human–animal interaction as a way of rethinking our subjectivity. In the Warsaw exhibition, the recognition of the commonality of men and animals but also the insurmountable differences between the species is evoked by filmed performances of Joseph Beuys's 'I Like America and America Likes Me' (1974) and Pilar Albarracin's 'She-Wolf' (2006). In both pieces the question of human–animal bond accompanies the exploration of broader themes of identity. In Beuys's work, for instance, the issue of relationality as the fundamental law of the ecosystem is intertwined with the question of American national identity. The latter is addressed by bringing together the New York performance space and the coyote, whose physical and symbolic presence in the urban interior refers to America's past, as well as to the evolutionary and adaptive processes which have been part and parcel of its definition as a nation.[39]

At the same time, both exhibitions hint at the loss of the bond between men and animals and at our conditioned indifference and blindness that prevent a more conscious consideration of the animal world and its contiguity with human life. In *After Darwin*, this issue is evoked, among others, in the farthest octagonal space where Ruth Padel's poems are juxtaposed with Gautier Deblonde's photographs of the Natural History Museum's Zoology store. Padel's poems respond to the formative moments in Darwin's life and to his passions: the loss of his mother, his zealous collecting, his work with Jenny the Orangutan. 'Likeness' explores the moment of their meeting in the London zoo in which Darwin's examination of the animal metamorphoses into the recognition of human and animal likeness:

> Man thinks himself,
> in his ignorance, a great work,
> and worthy a deity's glance.
>
> More humble – and true,
> I'd assert –
> to think him created, not bandbox new
> but slowly from this. From the animals.[40]

And yet, the repercussions of this revelation seem to have inspired only a hollow echo in a contemporary world which appears to have lost not only the consciousness of Darwin's evolutionary theory but also the link to the animal realm. What is our relation to the animal kin? How do we gain our knowledge about other species?

The ethics of (self-)observation

A corporeal sensitivity to the world can be regarded as a basis for a more ethical way of comprehending our surroundings and ourselves. The cultural anthropologist Thomas Csordas refers to such an engagement as 'somatic modes of attention' and argues for the supremacy of such 'multisensory engagement' over the predominant modes of strictly visual observation.[41] This method of observation is advocated by both aforementioned exhibitions which foreground the sensory and physical engagement with the works of art. According to Traci Warkentin, such preoccupation with the world can also be a basis for a more ethical interaction with other species. In 'Interspecies Etiquette: An Ethics of Paying Attention to Animals' (2010), Warkentin inspects the 'behavioural semantics' of human–animal encounter and calls for a 'praxis of attentiveness' as a way of approaching different species.[42] This 'ethical praxis of paying attention' requires more than simple observational skills, it 'involves one's whole bodily comportment and a recognition that embodiment is always in relation to social others, both animal and human'.[43] Such a novel, corporeally-based approach to the animal kin also requires an emotional engagement with the world.

Ecofeminists have criticised the cultural and historical emphasis on the supremacy of intellect over corporeality and have pointed out and admonished patriarchal sanctions on emotionality as a viable epistemological tool.[44] Warkentin regards this 'patriarchal legacy of valuing a masculinised, pure, detached reason over what has been constructed as 'feminized' embodied modes of knowing, such as emotion', as 'foster[ing] a diminution of situated and relational knowledge-making'.[45] To counteract this disembodied politics, ecofeminists have propagated 'an ethics of care' to which intersubjective relations are vital.[46] They have also recognised the significance of embodiment to the development and practice of ethics:

> [Ecofeminists] strongly suggest that embodiment enables the expression of ethical comportment towards others, while also providing a kind of empathic approximation of the experience of others in our midst, which can (and should) inform our responsive interactions with them.[47]

The exhibitory spaces of *All Creatures* and *After Darwin* privilege this entangled and embodied mode of vision as the epistemological tool in the exploration of the human–animal bond. This preoccupation, however, does not in itself guarantee a more ethical approach towards the world. It is rather through the activation of the meta-narrative of self-observation that these two exhibitions take up ethics as their major issue.

The question of ethics is evoked through the foregrounding of the problem of self-observation as crucial to any empathic engagement. Both exhibitions take up this issue with reference to individual embodiment and in relation to the question of display as a dehumanising factor that contributes to the severing of the human–animal bond, and which thus precludes any possibility of an ethical engagement. *After Darwin* pivots around the question of visitors' self-conscious attentiveness to their emotional responses. While the whole exhibition foregrounds the subject of embodiment and observation, one of the exhibited works, Tina Gonsalves's *Chameleon Project* (2008–10), stresses the social character of emotional expressions. Created at the interface of art, neuroscience and technology, the work interacts with the spectator's facial expression. Recognising and assessing the muscular changes in the faces of the spectators, it adapts the emotional expressions of the characters on screen, which thus seem to respond to the changes in the spectator's facial muscles. In this way, the installation generates and simultaneously explores the issue of 'emotional contagion':

> [It] strive[s] to create a feeling of empathy with the audience, drawing attention to how we innately and continuously synchronise with the facial expressions, voices, and postures of others by unconsciously infecting each other with our emotions. The later prototypes in the series attempt to implicate viewers in an emotional drama, using the audience's emotions to drive the emotional dialogue.[48]

While promoting this dialogical engagement, the project also foregrounds the question of display and empathic mimicry as crucial to interhuman and interspecies communication. This brings us closer to the recognition of the significance of the embodied self in the perception, cognition and interpretation of emotional states of others. Csordas proposes the method of cultural phenomenology as a way of approaching human–animal relations. Cultural phenomenology is 'concerned with synthesizing the immediacy of embodied experience with the multiplicity of cultural meaning' and goes beyond a simple attention to one's own body and the observational techniques that it initiates.[49] Summarising Csordas's program, Warkentin urges us to recognise that:

> [a]ttending with one's body goes far beyond just looking toward or gazing at; it means turning a full bodily engagement toward others, it means that our very own bodies can actually tell us something about others and about the world if we are attentive to it and them.[50]

Thus our attentiveness to the changing expression on the screen of the *Chameleon Project* not only tells us about the possible expression painted on our faces; it also promotes our interest in, and contemplation of, our own body and its external appearance: how do others react to the minuscule changes of our mimicry, and how does it influence their mood and frame of mind? What can our corporeal states tell us about the possible feelings of others, and how can they become the platform for an empathic intersubjective engagement?

The installation *Aping* at the *After Darwin* exhibition emphasises attentiveness to one's own emotional states and their expression as a way of a more empathic engagement with the animal kin. When accessed from within the exhibition space, the pane, which is also visible from the corridor of the Natural History Museum, becomes a display window in which we ourselves become the exhibit. Stimulated to mimic chimpanzee's expressions, we become aware of our status as objects of display. While this reversal of roles certainly addresses questions of the ethics and politics of display, it also points to a more important problem in this context: how do we read and interpret the expression of emotion of other people and other species? To what extent are they contiguous? In how far does our joint legacy make it possible for us to understand each other and to what extent is our bodily enmeshment in the world a key to this understanding? Finally, are we willing to reverse and project this situation? Are we willing to impersonate an animal? What do we feel when we do so? Those visitors who decided to imitate animal expressions while exposed to the intrigued glances of other spectators looked into the crowd in front of the window pane for familiar faces. Doesn't *Aping*, then, make us aware of our fundamental desire for empathy from others; for the recognition of our relatedness? Doesn't our attention to our corporeal frame allow us a more empathic relation to others?

Feeling with Darwin

By activating Darwin's entangled vision, the Warsaw exhibition *All Creatures Great and Small* and the London project *After Darwin: Contemporary Expressions* promote an intersubjective and embodied cognition as a way of comprehending nature and our place in it. The work that explores this main postulate of both exhibitions is Artur Malewski's two-piece sculpture *Hexakosioihexekontahexaphobia* (2008). The sculpture, which is part of the Warsaw exhibition, signals the significance of perception as both emotive and corporeal in character. The main theme of *Hexakosioihexekontahexaphobia* – the acquisition of knowledge – is reflected in the combination of the statue of Charles Darwin and the imagined pagan deity Baphomet. Iconographically modified, Malewski's Baphomet, 'a symbolical figure representing the First Master of the Magnum Opus, ... the Astral Light',[51] has filed horns and a burnt out candle instead of the torch of knowledge. In his ambivalent sagacity, Baphomet points towards the figure of Charles Darwin with the hand on which the verb 'solve' is inscribed. Darwin, who stands in a pile of plaster that must have fallen from his body, is covered by an abundance of bodily hair. He is blind. This blindness can be regarded as a sign of higher knowledge, of a revelation that is clear to him but stays incomprehensible to us. In this context, he is an archetypical blind seer.

Darwin's blindness acquires further meanings with reference to the major issues explored by the Warsaw exhibition. The combination of Darwin's bearded head with the sensorial and instinct-driven corporeality, as evoked by his hairy body, not only recycles the imagery that has accompanied the popularisation of his ideas for the last two centuries. It also links the question of knowledge to the physical, sensual, engagement with the world. Janet Browne comments on the presentation of Darwin in the late nineteenth century: 'virtually all that the public saw...were his beard, his hat, and his eyes...as a physical presence [he] had almost disappeared. All that was left was an intense impression of mind'.[52] Such portrayal linked Darwin to the depiction of philosophers, Old Testament prophets and Victorian patriarchs.[53] In the context of the contemporary pervasive iconisation of Darwin, his bearded head continues to function as a symbol of knowledge and intellect.[54] Yet, in Malewski's sculpture, that higher knowledge of the old, infirm man seems to refer to his corporeal integration in nature. The Biblical clay that has fallen off his body is also the clay that, in the Book of Genesis, is a material of the whole creation. The matter that links him to the world is also the foundation of his knowledge. Here, then, his cognition is not simply based on optical observation but, rather, accumulates in the recesses of his whole body. It is the phenomenological, lived-body experience that becomes the platform for the engagement with the world and for the acquisition of knowledge about it.

Darwin's body, iconographically linked to the world of lower animals through its simian appearance, becomes the temple of knowledge about the surrounding world. Darwin's physical being matters in the sense that it points out his sensory engagement and implies that the physical constitution of the observer is indispensable to his power of cognition. While the corporeality of the observer is restrictive in the sense that it does not allow any boundless intermingling with the world, it is also the only means of engagement with nature that surpasses the limitations of optical perception. Such a mode of cognition is entwined with the question of self-reflective attention to one's own body. It also privileges our relationship with animals as fundamental to a more ethical approach to nature. Thus, both exhibitions invite us not to see ourselves mirrored in the monolithic image of Darwin's pure mind, but, rather, encourage us to walk with the experiencing Darwin and explore with him 'endless *forms* most *beautiful* and most wonderful' by way of our corporeal and emotive engagement with nature.[55]

Notes

1 The Warsaw exhibition was hosted by Zachęta, one of the most established Polish exhibition spaces. Press material relating to the exhibition can be found at http://www.zacheta.art.pl/en/article/view/245/all-creatures-great-and-small (accessed 15 November 2011). Unfortunately, the exhibition was not accompanied by a catalogue.

2 An information leaflet about the project can be found at http://www.nhm.ac.uk/resources-rx/files/after-darwin-contemporary-expressions-leaflet-31923.pdf (accessed 14 July 2010). The London project was accompanied by a volume of essays edited by Bergit Arends, *Expressions: From Darwin to Contemporary Arts* (London: Natural History Museum, 2009).

3 Janet Browne, 'I could have retched all night: Charles Darwin and his Body', in Christopher Lawrence and Steven Shapin (eds), *Science Incarnate: Historical Embodiments of Natural Knowledge* (Chicago: University of Chicago Press, 1998), 240–87: 252.

4 Browne, 1998, 253.

5 Philip French, 'Creation', *Guardian*, 27 September 2009, http://www.guardian.co.uk/film/2009/sep/27/creation-review (accessed 5 July 2010).

6 Suzanne Gapps, 'Charles Darwin as an Icon', *Culture and Organization*, 12 (2006), 341–57: 364.

7 Charles Darwin, 'The Man and the Scientist', *The North American Review*, 146 (1888), 106-107, http://www.jstor.org/stable/25101403 (accessed 27 July 2010); and Francis Darwin (ed.), *The Life and Letters of Charles Darwin* (1896; New York: Ams Press, 1972).

8 George Levine, *Darwin Loves You: Natural Selection and the Re-Enchantment of the World* (Princeton and Oxford: Princeton University Press, 2006), 132.

9 Michael Ruse, 'The Changing Face of Darwinism', *Victorian Studies*, 45 (2003), 305–17: 306, 309.

10 On 17 November 2009, the site of the International Darwin Day Foundation (http:// www.darwinday.org/) listed 788 events scheduled in 46 countries, among them a number of private events such as 'Abe and Chuck's Birthday Party for Darwin' in Houston, Texas, USA. A few trends can be distinguished. Apart from the general Darwin Day/ bicentennial birthday, there were: 1) events that established Darwin's legacy and significance across academic disciplines, and particularly, his influence in the humanities, for example, 'Darwin's Reach. A celebration of Darwin's Legacy Across Academic Disciplines', at Hofstra University, USA; '150 Years of Evolution – Darwin's Impact on Humanities', at San Diego State University, USA; 2) events that highlighted Darwin's contemporary significance: 'Why Darwin Still Matters', at Pepperdine University, USA; '(Dis)Entangling Darwin: Cross-Disciplinary Reflections on the Man and his Legacy', at University of Porto, Portugal; 3) events that sought to bring together Darwin and religion, for example, 'Exploding the Myths, Science and Religion', at Beverley Baptist Church, UK; 'Did Darwin's Discovery Make God Redundant?', at Rutherglen, UK.

11 Randal Keynes, *Creation: The True Story of Charles Darwin* (2001 [published as *Annie's Box*]; London: John Murray, 2009), front matter.

12 Ralph Colp, *Darwin's Illness* (Gainesville: University Press of Florida, 2008).

13 'Darwin, My Ancestor', *BBC Radio 4*, http://www.bbc.co.uk/radio4/ darwin/my_ancestor.shtml (accessed 10 July 2010).

14 Janet Browne, 'Making Darwin: Biography and the Changing Representations of Charles Darwin', *Journal of Interdisciplinary History*, 40 (2010): 347–73: 373.

15 Browne, 2010, 372.

16 Bernard Lightman, 'The Visual Theology of Victorian Popularizers of Science: From Reverent Eye to Chemical Retina', *Isis*, 91 (2000), 651–80: 655. On the concept of 'entangled vision', see James Krasner, *The Entangled Eye: Visual Perception and the Representation of Nature in Post-Darwinian Narrative* (New York: Oxford University Press, 1992). For further discussion of these ideas, especially with reference to Amiel's *Creation*, see Pietrzak-Franger, 'Conversing with Ghosts: Or, the Ethics of Adaptation', in Pascal Nicklas and Oliver Linder (eds), *Adaptation and Cultural Appropriation* (Berlin: DeGruyter, 2012), 70-88.

17 Darwin, *The Expression of the Emotions in Man and Animals* (London: John Murray, 1872), 14.

18 *Ibid.*, 15.

19 Paul White, 'Darwin's Emotions: The Scientific Self and the Sentiment of Objectivity', *Isis*, 100 (2009), 815

20 White, 2009, 826, my emphasis.

21 Levine, 2006, 133.

22 Maurice Merleau-Ponty, *Phenomenology of Perception* (1945; London: Routledge, 2002).

23 David Kohn, 'The Aesthetic Construction of Darwin's Theory', in Alfred I. Tauber (ed.), *The Elusive Synthesis: Aesthetics and Science* (Kluwer: Kluwer Academic Publishers, 1996), 15.

24 *Ibid.*, 26.

25 *Ibid.*, 41.

26 Krasner, 1992, 34.

27 *Ibid.*, 35–6.

28 Gillian Beer, *Darwin's Plots: Evolutionary Narrative in Darwin, George Eliot and Nineteenth-Century Fiction* (1983; Cambridge: Cambridge University Press, 2009), 19.

29 Jed Mayer, 'The Expression of the Emotions in Man and Laboratory Animals', *Victorian Studies*, 50 (2008), 402.

30 Maria Brewińska, 'All Creatures Great and Small', Press Material (Warsaw: Zachęta National Gallery of Art, 2009), 1.

31 Brewińska, 2009, 1.

32 Bergit Arends, 'Curating the Exhibition After Darwin: Contemporary Expressions', in Bergit Arends (ed.), *Expressions: From Darwin to Contemporary Arts* (London: Natural History Museum, 2009), 29.

33 Anna Orzechowska-Barcz, W Kwestii Zwierząt, 'Wszystkie Stworzenia

Duże i Małe', *RecyklingSztuki*, 2 February 2010, http://recyklingidei.pl/orzechowska-barcz-w-kwestii-zwierzat (accessed 5 July 2010), my translation.

34 Marta Lisok, 'Boeuys and Kojot Obserwują przez Okno Ulice Nowego Jorku', *Dwutygodnik Strona Kultury*, 23 (2010), http://www.dwutygodnik.com/artykul/848-beuys-i-kojot-obserwuja-przez-okno-ulice-nowego-jorku.html (accessed 5 July 2010), my translation.

35 Lisok, 2010, my translation.

36 Lisok, 2010, my translation.

37 Diana Thater, 2003, qtd. Bergit Arends (ed), 'Becoming Gorilla', in *Diana Thater. Gorillagorillagorilla* (Köln: Walther König and Kunsthaus Graz, 2009), 51.

38 Arends, *Expressions*, 2009, 30.

39 David Levi Strauss, *Between Dog & Wolf: Essays on Art and Politics* (New York: Autonomedia, 1999).

40 Ruth Padel, 'On Darwin and the Expression of the Emotions in Man and Animals', in Arends, *Expressions*, 61.

41 Thomas J. Csordas, 'Somatic Modes of Attention', *Cultural Anthropology*, 8 (1993), 135–56: 138; Traci Warkentin, 'Interspecies Etiquette: An Ethics of Paying Attention to Animals', *Ethics and Environment*, 15 (2010), 109.

42 Warkentin, 2010, 103.

43 Warkentin, 2010, 102.

44 Elizabeth Grosz, 'Bodies and Knowledges: Feminism and the Crisis of Reason', in Linda Alcoff and Elizabeth Potter (eds), *Feminist Epistemologies* (New York: Routledge, 1993), 187.

45 Warkentin, 2010, 104.

46 Warkentin, 2010, 104.

47 Warkentin, 2010, 103.

48 Tina Gonsalves, 'The *Chameleon Project*', in Arends, *Expressions*, 70–73: 70.

49 Csordas, 1993, 143.

50 Warkentin, 2010, 110.

51 Eliphas Lévi, *The History of Magic*, trans. by A.E. Waite (1913; Boston: Red Wheel/Weiser, 2001), 211.

52 Browne, 1998, 279.

53 Browne, 1998, 274–75.

54 Gapps, 2006, 342.

55 Darwin, *On The Origin of Species* (1859; New York: PF Collier, 1909), 529.

8

Reckoning with the Emotions: Neurological Responses to the Theory of Evolution, 1870–1930

L.S. Jacyna

In 1994 the American neurologist, Antonio R. Damasio, drew attention to what he called 'Descartes' Error'. He summed up this error as a pervasive tendency in modern western thought – including the biomedical sciences – to insist on:

> the abyssal separation between body and mind, between the sizable, mechanically operated, infinitely divisible body stuff, on the one hand, and the unsizable, unidimensioned, unpushable, nondivisible mind stuff; the suggestion that reasoning, and moral judgment, and the suffering that comes from physical pain or emotional upheaval might exist separately from the body.[1]

More particularly, Damasio draws attention to the degree to which the conventional neuroscience that he was taught envisioned 'separate neural systems for reason and emotion'. In other words, 'the mechanisms of reason existed in a separate province of the mind, where emotion should not be allowed to intrude…'.[2] Damasio's own clinical experience indicated, on the contrary, that: 'feeling was an integral component of the machinery of reason'.[3] Based on his reading of the neurological literature – including the celebrated case of Phineas Gage[4]– as well as on his own clinical experience, Damasio therefore proposed an alternative view of the relations between reason and emotion: one that sees the two as interdependent; that insists that feeling is integral to thinking. 'Emotions and feelings', he declares, 'are not intruders in the bastion of reason at all: they may be enmeshed in its networks, for worse *and* for better'.[5]

This is not the place to review the detail of Damasio's argument. Suffice it to say that he insists on the need to understand the mechanisms of the human nervous system as the outcome of strategies elaborated over the course of evolutionary time. He adopts an explicitly Darwinian stance in maintaining that only those adaptations calculated to ensure the survival of the organism have been preserved in the course of human evolution. From this perspective, it can be seen that emotions developed from basic organismic needs and reactions at a relatively early stage. The parts of the brain that subserve what are usually regarded as purely 'rational' operations – notably the pre-frontal lobes – are of much more recent origin. The health and safety of the organism can, however, only be optimised when these neural systems interact and cooperate. In particular, Damasio argues that a loss of affect – the ability to *feel* – can undermine the organism's ability to make rational judgments.[6] This chapter does not seek to determine whether 'Descartes' error' *is* an error, nor whether if it is an error the error is *Descartes'*. Instead it addresses the question of how, in the period running roughly from 1870–1930, the emergent brain sciences treated the emotions and their relations with other departments of the mind. In particular, it considers the impact of evolutionary ideas on how the emotions were treated.

Placing the Emotions

The period under consideration here is of course 'Post-Darwinian'. However, for the purposes of this discussion Darwin is less important than some of his contemporaries. The impact of evolutionary ideas on neurology certainly *is* a major theme in the last decades of the nineteenth and the first of the twentieth centuries. But it is the evolution described by Herbert Spencer, rather than Darwin, that is more significant.

The period is also one in which the issue of *localisation* becomes predominant in clinical and experimental investigations of the nervous system. This was the era in which doctors, anatomists, and physiologists undertook the heroic task of mapping the brain: of identifying its functional topography – an undertaking that is sometimes referred to as the 'New Phrenology'.[7] Prominent among the functions to be assigned a 'seat' or 'centre' in the brain was muscular movement and sensation. But higher intellectual functions, notably language, were also localised. When, in 1890, the American psychologist William James reviewed these studies he nevertheless noticed a curious lacuna:

> the emotions have been so ignored in all these researches that one is tempted to suppose that if these investigators were asked for a theory of them in brain-terms, they would have to reply, either that they had as yet bestowed no thought upon the subject, or that they had found it so difficult to make distinct hypotheses that the matter lay among the problems of the future, only to be taken up after the simpler ones of the present should have been definitively solved.[8]

These remarks need to be placed in the context of James's more general discussion of the current state of investigation into the emotions. He noted at the outset that: 'the merely descriptive literature of the emotions is one of the most tedious parts of psychology'.[9] Having toiled through such classifications out of a sense of duty, 'I should as lief read verbal descriptions of the shapes of the rocks on a New Hampshire farm as toil through them again'.[10] Yet mere cataloguing of the emotions was the most frequent approach to the subject in the psychological literature.[11]

The problem with this way of dealing with the subject, James argued, was that it assumed that the emotions: 'are regarded too much as absolutely individual things'.[12] In this regard, psychology was stuck at the impasse that had stymied pre-evolutionary biology. So long as the emotions were: 'set down as so many eternal and sacred psychic entities, like the old immutable species in natural history', it was impossible for psychology to move beyond mere description to the higher scientific end of providing an *explanation* of these psychic phenomena. Once psychology was in a position to view the emotions: 'as products of more general causes (as 'species' are now regarded as products of heredity and variation), the mere distinguishing and cataloguing becomes of subsidiary importance'.[13] James's approach to the question of the emotions may therefore be seen as an aspect of his wider endeavour to apply the principles of Darwinian evolution to the study of the human mind.

Any scientific explanation of the emotions must, in James's view, be physiological in nature; that is, to explain feelings by reference to dynamic material processes occurring in the organism. In this regard, James may be viewed as the heir of the Scottish psychologist, Alexander Bain.[14] James had in 1884 provided such an explanation, which maintained that, contrary to the popular opinion, the psychic aspect of the emotion *succeeded* some bodily event, rather than preceding it.[15] In 1890 James saw no need to revise this theory – which came to be known as the 'James-Lange[16] Law' – even though he acknowledged that it had provoked much opposition from some psychologists.[17]

In an interesting variation on Damasio's argument, James averred that part of the resistance to his theory derived from a belief among the 'platonizers in psychology' that any such physiological account of the emotions would in some way debase the perceived value of these feelings.[18] An instance of such a defensive stance can be found in an 1872 article by James More written in direct response to Darwin's *Descent of Man* (1871) and *The Expression of the Emotions in Man and Animals* (1872). Contrary to Darwin, More insisted that there were fundamental differences between emotional manifestations in the human and in the mere 'brute'. In animals, the emotions were indeed: 'all of somatic and physical origin'.[19] In humans, however, emotion was modulated and controlled by the intellect. Thus a dog that was undergoing vivisection might, irrationally, lick: 'the hand that was inflicting upon him so much pain'. In contrast: 'the operator did not feel remorse, or, rather, did not allow his feeling of remorse to influence his conduct, as his reason told him that he performed the operation for a scientific purpose'.[20] More's conclusion was: 'the feelings and emotions of man give him quite a distinctive character, and seem to exclude him from the scale of mere animal being'.[21]

James sought to counter any such reservations by insisting on a distinction between the origins of an emotion and its worth within a system of human values:

> our emotions must always be *inwardly* what they are, whatever be the physiological ground of their apparition. If they are deep, pure, spiritual, worthy, spiritual facts on any conceivable theory of their physiological source, they remain no less deep, pure, spiritual, and worthy of regard on this present sensationalist theory. They carry their own inner measure of worth with them; and it is just as logical to use the present theory of the emotions for proving that sensationalist

> processes need not be vile and material, as to use their vileness and materiality as a proof that such a theory cannot be true.[22]

Nonetheless, even James was somewhat equivocal on this issue. He felt it necessary to draw a distinction between the 'coarser' and the 'subtler' emotions. It is worth pausing to dwell on the complex semantics of the word 'feeling', which in English can refer to a pinprick or a sense of nausea caused by dyspepsia, as well as to the heights of exaltation or the depths of despair. The former were more easily represented as the psychic consequences or concomitants of bodily events. But while the latter might appear to be purely intellectual in nature – 'and to borrow nothing from any reverberation surging up from the parts below the brain' – James maintained that in all but rare instances these finer feelings too possessed a physiological aspect.[23] The care with which he broached this subject was, however, indicative of its sensitivity. While the 'coarser' emotions might more readily be consigned to the realm of the corporeal, James recognised a cultural resistance to the suggestion that 'subtler' emotions, including the 'intellectual or moral rapture' that might accompany religious, ethical, or aesthetic experience, possessed a bodily aspect rather than being purely spiritual in nature.

On the other hand, he maintained that the theory of the emotions that he had propounded enjoyed, from a scientific point of view, a significant advantage. If 'the neural processes underlying emotional consciousness be what I have now sought to prove it, the physiology of the brain becomes a simpler matter than has been hitherto supposed'.[24] According to James's theory, it was possible to explain the emotions in terms of the 'sensational, associational, and motor elements' that neurologists employed in their accounts of other functions of the brain. James argued that one of two propositions regarding the neurology of the emotions must be true: 'Either separate and special centres, affected to them alone, are their brain-seat, or else they correspond to processes occurring in the motor and sensory centres already assigned, or in others like them, not yet known'.[25] The former hypothesis would necessitate a revision of the current neurological orthodoxy that the cerebral cortex was no more than a sensory-motor organ. But in James's theory: 'No new principles need to be invoked, nothing postulated beyond the ordinary reflex circuits, and the local centres admitted in one shape or another by all to exist'.[26] James's assertion that 'the emotions have been so ignored' by those researchers working within the dominant sensory-motor model reveals, however, a remarkable blind-spot in his knowledge of this literature. One of the most prominent and most rigorous proponents of this strictly sensori-motor notion of cortical function, the English physician John Hughlings Jackson, had, since the 1860s, devoted considerable attention to incorporating the emotions within his general understanding of the workings of the human nervous system.

John Hughlings Jackson on Emotional States

By the turn of the twentieth century investigators were actively exploring the nature of 'feeling' – in the full extent and complexity of that term – and seeking both to identify the neural mechanisms that subserved it and to understand how these interacted with the other functional units that composed the human nervous system. An interesting feature of the neuropsychological thought of this period is the way in which a *continuum* between these various forms of affect was identified.

It is worth dwelling briefly on the general features of the concept of localisation before considering how the emotions in particular were treated. To localise is to find a place for some function or attribute. It thus involves the demarcation and separation of various entities. These processes of separation often reveal a great deal about the classificatory systems of the individuals and communities involved. Moreover, to find a place for something is often also to establish the relations between that entity and others. These may be relations of contiguity, contrast, and perhaps most importantly, of *hierarchy*. Those engaged in the enterprise of cerebral localisation therefore more or less unwittingly reproduced many of the most basic values of their culture when they came to assign the various faculties of the human mind a place within the architecture of the nervous system.

To give just one example from a slightly earlier period than the one with which we are primarily concerned, in the first half of the nineteenth century the French physiologist Pierre Flourens performed some of the first ablation experiments designed to identify what functions the different parts of the brain performed. He concluded that the cerebral hemispheres as a whole were the seat of the 'higher' functions, including volition. Flourens refused, however, to allow that the hemispheres possessed any motor function directly; that role was delegated to centres lower in the cerebro-spinal axis. Flourens thus sought to maintain a demarcation in his physiology between the intellectual and the corporeal: a division that of course corresponded to key distinctions and value judgments within his culture.[27]

The central nervous system was thus endowed with an 'upper' and 'lower' axis, words which of course have an evaluative as well as topographic connotation. But the brain also has a left/right dimension. Parallel to the movement after 1860 to localise function on the surface of the brain was a growing recognition that, while anatomically symmetrical, the two hemispheres that composed the cerebrum were not functionally identical. One of the implications that came to be drawn from Paul Broca's identification of a language centre in the third frontal convolution of the brain was that in each individual one hemisphere was 'dominant'. The left hemisphere was seen as the seat of what were usually regarded as masculine functions: these were seen to be logical and analytical; while to the right hemisphere were relegated more feminine attributes, that is, those that were more intuitive and emotional in nature.[28] The emerging neurosciences can in this as in other instances be seen to have deployed commonplace cultural assumptions when seeking to understand the functional topography of the brain. It is worth noting that the traffic was not one way. The notion of the masculine left and feminine right hemispheres gained a remarkable currency in popular culture, so much so that in the second half of the twentieth century Michael Gazzaniga felt it necessary to publish a book that attempted to 'debunk' the 'left brain, right brain mania'.[29]

The tendency to bifurcate brain function into masculine/rational and feminine/affective components is also evident in the work of the British neurologist, John Hughlings Jackson, who was perhaps the most significant theorist of the functional organisation of the nervous system in the second half of the nineteenth century. What is most significant about Jackson's contribution to the novel understandings of the structure and function of the nervous system that emerged in the second half of the nineteenth century was his explicitly evolutionary stance.

One consequence of this evolutionary approach was that Jackson recognised no fundamental dichotomies in the functioning of the various levels of the human nervous system. The highest and lowest levels of nervous organisation were alike exclusively sensory-motor in their mode of operation. He insisted that: 'the very highest processes (the substrata of consciousness) are only the most multifold and complex of all sensori-motor processes …'.[30] Jackson recognised that by suggesting that muscles and viscera were represented in 'the "organ of mind", and in the very highest parts of that organ', he was transgressing established boundaries between the mental and the corporeal. Such a claim might at first glance appear 'extravagant'.[31] However, the theory of evolution, which he took to be axiomatic, demanded that the mode of function of the highest centres should display an identity with the workings of the more primitive forms of nervous organisation. As Jackson himself put it: 'I cannot conceive of what other "materials" the "organ of mind" can be composed, than of processes representing both movements and impressions'.[32]

The consciousness that accompanied the working of these highest centres was unitary. If 'analytically we speak of our highest mental states under the names of will, memory, emotion, and reason, we are only giving names to different aspects of consciousness'. Any system of faculty psychology – such as that advocated by Scottish Common Sense Philosophy[33] – was thus fundamentally flawed. Jackson even ventured to make a joke to underline the point: 'Taking "will", "memory" and "emotion" to be real independent entities or faculties reminds one of the old woman's remark, "how lucky it was that Adam called all the animals by their right names"'.[34] This jibe was as much applicable to proponents of a quasi-phrenological version of cerebral localisation as to faculty psychology. Likewise, contrary to the views of some localisationists, there were no 'separate centres for the will, nor for the memory, nor for reason, nor for emotion; all four arise simultaneously during activity of the highest cerebral centres….'.[35] Jackson reiterated this axiom in his 1884 lectures on 'Evolution and Dissolution of the Nervous System', insisting that: 'will, memory, reason, and emotion are simply artificially distinguished aspects of one thing, a state of consciousness'.[36]

Moreover, in 1875 Jackson came close to James's theory of the nature of the emotions when he declared: 'Emotional manifestations are wide and yet temporary bodily states'.[37] In particular, in 'emotional manifestations it is plain that the most automatic bodily processes are concerned'.[38] Elsewhere he demanded: 'how would anyone set about explaining what occurs physically during the emotional manifestations of fear, joy, etc., unless he assumed that the organic parts are represented in the highest centres?'[39] In a further anticipation of the James-Lange doctrine, Jackson in 1884 denied that an emotion could affect a bodily state. While repudiating any form of 'crude materialism', he maintained that the proper way of viewing the relationship was to recognise:

> The *physical basis* of what is the psychical state, an emotion, is of nervous arrangements of the highest centres especially *representing those parts of the body concerned in, and as they are concerned in, the manifestation of that emotion.* For example, the physical basis of anger (the emotion of combat) is of nervous arrangements representing the circulatory, respiratory and digestive systems, sphincters and other, no doubt all other, parts, animal and organic.[40]

The same set of organs might be involved in a range of emotions. What determined the particular emotion evoked was the representation of a certain combination of body parts involved in putting the organism in a particular posture. For instance, the emotion of anger accompanied a bodily state of readiness for combat. In an 1876 discussion of epilepsy, Jackson provided a definition of emotion that was consistent with these fundamental tenets about how the nervous substrate of consciousness must operate. In this paper he held that the substrate of consciousness were 'double'. Thus not only did we have the consciousness of say a rose, but: 'but *we* seem to know that *we* have it'.[41] Jackson designated the latter form of consciousness as 'subject' as opposed to 'object' consciousness.

The operations of the nervous centres underlying subject consciousness were those: 'most concerned in the wide bodily states which are the physical side of the emotions'.[42] He felt obliged to provide some definition of the term, one that relied implicitly upon the breadth of denotation that the word 'feeling' possessed. 'Feeling', in its elemental sense, signified the organism's primitive awareness of its environment; it was indeed in large part definitive of animal life. In Jackson's terms, '*Emotion* is the most general feeling: we are always in some state of feeling, just as much as we are always alive'. What were in common parlance known as the: 'emotions, meaning particular emotions, are variations of this constant state'. While emotion was dependent on the global state of the organism, 'each particular emotion is, so to speak, a special attitude of it'. Which emotion tended to predominate depended entirely upon what parts of the body experienced 'the leading excitement or depression'.[43] The emotions thus were not discrete entities to be classified in the way that James later derided; they formed a *continuum*. For example, Jackson maintained that fear constituted a degraded form of anger. The same bodily combination that would produce the latter psychic state in a 'strong man' would evoke the former in the 'weak man'.[44] Jackson used 'weak' and 'strong' to characterise different physiologies. Nonetheless, his choice of terms and example betrays tacit assumptions about the gendered nature of nervous performance.

Jackson acknowledged the influence of the ideas of Herbert Spencer on his own work.[45] James, in contrast, was repelled by what he saw as the deterministic aspect of Spencer's system. Spencer was of course the author of an ambitious 'synthetic philosophy' that attempted to encompass the workings of the entire physical, organic, psychological, and social worlds. His claims to scientific objectivity notwithstanding, Spencer had strong political commitments as well as sharing in many of the biases and assumptions of his time, class, and gender, which were insinuated into the view of cosmic evolution that he propounded. Spencer's concept of evolution was strongly teleological in nature.[46] The end point of the evolutionary process was the appearance of a rational, emotionally restrained, decidedly masculine individual: in fact, a middle-class Victorian gentleman, not dissimilar to Spencer himself.[47]

Similar cultural investments are also implicit in Jackson's writings. This is especially evident in some of the metaphors he used to explicate his notion of the workings of the nervous system. Jackson followed Spencer in assuming that in the course of evolution a succession of 'higher' – in the sense of more recent – centres were added to older parts of the nervous system. The use of the term 'adding on' deserves special attention because it embodies Jackson's theory of how the human nervous system had attained its present complexity in the course of evolutionary time: new centres had *supplemented* older ones, and in the process profoundly modified the workings of those more primitive nervous structures. They had not, however, *supplanted* the more ancient parts of the brain. When the nervous system functioned properly, the new centres exercised a supervisory and moderating role over the old. To use Jackson's own words, the 'adding on' of new units was also a 'keeping down' of the old. In particular, in the highly evolved nervous system, impulsive, automatic behaviour was replaced by more deliberate, flexible kinds of action.

The true nature of this keeping down was manifested when the more civilised parts of the brain were damaged by injury or disease. In such cases the process of evolution was reversed: there was, to use Jackson's term, 'dissolution' of nervous function. This involved not only a 'taking off' of the higher but also a 'letting go' of the lower.[48] Jackson used a political metaphor to illustrate the nature of this dissolution: 'If the governing body of this country were destroyed suddenly we should have two causes for lamentation: (1) The loss of the services of eminent men; and (2) the anarchy of the now uncontrolled people'.[49] It is apparent that the 'eminence' of the natural governors of society, as of the body, lay in their freedom from impulsive action. In contrast the 'people' – that is, those who needed to be governed – were slaves to their emotions, and, for that reason, inferior, more primitive, beings.

Jackson developed these dichotomies most fully in his discussion of the nature of the language deficits that occurred in cases of aphasia. He noted that in the typical case of aphasia the sufferer was not dumb; they still possessed a certain service of words, but of a severely restricted and imperfect kind; what Jackson called 'jargon'. There was, Jackson claimed, 'a loss of intellectual, the more voluntary language, with persistence of emotional, the more automatic language'.[50] Jackson proposed an explanation for the jargon that aphasics were wont to emit that relied upon the tenet that the right cerebral hemisphere was intrinsically more 'emotional' in character. In such cases, the right hemisphere was, he argued, acting in an abnormal or 'hurried' fashion. What, he asked: 'is the cause of this hurry? I believe it to be strong emotion'.[51] The aphasic was liable to be in an unusually heightened emotional state. This was indeed only an exaggerated form of the effect that powerful emotion could have upon speech even in the physiological state. 'Strong emotion', Jackson declared, 'tends to more automatic, inferior, utterance'.[52]

The degree to which this was manifest – and therefore the degree to which the case was liable to be characterised by marked jargon – depended, however, on the 'patient's temperament'. In one of the cases that Jackson cited in this regard, the aphasic had been 'a most excitable woman. Her husband used to remark on this'.[53] Aphasia thus served merely to highlight the, on Spencerian premises, more impulsive and therefore less evolved aspect of the female personality. A series of associations and dichotomies was thus simply assumed rather than asserted. The rational was, almost by definition, also the higher, most evolved, kind of language. It was also distinguished by being voluntary in nature. In contrast, emotional language was a rather grubby residuum that persisted when linguistic faculties were expunged. It was characterised by its 'automatic' character; it was, in other words, beyond rational control.

The most typical expression of this lower form of language was obscenity. Jackson noted that swearwords were among those most often left at the disposal of the aphasic. These were uttered with no regard to the context in which they were used. The obscenity was, in Jackson's view, bereft of intellectual content or discrimination: it was a merely a verbal ejaculation. Some utterances were, in Jackson's words, 'intellectually dead'. He maintained that:

> Although oaths differ from mere alterations of tone, in that they consist of *articulate words*, they are generally used in talking, not to express ideas, but to make up for by vigour what is wanting in precision of expression. They may, indeed, be considered as phrases which emotion has filched from the intellect, to express itself in more definite terms than it could by mere violence of tone or manner.[54]

It is worth dwelling on this rather prissy passage because it articulates a complex of attitudes about the nature of emotion and its relations to reason. Emotion is treated with both suspicion and disdain: it is represented as a disreputable entity that 'filches' words from reason. It is also apparent from Jackson's writings that part of the reason that 'common emotional language' is disparaged lies in its proximity to the corporeal; its rank within the hierarchy of linguistic performances was 'little above that of bodily starts…'.[55] In this disparaging tendency to see the emotional as somehow more closely bound to the body than the rational we can see in Jackson some traces of what Damasio calls 'Descartes' Error'. It also demonstrates how persistent this tendency to derogate the emotional was even in a pronounced evolutionist.

At the same time, however, a closer reading of Jackson's work reveals a more subtle view of the relations of reason and emotion than these stark dichotomies might suggest. He maintained that there was a continuum between the various forms of language. The highest 'emotional' utterance was scarcely distinguishable from the lowest rational expression. Moreover, in healthy individuals the subordinate, emotional, service of words was necessary to the voluntary exercise of rational speech. Although localisation was not one of Jackson's principal concerns, he did assign the two services of words a seat in each of the cerebral hemispheres. The voluntary service was, of course, located in the dominant hemisphere, while the other resided in the other half of the brain. It is interesting to note that Jackson resorted to Spencer's notion of the 'survival of the fittest' to try to give an insight into what went on in the subordinate hemisphere. There a struggle between competing images and impulses occurred, of which the 'fittest' fought its way to consciousness. This depiction of the seat of the emotions as a chaotic, indeed violent, locale where a state of nature obtained is itself indicative of an inherently suspicious view of the emotional life. Presumably matters were conducted in a more decorous and orderly way in the dominant hemisphere.

Jackson's Legacy

The general principles of nervous action that Jackson enunciated were to have a lasting effect on later work in both neurology and psychology. For example, they played an important part in the development of Sigmund Freud's theories of the dynamics of the human mind.[56] In Britain, Jackson's legacy was manifest in the work of a number of prominent neurological investigators around the turn of the twentieth century. Among the most significant of these was Henry Head who declared that fundamental principles derived from Jackson had been fundamental to his own thinking. Above all, Head had derived from Jackson the doctrine that the various centres of the nervous system formed a hierarchy in which some exercise higher functions than the others.

Between the 1905 and 1918 Head published a series of papers that argued for a novel understanding of the sensory nervous system. Head's studies encompassed the full range of feeling from cutaneous sensation to the processing of afferent impulses at the cerebral level.[57] These investigations began with a series of experiments that Head conducted in the early years of the twentieth century in conjunction with W.H.R. Rivers. For the purpose of these investigations Head underwent a surgical procedure in which two cutaneous nerves in his arm were severed leading to a loss of sensation in the limb. Over a period of years Head and Rivers subjected the arm to a battery of tests to chart the process by which feeling was gradually restored. The main conclusion at which Head and Rivers arrived was that these enquiries proved the existence of two separate forms of feeling, subserved by different nervous structures. These discrete structures underlay kinds of feeling that had emerged at different phylogenetic periods.

The first of these was what Head and Rivers called *protopathic* sensation. This was a primitive form of feeling characterised by sensations that were at once intensive in nature and extensive in distribution and distinguished by their 'all or nothing' quality. The second form of sensation Head called *epicritic*. This was much more discriminating and graduated than the protopathic and was a more recent evolutionary development.[58] Head revealed his debt to Jackson in his account of the relations between the two kinds of sensation. In the healthy subject, epicritic inhibited protopathic sensation, providing the organism with a finely modulated awareness of outside stimuli and avoiding the kind of massive, uncontrolled, reaction that characterised the more primitive form of feeling. It is easy to recast this in terms of a narrative of the gradual triumph of civilisation over savagery in human history – or of the eventual ascendancy of control and restraint over untrammelled impulse in individual development. It was only in cases of injury and disease that the protopathic was able to manifest itself in its stark, primitive, form.

All these were physiological processes, and Head stressed that most of them never entered consciousness. At the higher levels of the nervous system, however, some of these physiological responses were accompanied by psychic states. Head distinguished between two particularly important cerebral centres. One was the cerebral cortex, which, as we have already seen was the site on which many functions had been localised in the preceding decades. Head saw the cortex as the seat of what might be called the more discriminative aspects of sensation, such as spatial orientation. Of more interest for our present purposes, however, was the role he assigned to a set of sub-cortical ganglionic structures known as the thalamus. If the cortex was the locus of the epicritic aspect of sensation at the higher levels of the nervous system, then the thalamus possessed a distinctly protopathic aspect. In particular, it was responsible for endowing sensations with their peculiarly *affective* tone. At its most basic, this meant giving a particular feeling the quality of pleasure or pain, which would help determine its response to some outside stimulus. From these primitive instincts had evolved the full range of emotional responses. Under normal conditions the discriminatory and the affective aspects of sensation cooperated to orientate the organism's behaviour. When, however, the moderating influence of the cortical centres was weakened because of injury or disease, the responses of the individual took on a distinctly 'thalamic' character, with the affective, emotional aspects of sensation assuming a predominant, pathological character.[59]

We can thus see in Head's work something of the same value judgments that we have already noted. The emotional was equated, at both the physiological and psychological levels, with a lower, because more impulsive and less reflective, form of action. It was more 'primitive' in phylogenetic terms, but also tainted with the pejorative overtones of that word. The protopathic is seen to have its uses, to be necessary even to a thoroughly modern organism, but only when restrained and moderated by a more refined sensibility.

Head's ideas on the nature of sensation were deemed to have important implications in a variety of fields, including epistemology, anthropology, and aesthetics as well as physiology and psychology.[60] Among those who in the first half of the twentieth century took up and developed these ideas on the nature of sensation was Grafton Elliot Smith, Professor of Anatomy at University College London. He had collaborated directly with Head and provided materials for his researches. Smith had developed an ambitious project for a new kind of anatomy that would make use of clinical as well as morphological and comparative studies to seek to elucidate the nature of nervous function.

Smith later came to collaborate with George G. Campion, a psychologist and philosopher who maintained that recent developments in neurophysiology had major implications for some of the fundamental issues with which the philosophy of mind had traditionally been concerned. In particular, he held that the concept of the 'concept' had to be thoroughly revised in the light of current biological knowledge. But Campion was also concerned with understanding the place of the emotions within the overall mechanism of mind. In 1934, Smith and Campion joined forces to produce a book entitled: *The Neural Basis of Thought*. This book appeared in a series called 'Philosophy and Scientific Method', which was intended to make an educated lay readership acquainted with recent developments in the sciences. In this work, Smith and Campion accepted and developed the main outlines of Head's localisation of mental capacities in the brain. They agreed, in particular, that the thalamus and its associated structures were the seat of what they called the 'affective' aspects of mind, while the cortex was responsible for the 'cognitive' functions. The authors defined the affective factor as what had variously been described as: 'libido', 'love', 'interest', 'feeling', 'desire', 'liking', etc.[61] In his portion of the book Smith gave a detailed account of how these two centres had emerged in the course of mammalian evolution to meet the developing needs of the organism.

What is remarkable however about Smith and Campion's discussion of the relation of cortex to thalamus – and thus of reason to emotion – is the degree to which they insist on the harmonious nature of the relations between these two structures. In other words, the cortex and thalamus are seen to cooperate and to complement one another's functions. This is in sharp contrast with the antagonistic and hierarchical relations between reason and emotion found in previous authors. Smith and Campion did not speak of the cortex inhibiting or keeping down the thalamus. Their favoured metaphor was instead that of a *circulation* of impulses between these two centres.[62] Thalamus and cortex were thus seen as engaged in a process of constant consultation with respect on each side for the input of the other. The brain thus constituted a kind of liberal utopia. This was not, however, a celebration of liberal individualism. The emphasis was instead upon the *integration* of the organism into a cohesive whole through the action of a highly evolved nervous system.[63]

This attitude was fully consistent with their Darwinian – as opposed to a predominantly Spencerian – understanding of the mind and nervous system. If emotion was such a retrograde, destructive force, then it was unlikely that natural selection would have permitted it to survive as part of human psychological make-up.Smith and Campion strove instead to demonstrate how emotion was a valuable – indeed *essential* – component of the mind from a purely biological point of view. They argued that as well as the cognitive and affective faculties, the *conative* aspect of mind had to be recognised. In other words, all organisms, whether primitive or advanced, were always orientated towards action. They did not merely become aware of their environments thanks to impulses mediated by the nervous system; they responded to those stimuli in ways that were intended to ensure the survival of the individual and of the species.[64]

The affective aspect of thought played a crucial part in the process of priming the organism for action by giving an emotional colour to various stimuli. At its most basic, this affective element would help the organism decide whether a particular environmental stimulus was to be pursued – for example, because it was food or a potential sexual partner – or avoided because it was noxious or a predator. On this thinking, fear and desire were the most elemental of the emotions. The more complicated an organism became, so the range of its affective responses grew wider and more nuanced. From a biological point of view, however, the essential function of emotion remained the orientation of the organism towards its environment and as a necessary precondition for action.

This is very similar to the kind of view on the role of emotion in the shaping of human personality and action put forward by Damasio sixty years later. Damasio appears to have been unaware that many of his ideas had thus been anticipated. This raises interesting questions about why what might be called consensual models of neural functioning seem to have had less appeal and currency than those which portray the workings of the brain in terms of conflict and constant tension. Is it simply that the latter are more dramatic and exciting?

A somewhat more sophisticated historical explanation would seek to place developments in neurology in the context of more general intellectual and cultural trends. It would, for instance, seek to locate the views of 'neuropsychologists', such as Smith and Campion, against the background of the pervasive trend towards biological holism in the interwar period.[65] This in its turn may be seen as a somewhat forlorn attempt by intellectuals to assert the values of harmony and unity in a world where discord and division appeared to be in the ascendant. For example, Campion, in a 1923 book entitled *Elements in Thought and Emotion,* moved easily from a discussion of the individual mind to pronouncements upon social psychology. He was concerned with the harmful effects of certain 'obsessions' that could become engrained in the psyche of a population leading to fanaticism and anarchy. As an example of such social pathology Campion cited the example of Marxism.[66] He also discussed at length the mental state of the German nation at the outset of World War 1: in this case a whole people had been dominated by a single emotion – namely, hatred. Campion maintained that this morbid – and in its consequences calamitous – state of affairs had arisen because of the failures of a system of education that encouraged both an excessive self-love and a corresponding extreme antipathy for other nations.[67] Sadly the history of the decades that followed the publication of Smith and Campion's ideas on the neural basis of thought appeared to vindicate the Jacksonian view of human nature as a seething mass of passions barely restrained by a fragile reason rather than their own more optimistic vision.

Notes

1 Antonio R. Damasio, *Descartes' Error: Emotion, Reason, and the Human Brain* (New York: G.P. Putnam, 1994), 249–50.

2 *Ibid.*, xi.

3 *Ibid.*, xii.

4 Gage was the American railway worker who in 1848 suffered a serious brain injury when a tamping rod passed through his skull. Because Gage survived this accident, his case provided the opportunity for observations and speculations on the effect a lesion of this kind might have on a person's character.

5 Damasio, 1994, xii, emphasis in original.

6 *Ibid.*, 128.

7 Robert M. Young, *Mind, Brain and Adaptation in the Nineteenth Century: Cerebral Localization and its Biological Context from Gall to Ferrier* (New York: Oxford University Press, 1990).

8 William James, *The Principles of Psychology*, 2 vols (London: Macmillan, 1890), ii. 472.

9 *Ibid.*, ii. 448.

10 *Ibid.*, ii. 448.

11 Such a classificatory approach to the subject was compatible with an explicitly evolutionary understanding of the emotions: see Charles Mercier, *The Nervous System and the Mind: A Treatise on the Dynamics of the Human Organism* (London: Macmillan, 1888), chapters 12–14.

12 James, 1890, ii. 449.

13 *Ibid.*, ii. 449.

14 See Jill Matus, 'Emergent Theories of Victorian Mind Shock: From War and Railway Accident to Nerves, Electricity and Emotion', in Anne Stiles (ed.), *Neurology and Literature, 1860–1920* (London: Palgrave, 2007), 175–6.

15 William James, 'What is an Emotion?', *Mind*, 9 (1884), 188–205.

16 Carl Lange (1834–1900) was a Danish neurologist who wrote on a number of psychological topics.

17 For a discussion of the response to the James–Lange theory, see: Thomas Dixon, *From Passions to Emotions: The Creation of a Secular Psychological Category* (Cambridge: Cambridge University Press, 2003), chapter 7.

18 James, 1890, ii. 453.

19 James More, 'The Emotions and the Psychology of Darwin', *Lancet*, 2 (1872), 112.

20 *Ibid.*, 113.

21 *Ibid.*, 114.

22 James, 1890, ii. 453, emphasis in original.

23 *Ibid.*, ii. 470–1.

24 *Ibid.*, ii. 472–3.

25 *Ibid.*, ii. 473.

26 *Ibid.*, ii. 474.

27 Edwin Clarke and L.S. Jacyna, *Nineteenth-Century Origins of Neuroscientific Concepts* (Berkeley: University of California Press, 1987), 244–85.

28 Anne Harrington, *Medicine, Mind, and the Double Brain* (Princeton: Princeton University Press, 1987).

29 Michael S. Gazzaniga, *The Social Brain: Discovering the Networks of the Mind* (New York: Basic Books, 1985), chapter 4.

30 John Hughlings Jackson, 'On the Anatomical and Physiological Locali-

sation of Movements in the Brain' (1875), in James Taylor, Gordon Holmes and F.M.R. Walshe (eds), *Selected Writings of John Hughlings Jackson*, 2 vols (London: Hodder and Stoughton, 1931–2), i. 46.

31 *Ibid.*, i. 47.

32 *Ibid.*, i. 48.

33 A system of philosophy originating in Enlightenment Scotland chiefly associated with Thomas Reid (1710–1796) that enjoyed considerable popularity in the nineteenth century, especially in America.

34 John Hughlings Jackson, 'On Affections of Speech from Disease of the Brain' (1878–9), in Taylor et al, 1931–2, ii. 154–204: 199.

35 John Hughlings Jackson, 'Lectures on the Diagnosis of Epilepsy' (1879), in Taylor et al, 1931–2, i. 276–307: 289.

36 John Hughlings Jackson, 'Evolution and Dissolution of the Nervous System' (1884), in Taylor et al, 1931–2, ii. 66.

37 Jackson, 1875, i. 47.

38 John Hughlings Jackson, 'The Physiological Aspects of Education', (1872), in Taylor et al, 1931–2, ii. 266.

39 Jackson, 1879, i. 291.

40 John Hughlings Jackson, 'Remarks on Evolution and Dissolution of the Nervous System' (1887), in Taylor et al, 1931–2, ii. 106, emphasis in original.

41 John Hughlings Jackson, 'On the Scientific and Empirical Investigation of Epilepsies' (1874–6), in Taylor et al, 1931–2, i. 243, emphasis in original.

42 Jackson, 1874–6, i. 243–4.

43 *Ibid.*, i. 244, note, emphasis in original.

44 Jackson, 1887, ii. 107.

45 On Jackson's debt to Spencer, see C.U.M. Smith, 'Evolution and the Problem of Mind: Part II. John Hughlings Jackson', *Journal of the History of Biology*, 15 (1982), 241–62.

46 *Too* teleological for James who in 1878 alleged that: 'Spencer merely takes sides with the [telos] he happens to prefer', William James, *Essays in Philosophy* (Cambridge, MA: Harvard University Press, 1978), 18.

47 L.S. Jacyna, *Lost Words: Narratives of Language and the Brain, 1825–1926* (Princeton: Princeton University Press, 2000), 134–5.

48 Jackson, 1884, ii. 58.

49 *Ibid.*, ii. 58.

50 *Ibid.*, ii. 49.

51 Jackson, 1878–9, 2, 194.

52 *Ibid.*, ii. 196.

53 *Ibid.*, ii. 195.

54 John Hughlings Jackson, 'Loss of Speech: Its Association with Valvular Disease of the Heart and with Hemiplegia on the Right Side. – Defects of Smell. – Defects of Speech in Chorea. – Arterial lesions in Epilepsy' (1864), in Henry Head, 'Reprint of some of Dr Hughlings Jackson's Papers on Affections of Speech', *Brain*, 38 (1915), 40, emphasis in original.

55 John Hughlings Jackson, 'On Affections of Speech from Disease of the Brain' (1879–80), in Taylor et al, 1931–2, ii. 178.

56 Frank J. Sulloway, *Freud, Biologist of the Mind: Beyond the Psychoanalytic Legend* (London: Fontana, 1980), 270–1.

57 W.H.R. Rivers and Henry Head, 'A Human Experiment in Nerve Division', *Brain*, 31 (1908), 323–450.

58 L.S. Jacyna, *Medicine and Modernism: A Biography of Sir Henry Head* (London: Pickering and Chatto, 2008), 125–30.

59 *Ibid.*, 131–7.

60 *Ibid.*, 144–51.

61 George G. Campion and Grafton Elliot Smith, *The Neural Basis of Thought* (London: K. Paul, Trench, Trubner & Co., Ltd.; New York: Harcourt, Brace and Co., 1934), 103–5.

62 *Ibid.*, 107–9.

63 *Ibid.*, 126.

64 *Ibid.*, 125–6.

65 Christopher Lawrence and George Weisz (eds), *Greater than the Parts: Holism in Biomedicine, 1920–1950* (Oxford: Oxford University Press, 1998).

66 George G. Campion, *Elements in Thought and Emotion: an Essay on Education, Epistemology, and the Psycho-Neural Problem* (London: University of London Press; 1923), 5.

67 *Ibid.*, 74.

9

Darwin's Changing *Expression* and the Making of the Modern State

Rhodri Hayward

One of the more depressing political developments in twenty-first century Britain has been the government's embrace of a happiness agenda. From the beginning of the new century there has been a rhetorical consensus between leaders of the Labour and Conservative parties that the state's pursuit of 'gross domestic product' should be supplemented by a search for 'gross national hap piness'.[1] In 1999, the prime minister, Tony Blair promised a new era of 'lifestyle politics' arguing that health and education policy would now be benchmarked against 'quality of life indicators'. He noted 'that money isn't everything but in the past, governments have seemed to forget this'.[2] In 2006, David Cameron, the newly elected leader of the Conservative party, warmly embraced this philosophy. In an interview with the BBC he claimed that any future government must concentrate on 'not just what is good for putting money in people's pockets but what is good for putting joy in people's hearts'.[3] Four years later, as prime minister of the coalition government, Cameron formally commissioned the creation of a happiness index by the Office for National Statistics.[4]

These schemes of psychological state building must seem far distant from the anthropological and biological speculations Darwin engaged in when he wrote *The Expression of the Emotions in Man and Animals* in 1872. Darwin's work largely eschews reference to politics or morality and instead presents a natural history of the passions in which universal forms of emotional expression are identified and attributed to evolutionary accidents and adaptations. Yet it is this work, I will argue, that has provided the basis for our current political programmes. Darwin's arguments and their subsequent reconstruction in the writings of a wide range of thinkers including Walter Cannon, Konrad Lorenz, Silvan Tomkins, John Bowlby and Paul Ekman have made possible a kind of 'prehistoric utopianism' in which political decisions are grounded through reference to two temporal registers. They look to the complexities of our current situation while at the same time referencing human needs developed in the Pleistocene era – a time that evolutionary psychologists designate the environment of evolutionary adaptedness (EEA).

The main sources for both Cameron's and Blair's enthusiasm for emotional government have been the work of Richard Layard at the London School of Economics, and the economist Amartya Sen at Trinity College, Cambridge.[5] These are the best-known of a number of economists who have argued that the decades of sustained economic growth witnessed since the Second World War have been bought at a high psychological cost.[6] Drawing on a mixture of epidemiological sources including obesity reports, meta-analyses of anxiety estimates, cohort studies of depression and World Health Organization (WHO) surveys of the global burden of disease, they argue that the last four decades have witnessed a dramatic growth in individual misery resulting in clinical depression rapidly becoming the major cause of ill health in the Western world.[7] This miserable increase is attributed to the ongoing frustration of our primeval needs and expectations.

My purpose here is not to make cheap political points, nor is it to question the fundamental assumptions of the pursuit of happiness.[8] Rather it is practical. I want to examine how such political programmes have been made thinkable, and explore the relationship between these programmes and the model of the affective life that Darwin articulated. This relationship is not clear cut. The objective evidence upon which the politics of happiness is grounded is largely drawn, as we have seen, from psychopathology and psychosomatic medicine – subjects which are largely absent from Darwin's work. It relies on a faith that the signs collected through surveys of sickness, questionnaire methods, investigations of obesity, alcohol intake or cigarette smoking can be reliably read as expressions of a troubled inner state. And it rests upon a conviction that these inner states can be unified into a meaningful pattern that somehow provides a physical or psychological commentary on our current state of affairs. The symptoms presented cannot be attributed to idiosyncratic habits or eccentric responses to the vagaries of everyday life – such as a sick hamster or difficulties with the milkman – rather they have to be united and understood as a universal response to a shared situation. This is where Darwin's arguments provide a crucial foundation for the new politics. They allow the evanescent emotions we experience day-to-day to be brought together and seen as meaningful manifestations of a common condition. While our lives may be complex and fragmented, our responses are united through their common ancestry.[9]

In descriptions of the pathological effects of modern forms of social organisation the continued reference to evolutionary explanations introduces a new form of the 'state of nature' argument into political discourse. This argument, closely associated with the political philosophy of Hobbes and Locke, is transformed through the new evolutionary psychologies into a kind of prehistoric utopianism. The body's expressed emotions provide a commentary not simply on the way things are but how they have been. It is this tension between the two temporalities that sustains the critique of modern life.

This somatic commentary is itself open to negotiation. Darwin's arguments, as we shall see, raise hermeneutic questions about the boundaries of expression and the difficulty of discriminating between those physiological and anatomical changes that can be said to indicate emotion and those which do not. Darwin's theory, I argue, makes possible a very fluid model of the psychosomatic and is capable of sustaining a wide variety of narratives about the changing emotional state of the population. Although it suggests that our emotional responses have been scripted at some far distant time, there is little agreement on the content of those scripts and the elements that comprise them. This chapter shows how those scripts have been rewritten as different editions of Darwin's *Expression* have been produced. In these editions the emotions are read variously as expressions of the racial past, individual memory, communicative problems and the psychosocial environment. These different understandings create different political worlds.

Darwin's Original *Expression*

Darwin scholars agree that *Expression* is a problematic work.[10] Whereas in *The Descent of Man* (1871), Darwin had shown how human behaviour was adaptive and served a wider biological function, in the *Expression*, the emotions simply appear as vestigial habits unadapted to the contemporary environment. They are the embodied memories of past actions acquired and transmitted through some obscure process of Lamarckian inheritance.[11] Moreover it is difficult to see what they actually express.[12] Darwin shies away from the idea that emotions might serve as forms of non-verbal communication and instead depicts them as simple physiological routines initiated by specific environmental triggers. As Darwin noted in his conclusion to the *Expression*:

> The power of communication between members of the same tribe by means of language … is much aided by the expressive movements of the face and body … Nevertheless there are no grounds, as far as I can discover, for believing that any muscle has been developed or even modified exclusively for the sake of expression.[13]

This reluctance to ascribe any communicative value to the emotions arises in part, as Richard Burkhardt and others have noted, from Darwin's engagement with the claim, advanced in natural theology, that facial expression was a God-given endowment to promote human dialogue.[14] The position was exemplified in Charles Bell's *Anatomy of Expression* (1844). Bell claimed that 'in man there seems to be a special apparatus, for the purpose of enabling him to communicate with his fellow-creatures, by that natural language, which is read in the changes of countenance'.[15] Darwin resisted the idea that the human facial muscles were somehow especially designed or adapted for the purpose of communication and insisted instead that such expressions should be understood as vestigial inheritances. The argument created an oddly denuded model of human emotion. Emotional expressions operated neither as aids to communication nor encouragements to action.

In his insistence that emotional expressions served no local purpose, Darwin's work undermined design arguments while at the same time emphasising the psychological unity of mankind. It could be read, as James Moore, Adrian Desmond and Greg Radick have forcefully argued, as a further demonstration of his commitment to the 'sacred cause' of common origins.[16] It reiterated the thesis of *Descent*, arguing that a common repertoire of expressions is shared by savage and civilised races across the globe. As Darwin wrote: 'the many points of close similarity in the various races were due to the inheritance from a single parent-form, which had already assumed a human character'.[17]

Although Darwin recognised that the rules on emotional expression were culturally specific, he insisted that the form the emotions took was bound up with the common architecture of the body. Darwin identified three main principles that governed their expression. The first mechanism (or 'principle', to use Darwin's term) is drawn from Alexander Bain's *The Senses and the Intellect* (1855). Here emotional reactions are presented as 'serviceable associated habits' – a kind of ingrained bodily response, triggered by events that had once served the purposes of our ancestors. As he noted 'there is a tendency through the force of habit and association for the same movements to be performed though they may not be of the least use'.[18] Darwin gave the example of indignation in which the erect head, downward frown, tensed shoulders and clenched fists serve to indicate the readiness to fight and remove the source of irritation.[19] And in the case of sadness, which would become central to psychological politics, the downward mouth and furrowed brow (now held up as the omega sign of melancholia) were understood as indications of the suppressed scream.[20]

The second principle, 'antithesis', is based upon the idea that these 'serviceable associated habits' would set up corresponding links between symmetrically opposed groups of emotions and behaviours. Anger leads to the tensing of the shoulder muscles while resignation leads to their collapse. As Darwin notes,

> The helpless man unconsciously contracts the muscles of the forehead which are antagonistic to those that cause a frown, and thus raises his eyebrows; at the same time he relaxes the muscles about the mouth so that the jaw drops. The antithesis is complete in every detail, not only in the movements of the features, but in the position of the limbs and in the attitude of the whole body…As the helpless or apologetic man often wishes to show his state of mind, he then acts in a conspicuous or demonstrative manner'.[21]

Although Darwin, through his dispute with Bell, wished to downplay the communicative aspects of emotions and suggests that such behaviour may have evolved through a general antagonism to expressions created by 'serviceable associated habits', he concedes that since

> the power of intercommunication is certainly of high service to many animals, there is no *a priori* improbability in the supposition, that gestures manifestly of an opposite nature to those by which certain feelings are already expressed, should at first have been voluntarily employed under the influence of opposite states of feeling.[22]

As we will see, the tension around the communicative work of the emotions was to have a lasting effect on the politics of emotion.

Darwin's third principle of expression eschewed reference to will and habit and looked instead to the anatomy of the cerebro-spinal system. This mechanism was grounded in the hydraulic models of nervous energy that dominated late nineteenth-century medicine. Drawing on the medical memoirs of his second cousin, Henry Holland, and Herbert Spencer's 'Physiology of Laughter', Darwin claimed that

> it is an unquestionable truth that, at any moment, the existing quantity of nerve force which in an inscrutable way produces in us the state we call feeling, must expend itself in some direction – must generate an equivalent manifestation of force somewhere; so that when the cerebro-spinal system is highly excited and nerve force is liberated in excess, it may be expended in intense sensations, active thought, violent movements, or increased activity of the glands.[23]

Here, the common architecture of the nervous system, rather than the common responses of ancestors, creates shared forms of embodied feeling and action.

This idea of an explosion of nerve force was central to nineteenth-century understandings of psychosomatic phenomena, but this was not an avenue that Darwin wished to explore. Although he notes the dramatic physical effects that might accompany strong outbursts of emotion (such as severe trembling or one's hair going white from shock) he dismisses such phenomena as 'obscure', saying that any discussion only allows us 'to clearly perceive our ignorance'.[24] This absence is surprising since connections between emotional excess, physical suffering and mental illness were well-established in nineteenth-century England. Leading physicians and alienists, including Darwin's own father, had warned against the dangers of overindulgence in emotion and morbid introspection.[25] Darwin was well aware of these arguments. He had himself suffered repeated bouts of depression and (it seems) psychosomatic illness following his father's death in November 1848, and his own doctors had often reminded him of the relationship between his chronic dyspepsia and his anxiety.[26] In his attempts to master his sadness, he employed his own prototypical theories of psychosomatic substitution, celebrating his bouts of eczema for the relief they offered from his internal suffering.[27] At one level, as Janet Browne's work suggests, Darwin's life can be read as a miniature demonstration of the political force of emotions. Negotiations around the strength and significance of his purportedly psychosomatic illnesses structured the domestic life of the Darwin family.[28]

Although Darwin's work failed to launch a coherent programme of research into the medical implications of the emotions, his three principles of explanation allow the connection between illness and environment to be imagined in different ways. Across the twentieth century aspects of Darwin's argument were taken up by competing movements in psychosomatic medicine and used to ground different models of the expressive nature of illness. Against Darwin's best intentions the physiological concomitants of emotions became communicative. They were taken as indices of different aspects of life and made possible different forms of social organisation.

Freud's *Expression*: Emotions and the Language of the Unconscious

Although British alienists and neurologists paid lip service to Darwin's ideas in their studies of melancholia and other emotional pathologies, it was in Freud's early essays on hysteria that the primitive origins of emotion was most closely explored.[29] Although Freud closely allied himself with the Darwinian project, his own work was to cut across the grain of Darwin's argument and resurrect the idea of emotional expression as a form of symbolic communication. This transformation is clearly visible in his *Studies of Hysteria,* co-authored with Josef

Breuer in 1895. In their analyses of 'Emmy von M.' (Fanny Moser) and 'Cäcilie M.' (Anna von Lieben), Freud and Breuer drew upon the concepts of serviceable habit and the idea of a overflow of nerve force to argue that physical illnesses (M.'s sudden spasms and gastritis) originated in a need to maintain internal equilibrium: an equilibrium disrupted by the tension between her external environment and her instinctual needs. Citing the feelings of cardiac pain and discomfort in the throat following an insult, Freud argued that such sensations were not simple embodiments of popular sayings ('a stab in the heart', 'swallowing an insult') but rather 'all these [neurotic] sensations and innervations belong to the field of "The Expression of the Emotions", which, as Darwin [1872] has taught us, consists of actions which originally had a meaning and served a purpose'.[30] In Freud and Breuer's reconstruction, this process of internal regulation is overladen by a new mechanism of hysterical conversion in which the old 'serviceable associated habits' are exploited as a symbolic language of protest.

Thus although Freud allied himself with Darwin, the mechanism of hysterical conversion recast the emotions as a privileged form of communication. In his work, emotions collapsed different registers of time from those Darwin had described. Thus while he joined Darwin in his recognition of the phylogenetic origins of emotional expressions, he saw their selection and use as being determined in part by an ontogenetic project to maintain the internal equilibrium of psychic energy. This internal equilibrium was not simply disrupted by present tensions (as Darwin had observed) but past traumas. Unshackled from their direct relationship with primitive adaptations through the mechanism of the unconscious, emotional expression, in this account, took on new forms. In the work of Freud's followers, especially Sandor Ferenczi and Felix Deutsch, the vocabulary of this symbolic language was extended from the musculo-skeletal system to include physiological disturbances.[31] In this process of extension the potential signs of emotions multiplied and the inner life became visible to the state.

Beadnell's *Expression*: Endocrinology, Ancestral Physiology and Psychosomatic Medicine

In 1942 a new abridged edition of the *Expression* appeared, edited by

C.M. Beadnell, the president of Rationalist Press Association. Beadnell was an enthusiastic Darwinian, publishing his own short essays on the phylogenetic bases of kissing, laughter and prayer, but he regretted that Darwin's own researches into the origins of the emotions had been hampered by his ignorance of endocrine mechanisms. 'The astonishing facts that the innermost moods of the mind – love, hate, fear, rage, etc.' Beadnell claimed, "together with their outermost manifestations the muscular contractions of the limbs and face, are in great measure dependent upon the functioning of the chemical messengers or hormones...would *have rejoiced Darwin's heart*'.[32] They demonstrated that emotional expression was rooted outside consciousness in the actions of the sympathetic nervous system.

The discovery of endocrine mechanisms opened up new debates over the communicative role of emotion and disease. Against Freudian ideas of hysterical conversion and symbolic expression, American clinicians and physiologists, including Walter Cannon and George Crile, argued that psychosomatic illnesses might simply arise as the physiological concomitants of alarm reactions.[33] Drawing upon their experiences in treating cases of shock during the First World War, Crile and Cannon argued that many forms of illness were simply the pathological consequences of inherited responses. The emotions serve, in their account (as in Darwin's), as a form of mnemonic recalling ancestral encounters and primitive decisions. But the disjuncture between this early environment and the demands of contemporary life had a high physiological cost. As Crile explained:

> man has reached his present status of civilization with brutish organs…And now, though sitting at this desk in command of the complicated machinery of civilization, when he fears a business catastrophe his fear is manifested in the terms of his ancestral physical battle in the struggle for existence. He cannot fear intellectually, he cannot fear dispassionately; he fears with all his organs and the same organs are stimulated and inhibited as if it were a physical battle with teeth and claws.[34]

This combination of ancestral excitement and contemporary repression (Crile's 'auto-captivity' of modern man) generated new forms of disease. In these models of morbid excitement, later extended in the stress models of Hans Selye, forms of psychosomatic affliction now came to be seen as symbolic commentaries on the nature of the external environment. Just as the pathological organisation of the laboratory experiment was revealed in the adrenal lesions of the rats that Selye dissected, so the pathological organisation of society was revealed by patterns of psychosomatic illness distributed across the population.[35] 'Psychosomatics', as it was argued, could 'be considered an aspect of the study of response systems acquired by *Homo Sapiens* during domestication'.[36]

Psychosomatic medicine existed at the uneasy junction between Freudian theories of emotion as a form of symbolic communication and physiological ideas of emotion as a form of adaptive response. It supported different readings of the emotions – as indices of frustrated desire or social organisation – and it vastly multiplied the number of signs of emotion. Emotions were not simply expressed in accordance with the principles identified by Darwin; rather, expressions were generated through the choices of the unconscious or the collateral action of the hormones. By the end of the 1930s a vast range of illnesses from diabetes through to cancer could be read as forms of emotional expression.[37] It was a transformation that made visible the emotional life of populations. What had previously passed fleetingly in the gestures and grimaces of everyday encounters was now collated and recorded in sickness claims and national insurance returns.[38] The trace of the emotions now extended from the nervous system through the body and into the bureaucratic machinery of the welfare state.

By the time of the Second World War, this vast extension of the emotions and their significance would lead to many critical commentators disputing their existence. In the writings of psychological constructionists such as Luther Lee Bernard, Knight Dunlap, and Elizabeth Duffy, emotion was depicted as a strategic term used to connect perceptions and bodily processes to specific programmes of activity.[39] By 1941, William Hunt could conclude that emotion terms were simply a useful way of describing certain terms: 'We are no longer so apt to view emotion as a 'thing' *sui generis*, unique and separable form other psychological phenomena'.[40] It was a construction used for the purposes of communication.

Mead's *Expression*: Emotions, Cybernetics and Communication Networks

In 1955, the anthropologist, Margaret Mead, published a new edition of the *Expression*. She presented Darwin's work as a prototypical contribution to the new discipline of 'kinesis' (the study of body language).[41] Where Beadnell had emphasised the hormones, Mead argued that the 'one interpretive tool which Darwin lacked and which we have is the idea of cybernetics'. Cybernetics offered a new perspective on emotional expression,

> so that instead of considering the internal conditions within the organism, which lead to the curled lip or the distended nostril, one considers these together with the perception of other organisms of the same type, seeing each piece of behavior as part of an intra or inter-species system of communication. [42]

Drawing upon her anthropological work on the relativity of emotional expressions and behaviourist arguments on the conditioned basis of emotional expressions, Mead saw emotions as being determined by the culture: 'When we substitute the word *communication* for Darwin's word expression, each of his questions, asked with such vigor and acumen...can be asked anew'.[43] Mead's position collapsed her own anthropological researches on the cultural coding of emotions into behaviourist claims on the conditioned basis of expression.[44] Emotion was determined and sustained its role in networks of communication.

This elision of expression and communication was a central tenet of the cybernetics movement which had emerged in the USA and the UK at the end of the Second World War. Mead had been involved in the movement's early discussions (promoted by the Macy Foundation) and her reading of Darwin seems to reflect the position of her colleagues – particularly that of her ex-husband, the anthropologist, Gregory Bateson, and his colleague, the psychiatrist, Jürgen Ruesch.[45] Working together at the Langley Porter Neuropsychiatric Clinic in California, Ruesch and Bateson studied psychosomatic disease in recent immigrants, believing that rates of disturbance could be taken as an index of their social dislocation.[46] This environmentalist programme was soon superceded, however. As Ruesch admitted, 'it soon became evident that the stress associated with culture change was related to difficulties in communication'.[47]

Together with Bateson, Ruesch developed a cybernetic model of psychosomatic distress in which physical disturbances were attributed to failures of communication. They argued that the social field comprised the unity of all its members and that internal equilibrium could only be maintained through the free flow of information. When this flow was disrupted the undischarged information was transmitted through the bodily organs – hence two men speaking at cross purposes would become increasingly agitated in their movements as their failures of communication led to them seeking new forms of expression.[48] One can see here how the psychical energy (described in Darwin's third principle) was transformed into information. The cure for this cybernetic psychosomatic disturbance was a form of psychotherapeutic intervention, for health relied upon the restoration of open systems of communication. For Ruesch 'communication' was 'the link that makes the difference between life and death'.[49]

A few years later, writing with the poet and artist, Weldon Kees, Ruesch noted how:

> Darwin claimed that certain emotional states are associated with *reflex-like motions* – for example, rage with gritting teeth or fright with recoiling. His theory – that many of the expressive gestures observed in men constitute residuals of the lower levels in phylogenetic development – may or may not be correct. The important issue is that, for purposes of communication, people make the assumption that feelings are linked with certain expressive movements and that these movements escape voluntary control.[50]

As they explained, these movements included all the 'glandular, vascular and intestinal manifestations….' From Ruesch's perspective every aspect of human behaviour became a form of communication

And here it seems our story has come full circle. The emotions that Darwin attempted to wrest from Bell and treat not as expressions but as ancillary ancestral reactions now became a sign of the failure of communication and a means to restore a failing dialogue. It was a triumph of the sign over the signified.

Lorenz's *Expression*: Ritualisation and the Incarnation of the Social System

In 1965 a new edition of the *Expression* was published with an introduction by Konrad Lorenz. Lorenz, alongside Niko Tinbergen, was a pioneer of the new science of ethology and he celebrated Darwin's anticipation of his own approach.[51] Darwin's arguments underlined the biological basis of behaviour undermining the claims of behaviourism and learning theory.[52] As Lorenz noted: 'The fact, which is still ignored by many psychologists, is quite simply that behaviour patterns are just as conservatively and reliably characters of species as are the forms of bone, teeth or any other bodily structures'.[53] Where Beadnell and Mead had celebrated the *Expression*'s anticipation of endocrinology and cybernetics, Lorenz paid tribute to Darwin's recognition of the historical basis of emotional interaction. As Lorenz observed, 'The conservative persistence of behaviour patterns, even after they have outlived, in the evolution of a species, their original function, is exactly the same as that of organs; in other words they can become 'vestigial' or 'rudimentary' in just the same way as the latter can'.[54]

The vestigial routines, described in the ethological work of Lorenz and Niko Tinbergen, were rather different from the autonomic responses described in the interwar period by Cannon and Crile. They were not simple emergency reactions, but complex performances arising from the intricacies and frustrations of social interaction. Where once the 'fight or flight' response mediated by the autonomic nervous system, had provided the key to understanding psychological and psychosomatic illness, now ethologists turned to the central nervous system in an attempt to uncover the innate patterns of emotional behaviour. In their writings and those of a new generation of psychologists, gathered around the affect theorist Silvan Tomkins, emotions were reimagined as integrated programmes. They were coordinated acts of physiological mobilisation, hardwired into the subcortical regions and coded through evolutionary pressure. They were biological universals yet retained a communicative function. For the ethologists, the ritulisation of affects into meaningful displays would form a crucial element of group selection.[55]

The development of these new models encouraged psychiatrists to turn away from the study of alarm conditions, such as anxiety, and instead explore more nuanced experiences such as loss and humiliation.[56] In studies of such varied groups as American veterans of the Korean War, hospitalised children, widows in South London, and rhesus monkeys subjected to maternal deprivation and isolation, consistent connections were drawn between the loss of a loved object, learned helplessness and depression.[57]

Again this connection was explicated through reference to Darwin's *Expression*. Colin Murray Parkes, who, alongside John Bowlby, pioneered psychiatric investigations into grief, noted that:

> When, in 1959, I first reviewed the scientific literature on loss and grief I was struck by the absence of any reference in it to the common observation that animals, in their reaction to loss, show many of the features that are evident in human beings. One of the few people who have made this point is Charles Darwin who, in *The Expression of the Emotions in Man and Animals* (1872), described the way in which sorrow is expressed by animals, young children, and adult human beings. His work caused me to formulate a 'biological theory of grief' which has been developed over the last decade but has not required any major modification.[58]

Drawing on Lorenz, Tinbergen, and Darwin, Parkes and Bowlby described grief as a social activity. As Parkes explained:

> The facial expression that Darwin describes, coupled with the restless searching and verbal requests for help, is sufficient to excite the serious concern and cooperation of others. The suppressed cry has thus become part of the ritualized social signal system which is still only partially under the control of the will but which enables the separated individual to evoke and direct the help of others in an orderly and constructive manner. Only when searching is useless and reunion impossible, as in the statistically infrequent event of loss by death, does the involuntary expression of grief lose its utility.[59]

By the end of the 1960s, the withdrawal behaviour associated with prolonged grief and depression was being read as a strategy for maintaining group hierarchy and social cohesion.[60] The emotions incarnated the individual's relation to the wider society. They could be seen as a way of defusing intra-group challenges, with the weak and oppressed retreating from active engagement rather than embarking upon a costly challenge to the status quo. As the Maudsley psychiatrist John Price argued in 1969:

> Comparative ethologists have shown that, in most vertebrate species, disputes between individuals are settled by ritual fighting in which neither of the contestants is seriously harmed. If there is no physical incapacity or physical pain, what induces the loser in a ritual combat to yield the prize to the winner? One possibility is that ritual yielding is subserved by mental incapacity and mental pain.[61]

Mental pain encouraged the isolation of the weak and preserved social stability. As Price would later note, if those 'hen-pecked animals' such as low ranking Hamadryas baboons

> could talk and were allowed access, they would be queuing up at the Emergency Clinic to complain of anxiety, depression and widespread aches and pains. It seemed obvious that if you had to have social hierarchies, you had to have behaviour appropriate to those positions in the hierarchy that individuals came to adopt. [62]

In this account, emotions are no longer simple rehearsals of prehistoric encounters. Rather they incarnate the structure and hierarchy of primitive society.[63] 'Emotion', it was argued, 'lies at the core of the social process'.[64]

Ekman's *Expression*: Affect Programmes and the Politics of the EEA

In 1998 Paul Ekman published his 'third' edition of the *Expression*. Heavily annotated and with an extensive scholarly apparatus, Ekman's edition drew out the biological stability of affect programmes emphasised by Lorenz and Tomkins as well as the communicative function described by Mead. But where Mead had seen communicative signals as local achievement produced through relational networks, Ekman argued that the patterns of emotional expression were universal and determined in the EEA.[65] In many ways his redescription of Darwin can be read as a long argument with Mead, Bateson and Ruesch. His life and theirs were, for a time, closely connected. In 1956, the year after Mead's edition of the *Expression* was published, Ekman had joined Ruesch as graduate clinical psychologist at Langley Porter. In his early work he had followed Ruesch in trying to use non-verbal behaviour to benchmark clinical encounters but soon abandoned the idea that facial expression and embodied gestures were excess forms of communication.[66] He saw the emotions instead as hardwired physiological responses in which the various forms of somatic expression formed only a component part.

Ekman's work rested upon a reassertion of the signified. Non-verbal behaviour was meaningful, not for its role in the flow of information, but because it was rooted in discrete physiological states. As Ekman recalled, his position, drawn from that of Silvan Tomkins, earned the ire of Bateson and Mead:

> Gregory, like Margaret, objected to the use of the word expression. I was being misled by Darwin. Use of the word expression directed attention away from the role of facial movements as communicative signals. It was a mistake to consider expressions as tied to internal sensations and physiological activity; they were tied to the back-and-forth flow of conversation.[67]

Forty years later, in his afterword to the *Expression*, Ekman penned an imaginary letter to his old mentor, Bateson. He argued that their dispute had lain around the question of agency. Bateson had understood expressions as intentional forms of communication while Ekman saw them as unintended signals produced by internal physiological changes. Intentionality, in Ekman's account, was displaced to an earlier state of evolution:

> They inform us that something important is happening inside the person who has the emotion. Those internal changes are preparing the person to deal with an important event, most often some interpersonal encounter, in a way that has been adaptive in the past. The past refers to our past as individuals and what has been adaptive in the history of our species.[68]

The past imagined in Ekman's work was quite different from the past constructed by Darwin. In place of the atomised encounters and confrontations held up in the *Expression* as the origins of serviceable associated habits, Ekman allied himself with a tradition of ethological and psychiatric work that described the EEA as a complex and hierarchical primitive society. Social life was negotiated through the development of complex rituals: rituals designed to avoid social conflict and its attendant biological costs. Such rituals persisted in the emotional responses of primates. At one level, Ekman had accepted Mead's cybernetic understanding of emotion as a communication system, but it was a communication system relegated to the environment of evolutionary adaptation.

Afterword

Reflecting on the social implications of his work in 2004, Ekman noted:

> There's an enormous hunger in our country – I can't really say of other countries; I suspect it's no different in any literate culture – to try to improve emotional life. The emotion system wasn't designed, I don't believe, it didn't evolve for [self-]consciousness to play much of a role. It evolved in a period in our history on this planet when we spent our lifetime like these people in the Stone Age culture did. They lived in the same village from birth to death, the same 150 people. That's all they encountered. We live in entirely different circumstances now. If you would ask a person in this Stone Age culture, 'Can we go to the next village on Saturday?' they said, 'How can I know today? Ask me on Saturday'. I tried that when I came back. People would invite me to dinner ten days [in advance], and I'd say, 'I'll let you know in ten days. How can I know today what I'm going to want to do in ten days?' You don't get away with that in our [culture].[69]

Whatever one's feelings about Ekman's reliability as a dinner guest he provides a good example of the conflicting temporalities inherent in modern theories of emotion. It is these multiple temporalities that allow emotions to become political objects. The multifarious causes of misery in everyday life are united in the primitive scripts of our affect programmes traced out in our neural architecture.[70] These scripts encode a particular version of the state of nature: a state very different from the one described in the *Expression.*

The primitive society Darwin found incarnated in the serviceable associated habits was one of face-to-face encounters and immediate responses. In place of this denuded conception of social interaction, evolutionary psychologists find hardwired into our emotional programmes the traces of a complex hierarchical society organised around common ideals of justice and fairness. Although we may differ in our values and political opinions about the proper organisation of society and the distribution of resources, our bodily and psychological illnesses trace out a shared understanding of rank and hierarchy that emerged in the EEA.

The shape of our illnesses and discontents, recorded by epidemiologists, describes the values of EEA while at the same time critiquing contemporary arrangements. In the writings of Oliver James, Richard Layard, Michael Marmot and Richard Wilkinson, the growth of sadness and illness across the population is attributed to maladaptive episodes of social comparison.[71] Assessments of our status in contemporary society generate pathological traces across our bodies and our behaviour as emotional programmes of primitive withdrawal and learned helplessness are initiated.[72] The language of Darwinian emotion, extended through the hormonal systems, communicative networks, and modular programmes of Cannon, Mead, Lorenz, Ekman and others, renders meaningful a vast number of objects and activities. Almost every aspect of our bodily health and personal behaviour can be read as psychosomatic markers revealing both terrain of the new politics and the primitive politics of the EEA.

Reviewing Ekman's edition of the *Expression* in 1998, the philosopher of science Ian Hacking suggested that such works represented a new folk psychology in the making.[73] But, looking at current state programmes, it is the clear that the significance of evolutionary emotion theories goes much further than this. The 'obscure subject' of the psychosomatic that Darwin failed to contain, makes every aspect of our lived lives expressive. These expressions have gone beyond the simple embodiments of past encounters that Darwin described, these affect programmes instead map out the shape of a lost polity. The materials that we use to construct the inner life now provide the substance for a new kind of politics.

Notes

1 Sara Ahmed, 'The Happiness Turn', *New Formations*, 63 (2007), 7–16.
2 Tony Blair, 'Foreword', *A Better Quality of Life* (London: HMSO [Department for Environment, Food and Rural Affairs], 1999).
3 Mark Easton, 'The Politics of Happiness', *The Happiness Formula*, http://news.bbc.co.uk/1/hi/programmes/happiness_formula/4809828.stm (accessed 17 September 2007).
4 Allegra Stratton, 'Happiness Index to Gauge Britain's National Mood', *Guardian*, 15 November 2010, 1a.
5 Richard Layard, *Happiness: Lessons from a New Science* (London: Penguin, 2006); and Amartya Sen, *Inequality Reexamined* (Oxford: Oxford University Press, 1992).
6 See also Richard.A. Easterlin, 'Does Economic Growth Improve the Human Lot? Some Empirical Evidence', in Paul A. David and Melvin W. Reder (eds), *Nations and Households in Economic Growth: Essays in Honor of Moses Abramowitz* (New York: Academic Press, 1974), 89–125; Robert E. Lane, *The Loss of Happiness in Market Democracies* (London: Yale University Press, 2000),19–35, 347–8; David G. Blanchflower and Andrew J. Oswald, *Well-being over Time in Britain and the USA* (Warwick Economic Research Papers, 616, 2001); Avner Offer, *The Challenge of Affluence* (Oxford: Oxford University Press, 2006), 347–56; and Derek Bok, *The Politics of Happiness: What Government Can Learn from the New Research on Well-Being* (Princeton NJ: Princeton University Press, 2010), 63–74.
7 Eric Fombonne, 'Clinical Depression: Time Trends and Possible Explanatory Mechanisms', in Michael Rutter and David J. Smith (eds), *Psychosocial Disorders in Young People: Time Trends and their Causes* (Chichester: John Wiley, 1995), 616–85; Layard, 2006, 35–6; Jean M. Twenge, 'The Age of Anxiety? Birth Cohort Change in Anxiety and Neuroticism, 1952–1993', *Personality and Individual Differences*, 30 (2000), 1007–21; Jean M. Twenge, 'Self-Esteem and Socio-economic Status: A Meta-analytic Review', *Personality and Social Psychology Review*, 6 (2002), 59–71; and Office for National Statistics, *Survey of Psychiatric Morbidity Among Adults Living in Private Households* (London: HMSO, 2000).
8 Ahmed, 2007; Barbara Ehrenreich, *Smile or Die: How Positive Thinking Fooled America and the World* (London: Granta Publications, 2009); Frank Furedi, 'Why the Politics of Happiness Makes Me Mad', *Spiked*, 23 May

2006, http://www.spiked-online.com/index.php?/site/article/311/ (accessed 22 July 2012); and Adam Phillips, 'Over the Moon', *Guardian,* 4 September 2010, http://www.guardian.co.uk/books/2010/sep/04/adam-phillips-the-happiness-myth (accessed 22 July 2012).

9 Randolph M. Nesse, 'Natural Selection and the Elusiveness of Happiness', *Philosophical Transactions of the Royal Society B*, 359 (2004), 1333–47.

10 Gregory Radick, 'Darwin's Puzzling *Expression*', *Comptes Rendus Biologies,* 333 (2010), 181–7; Janet Browne, 'Darwin and the Expression of the Emotions', in David Kohn (ed.), *The Darwinian Heritage* (Princeton: Princeton University Press, 1985), 307–26; Alan J. Fridlund, 'Darwin's Anti-Darwinism in *The Expression of the Emotions in Man and Animals*', in K.T. Strongman (ed.), *International Review of Studies of Emotion*, 2 vols (Chichester: Wiley, 1992), ii. 117–37.

11 William Montgomery, 'Charles Darwin's Thought on Expressive Mechanisms in Evolution', in Gail Zivin (ed.), *The Development of Expressive Behavior: Biology Environment Interactions* (London: Academic Press, 1985), 27–50; and Fridlund, 1992.

12 Thomas Dixon, *From Passions to Emotions* (Cambridge: Cambridge University Press, 2003), 165–6.

13 Charles Darwin, *The Expression of the Emotions in Man and Animals*, ed. by Joe Cain and Sharon Messenger (1872; London: Penguin Books, 2009), 324–5. All subsequent references to *Expression* are to this edition.

14 Richard W. Burkhardt, 'Darwin on Animal Behavior and Evolution', in Kohn, 1985, 327–69; and Dixon, 2003, 168–71.

15 Charles Bell, *The Anatomy and Philosophy of Expression as Connected with the Fine Arts* (1844; London: George Bell and Sons, 1877), 113.

16 Adrian Desmond and James Moore, *Darwin's Sacred Cause: Race, Slavery and the Quest for Human Origins* (London: Allen Lane, 2009); and Radick, 2010.

17 Darwin, 1872, 361.

18 *Ibid.,* 38.

19 *Ibid.,* 42.

20 Edward Shorter, 'Darwin's Contribution to Psychiatry', *British Journal of Psychiatry*, 195 (1995), 473–4.

21 Darwin, 1872, 272.

22 *Ibid.,* 66–8, 63.

23 *Ibid.,* 21, 67, 71; Hebert Spencer, 'The Physiology of Laughter', in

Essays Scientific, Political, Speculative, 3 vols (1855; London: Williams and Norgate, 1863), ii. 458; and Henry Holland, *Medical Notes and Reflections* (London: Orme, Brown, Green, Longman and Co., 1839).

24 Darwin, 1872, 69.

25 Michael J. Clark, '"Morbid Introspection", Unsoundness of Mind, and British Psychological Medicine, c.1830–1900', in William F. Bynum, Roy Porter and Michael Shepherd (eds), *The Anatomy of Madness: Essays in the History of Psychiatry*, 3 vols (London: Tavistock Publications, 1985), iii. 71–101.

26 John Bowlby, *Charles Darwin. A Biography* (London: Hutchinson, 1990), 278–89, 372–4; Ralph Colp, *Darwin's Illness* (Gainesville: University Press of Florida, 2008); Janet Browne, 'I Could Have Retched All Night: Charles Darwin and His Body', in Christopher Lawrence and Steve Shapin (eds), *Science Incarnate: Historical Embodiments of Natural Knowledge* (Chicago: University of Chicago Press, 1998), 246; and Edward J. Kempf, *Psychopathology* (London: Henry Kimpton, 1920).

27 Bowlby, 1990, 372–3; Colp, 2008, 86–7; and Adrian Desmond and James Moore, *Darwin* (London: Michael Joseph, 1991), 517.

28 Browne, 1998.

29 George M. Robertson, 'Melancholia, from the Physiological and Evolutionary Point of View', *Journal of Mental Science*, 36 (1890), 53–67; George M. Robertson, 'Morison Lectures – The Differentiation of Melancholia, the Depressive Phase of Manic-depressive Insanity', *Journal of Mental Science*, 57 (1911), 415–58; John Macpherson, 'Mania and Melancholia', *Journal of Mental Science*, 36 (1891), 212–25; Charles Mercier, *Conduct and Its Disorders, Biologically Considered* (London: Macmillan, 1911); Frank J. Sulloway, *Freud: Biologist of the Mind* (London: Burnet Books, 1979); and Lucille B. Ritvo, *Darwin's Influence on Freud* (London: Yale University Press, 1990).

30 Josef Breuer and Sigmund Freud, *Studies on Hysteria*, in James Strachey (ed.), *The Standard Edition of the Complete Works of Sigmund Freud* (1893–5; London: The Hogarth Press, 1955), 1–319:181, also 91.

31 Sandor Ferenczi, *Further Contributions to the Theory and Technique of Psycho-Analysis* (London: Hogarth Press, 1929); and Felix Deutsch, *Applied Psychoanalysis* (New York: Grune and Stratton, 1949).

32 C.M. Beadnell, *The Origin of the Kiss* (London: Watts, 1942); and C.M. Beadnell (ed.), 'Foreword', in C. Darwin, *The Expression of the Emotions in Man and Animals* (London: Watts, 1934), v, emphasis in original.

33 Walter B. Cannon, *Bodily Changes in Pain, Hunger, Fear and Rage: An Account of Recent Researches in the Function of Emotional Excitement* (New York and London: D. Appleton and Co., 1929), 9–18, 70–9; Walter B. Cannon, *The Wisdom of the Body* (London: Kegan Paul, Trench, Trübner, 1932); George W. Crile, *The Origin and Nature of the Emotions* (Philadelphia and London: W.B. Saunders, 1915); W. Whately Smith, *The Measurement of Emotion* (London: Kegan Paul, Trench, Trübner, 1922), 22–4.

34 Crile, 1915, 61–2.

35 Hans Selye, 'The General Adaptation Syndrome and the Diseases of Adaptation', *Journal of Clinical Endocrinology*, 6 (1946), 117–230; Mark Jackson, *The Age of Stress* (Oxford: Oxford University Press, 2013).

36 M.F. Ashley Montagu, 'Contributions of Anthropology to Psychosomatic Medicine', *The American Journal of Psychiatry*, 112 (1956), 979; Jules Henry, 'Anthropology and Psychosomatics', *Psychosomatic Medicine*, 11 (1949), 221.

37 Erich Wittkower, 'Studies of the Influence of the Emotions on the Functions of the Organs', *Journal of Mental Science*, 81 (1935), 533–682; and H.F. Dunbar, *Emotions and Bodily Changes* (1935; New York: Columbia University Press, 1954).

38 James L. Halliday, *Psychosocial Medicine: A Study of the Sick Society* (London: Heinemann Medical, 1948); and Rhodri Hayward, 'Enduring Emotions: James L. Halliday and the Invention of the Psychosocial', *Isis*, 100 (2009) 827–51.

39 Luther Lee Bernard, 'Instincts and the Psychoanalysts', *Journal of Abnormal Psychology and Social Psychology*, 27 (1922), 350–66; Knight Dunlap, 'Are Emotions Teleological Constructs?', *American Journal of Psychology*, 44 (1932), 572–6; and Elizabeth Duffy, 'Is Emotion a Mere Term of Convenience?', *Psychological Review*, 41 (1934), 103–4.

40 William A. Hunt, 'Recent Developments in the Field of Emotion', *Psychological Bulletin*, 38 (1941), 249–76: 249. For an overview, see Maria Gendron and Lisa Feldman Barrett, 'Reconstructing the Past: A Century of Ideas about Emotion in Psychology', *Emotion Review*, 1 (2009), 316–339.

41 Margaret Mead, 'Preface', in Darwin, *The Expression of the Emotions in Man and Animals* (1955; New York: Greenwood Press, 1969). On kinesis, see Ray L. Birdwhistell, *Kinesics and Context* (London: Penguin, 1971); Martha C. Davis, 'Film Projectors as Microscopes: Ray L.

Birdwhistell and the Microanalysis of Interaction (1955–1975)', *Visual Anthropology*, 17 (2001–02), 39–49.

42 Mead, 1955, vi.

43 *Ibid.*, vi, emphasis in original.

44 C. Landis, 'The Interpretation of Facial Expression in Emotion', *The Journal of General Psychology*, 2 (1929), 59–72; Margaret Mead, 'The Concept of Culture and the Psychosomatic Approach', *Psychiatry*, 10 (1949), 57–76.

45 Steve J. Heims, *The Cybernetics Group* (Cambridge, MA: MIT Press, 1991); Mary Catherine Bateson, *With a Daughter's Eye: A Memoir of Margaret Mead and Gregory Bateson* (New York: Harper Collins, 1991).

46 Jürgen Ruesch, C. Christiansen, S. Dewees, R.E. Harris, A. Jacobson and M.B. Loeb, *Duodenal Ulcer: A Sociopsychological Study of Naval Enlisted Personnel and Civilians* (Berkeley: University of California Press, 1948); Jürgen Ruesch, *Langley Porter Institute and Psychiatry in Northern California, 1943–1975* (San Francisco, CA: Friends of the Langley Porter, 1978).

47 Jürgen Ruesch, 'Communication and Mental Illness: A Psychiatric Approach', in Jürgen Ruesch and Gregory Bateson, *Communication: The Social Matrix of Psychiatry* (1951; New York, Norton, 2008), 79.

48 Ruesch, 1951, 81–93; Jürgen Ruesch, 'Psychosomatic Medicine and the Behavioral Sciences', *Psychosomatic Medicine*, 23 (1961), 277–85.

49 Ruesch, 1961, 281.

50 Jürgen Ruesch and Weldon Kees, *Non-verbal Communication: Notes on the Visual Reception of Human Relations* (Berkeley, CA: University of California Press, 1970), 36, emphasis in original.

51 Richard W. Burkhardt, *Patterns of Behaviour: Konrad Lorenz, Niko Tinbergen and the Founding of Ethology* (Chicago: University of Chicago Press, 2005).

52 Burkhardt, 2005, 398–407, 427–30.

53 Konrad Lorenz, 'Preface', in Charles Darwin, *The Expression of the Emotions in Man and Animals* (1872; Chicago: University of Chicago Press, 1965), xii.

54 *Ibid.*, xii.

55 Konrad Lorenz, 'Evolution of Ritualization in Biological and Cultural Spheres', *Philosophical Transactions of the Royal Society of London Series B*, 251 (1966), 273–84; Niko Tinbergen, *Social Behavior in Animals* (London: Methuen, 1965); Silvan S. Tomkins, *Affect, Imagery, Consciousness*, 2 vols; i., *The Negative Effects* (London: Tavistock, 1963); Silvan S. Tom-

kins, 'Affect and the Psychology of Knowledge', in Silvan S. Tomkins and Carroll E. Izard (eds), *Affect, Cognition and Personality* (New York: Springer, 1964) 72–97; Eve Kosofsky Sedgwick and Adam Frank, *Shame and Its Sisters: A Silvan Tomkins Reader* (New Durham, NC: Duke University Press, 1995); I. Alexander and Silvan S. Tomkins, 'The Heart of the Matter', in Gregory A. Kimble, C. Alan Boneau and Michael Wertheimer, *Portraits of Pioneers in Psychology*, 6 vols (Mahwah, NJ: Lawrence Earlbaum, 1996), ii. 295–313; and John R. Durant, 'Innate Character in Man and Animals: A Perspective on the Origins of Ethology', in Charles Webster (ed.), *Biology, Medicine and Society, 1840–1940* (Cambridge: Cambridge University Press, 2002), 182–92.

56 George L. Engel, 'Anxiety and Depression Withdrawal: The Primary Affects of Un-pleasure', *International Journal of Psycho-Analysis*, 43 (1962), 89–97; John D. Adamson and A. H. Schmale, 'Object Loss, Giving Up, and the Onset of Psychiatric Disease', *Psychosomatic Medicine*, 27 (1965), 557–77; and Edward Bibring, 'The Mechanism of Depression', in Phyllis Greenacre (ed.), *Affective Disorders* (New York: International Universities Press, 1963).

57 Bill Seay and Harry F. Harlow, 'Maternal Separation in the Rhesus Monkey', *Journal of Nervous and Mental Disease*, 140 (1965), 434–41; Colin Murray Parkes, *Bereavement: Studies of Grief in Adult Life* (1972; Harmondsworth: Penguin, 1975); William R. Miller and Martin E. Seligman, 'Depression and Learned Helplessness in Man', *Journal of Abnormal Psychology*, 84 (1975), 228–58; Frederick T. Melges and John Bowlby, 'Types of Hopelessness in Psychopathological Processes', *Archives of General Psychiatry*, 20 (1969), 690–99; James Robertson, *Young Children in Hospital* (London: Tavistock, 1958); and John Bowlby, 'Childhood Mourning and Its Implications for Psychiatry', *American Journal of Psychiatry*, 118 (1961), 481.

58 Parkes, 1975, 44.

59 *Ibid.*, 60. For later statements of this model, see Randolph M. Nesse, 'An Evolutionary Perspective on Psychiatry', *Current Opinion in Psychiatry*, 25 (1984), 575–80; and Nesse, 2004.

60 Melges and Bowlby, 1969.

61 John S. Price, 'The Ritualization of Agonistic Behaviour as a Determinant of the Variation along the Neuroticism/Stability Dimension of Personality', *Proceedings of the Royal Society of Medicine*, 62 (1969), 1107–10: 1107; and John S. Price, 'Hypothesis: the Dominance

Hierarchy and the Evolution of Mental Illness', *Lancet*, 2 (1967), 243–6.

62 John S. Price, Russell Gardner, Daniel R. Wilson, Leon Sloman, Peter Rohde and Mark Erickson, 'Territory, Rank and Mental Health: The History of an Idea', *Evolutionary Psychology*, 5 (2007), 531–54: 532; and Peter Rohde, 'The Relevance of Hierarchies, Territories, Defeat for Depression in Humans: Hypotheses and Clinical Predictions', *Journal of Affective Disorders*, 65 (2001), 221–30.

63 On these primitive hierarchies see S. Zuckerman, *The Social Life of Monkeys and Apes* (London: Kegan Paul, Trench, Trubner, 1932); Pieter R. Adraiens and Andreas De Block, 'The Evolutionary Turn in Psychiatry: A Historical Overview', *History of Psychiatry*, 21 (2010), 131–43; Donna Haraway, 'Animal Sociology and the Natural Economy of the Body Politic, Part II: The Past is a Contested Zone', *Signs*, 4 (1984), 37–60; Lane, 2000, 44–8; and Price, 2007.

64 David A. Hamburg, 'Emotions in the Perspective of Human Evolution', in Peter H. Knapp (ed.), *The Expression of the Emotions in Man* (New York: International Universities Press, 1963), 316.

65 Paul Ekman (ed.), *Darwin and Facial Expression: A Century of Research in Review* (New York: Academic Press, 1973); and Paul Ekman, Wallace V. Friesen, and Phoebe Ellsworth, *Emotion in the Human Face: Guidelines for Research and a Review of the Findings* (New York: Pergmanon Press, 1972).

66 Paul Ekman, 'Body Position, Facial Expression and Verbal Behaviour During Interviews', *Abnormal and Social Psychology*, 68 (1964), 295–301.

67 Paul Ekman, 'Afterword', in Darwin, *The Expression of the Emotions in Man and Animals, with Introduction, Afterword and Commentaries by Paul Ekman*, 4th edn (Oxford: Oxford University Press, 2008), 372; Paul Ekman, 'A Life's Pursuit', in Thomas A. Sebeok and Jean U. Sebeok (eds), *The Semiotic Web 1986* (Berlin: Mouton de Gruyter, 1987), 13,19.

68 Ekman, 2008, 372.

69 'Face to Face: The Science of Reading Faces. Conversation with Paul Ekman, Professor of Psychology, UCSF Medical School', by Harry Krelser, 2004, http://globetrotter.berkeley.edu/people4/Ekman/ekman-con2.html (accessed 11 September 2009), 5.

70 Nesse, 2004.

71 Oliver James, 'Death by a Thousand Social Comparisons – The Evolution of Subordination and Depression', in *Britain on the Couch: Treating a Low-Serotonin Society* (London: Arrow Books, 1998), 42–68; Michael

Marmot, *Status Syndrome* (London: Bloomsbury, 2005), 89–97, 156–75; Richard Wilkinson and Kate Pickett, *The Spirit Level: Why Equality is Better for Everyone* (London: Penguin Books, 2010), 197–211.

72 Robert M. Sapolsky, 'The Physiology and Pathophysiology of Unhappiness', in Daniel Kahnemann, Ed Diener and Norbert Schwarz, *Wellbeing: The Foundations of Hedonic Psychology* (New York: Russel Sage Foundation, 2004), 465.

73 Ian Hacking, 'By What Links Are the Organs Excited?', *Times Literary Supplement*, 17 July 1998), 11–12.

10

Calling the Wild: Selection, Domestication, and Species

Harriet Ritvo

In *The Variation of Animals and Plants under Domestication* Charles Darwin developed a theme to which he had already accorded great rhetorical and evidentiary significance. A decade earlier, he had begun *On the Origin of Species* with a description of artificial selection as practised by farmers, stock breeders, and pet fanciers, thus using a reassuringly homely example – one recognisable by the general public as well as by members of the scientific community – to introduce the most innovative component of his evolutionary theory. In addition, domesticated animals and plants, because they were numerous and available for constant observation, provided a readily available body of evidence.

Reassuring as it was, the analogy between natural and artificial selection was far from perfect. The point of Darwin's analogy was to make the idea of natural selection seem plausible by emphasising the efficiency and shaping power even of artificial selection. He devoted special attention to domesticated pigeons in *Variation* (see Figures 10.1 and 10.2), allotting two entire chapters to them, while pigs, cattle, sheep, and goats had to share a single chapter, as did ducks, geese, peacocks, turkeys, guinea fowls, canaries, goldfish, bees, and silk moths.[1] He noted, for example, that some of the prize birds bred by London pigeon fanciers diverged so strikingly in size, plumage, beak shape, flying technique, vocalisations, bone structure, and many other attributes, that if they had been presented to an ornithologist as wild specimens, they would unquestionably have been assigned to distinct species, perhaps even distinct genera. Darwin argued that if the relatively brief and constrained selective efforts of human breeders had produced such impressive results, it was likely that the more protracted and thoroughgoing efforts of nature would work still more efficaciously.

Figure 10.1: English Pouter. From Charles Darwin, *Variation of Animals and Plants under Domestication,* 2 vols (London: John Murray, 1868), i. 137.

But as Darwin acknowledged, there were some fairly obvious reasons why the two processes might diverge. The superior power of natural selection – 'Man can act only on external and visible characters: nature…can act on…the whole machinery of life. Man selects only for his own good; Nature only for that of the being which she tends' – might constitute a difference of kind rather than of degree, as might the much greater stretches of time available for natural selection.[2] Further, although the mechanism of the two processes appeared superficially similar, their outcomes tended to be rather different. Natural selection produced a constantly increasing and diversifying variety of forms; it never reversed or exactly repeated itself. Anyone familiar with artificial selection would have realised that, although new breeds were constantly being developed and although neither improved wheat nor improved cattle showed much tendency to revert to the condition of their aboriginal wild ancestors, the strains produced by human selection were neither as prolific nor as durable as those produced by nature. Indeed, the animals and plants celebrated as the noblest achievements of the breeder's art were especially liable to delicacy and infertility. Highly bred strains, long isolated from others

Figure 10.2: English Fantail. From Charles Darwin, *Variation of Animals and Plants under Domestication,* 2 vols (London: John Murray, 1868), i. 147.

of their species to preserve their genealogical purity, far from serving as a springboard for future variation, often had to be revivified with infusions of less rarefied blood. Yet, paradoxically, any relaxation of reproductive boundaries threatened subsidence into the common run of conspecifics.

At least in part, the disjunction between these two versions of selection reflects a dichotomy between the wild and the domesticated that has operated powerfully within both scientific and general culture, although it has not normally been the subject of much reflection. With regard to animals especially (as opposed to plants) it has tended to be taken for granted.

Wild Inclinations

When Byron wrote that 'the Assyrian came down like the wolf on the fold' ('The Destruction of Sennacherib', 1815), his audience had no trouble understanding the simile or feeling its force, despite the fact that wolves had not threatened most British flocks since the Wars of the Roses. Almost two centuries later, expressions such as 'the wolf is at the door' remain evocative, though the Anglophone experience of wolves has diminished still further. For most of us, they are only to be encountered (if at all) in zoos, or in establishments like Wolf Hollow, which is located in Ipswich, just north of Boston in North America, where a pack of grey wolves lives a sheltered suburban existence behind a high chain link fence. Their Massachusetts captivity has produced some modification of their nomadic habits and their fierce independent dispositions. (The pack was established twenty years ago with pups, so that only inherent inclinations needed to be modified, not confirmed behaviours.) Their relationship with their caretakers seems affectionate and playful, sometimes even engagingly doglike – so much so that visitors need to be warned that it would be very dangerous for strangers to presume on this superficial affability. The animals themselves give occasional indications that they retain the capacities of their free-roaming relatives – that though apparently reconciled to confinement, they are far from tame. When large loud vehicles rumble past on nearby Route 133, the wolves tend to howl. And despite their secure enclosure within the built-up landscape of North American sprawl, their calls evoke the eerie menace that has immemorially echoed through the wild woods of fairy tale and fable. At least within the controlled setting of Wolf Hollow, this frisson of fear is clearly attractive. The website howls when you open it, and visitors are invited to howl with the wolves before they leave the sanctuary.[3]

The symbolic resonance of large ferocious wild animals – the traditional representatives of what seems most threatening about the natural world – has thus proved much more durable than their physical presence. Indeed, their absence has often had equal and opposite figurative force. Thus the extermination of wolves in Great Britain, along with such other unruly creatures as bears and wild boars, was routinely adduced as evidence of the triumph of insular (as opposed to continental) civilisation in the early modern period. As they dispersed around the globe, British settlers and colonisers set themselves parallel physical and metaphorical challenges, conflating the elimination of dangerous animals with the imposition of political and military order. In North America, hunters could claim bounties for killing wolves from the seventeenth century into the twentieth, although by the latter period wolves had abandoned most of their historic range, persisting only in remote mountains, forests, and tundras. In Africa and (especially) Asia, imperial officials celebrated the 'extermination of wild beasts' as one of 'the undoubted advantages … derived from British rule'.[4]

Very occasionally, large aggressive predators could symbolise help rather than hindrance. They served as totems for people whose own inclinations were conventionally wolfish or leonine. And alongside the legendary and historical accounts of big bad wolves existed a minority tradition that emphasised cooperation rather than competition. From this perspective the similarities of wolf society to that of humans implicitly opened the possibility of individual exchange and adoption. A slender line of imagined lupine nurturers ran from the foster mother of Romulus and Remus to Akela, who protects and mentors Mowgli in *The Jungle Book* (1894). But in this way, as in others, Rudyard Kipling's animal polity looked toward the past rather than the future. By the late nineteenth century, human opinions of wolves and their ilk had indeed become noticeably mixed. The cause of this amelioration, however, was not an altered understanding of lupine character or an increased appreciation of the possibilities of anthropo-lupine cooperation, but rather a revised estimation of the very qualities that had made wolves traditional objects of fear and loathing.

The shift in European aesthetic sensibility that transformed rugged mountains into objects of admiration rather than disgust is a commonplace of the history of aesthetics. For example, in the early eighteenth century, even the relatively modest heights of what was to become known as the English Lake District impressed Daniel Defoe as 'eminent only for being the wildest, most barren and frightful of any that I have passed over in England, or even in Wales itself'.[5] The increasingly Romantic tourists who followed him gradually learned to appreciate this harsh, dramatic landscape, so that a century later the noted literary opium eater Thomas De Quincey could characterise the vistas that had horrified Defoe as a 'paradise of virgin beauty'.[6] Of course, this altered perception had complex roots, but it is suggestive that it coincided with improvements in transportation and other aspects of the infrastructure of tourism. As economic and technological developments made the world seem safer and more comfortable, it became possible to experience some of its extremes as thrilling rather than terrifying. Or, to put it another way, as nature began to seem a less overwhelming opponent, the valence of its traditional symbols began to change. Ultimately (much later, after their population numbers and geographic ranges had been radically reduced) even wild predators began to benefit from this reevaluation. The ferocity and danger associated with wolves and their figurative ilk became a source of glamour, evoking admiration and sympathy from a wide range of people who were unlikely ever to encounter them. As representatives of the unsettled landscapes in which they had managed to survive, they inspired nostalgia rather than antagonism.

Symbolic shifts were supplemented by shifts in scientific understanding, which redefined high-end predators as a necessary element of many natural ecosystems. Late nineteenth-century attempts at wild animal protection were modelled on the hunting preserves of European and Asian elites. Thus the immediate antecedents of modern wildlife sanctuaries and national parks were designed to protect individual species that were identified as both desirable (whether intrinsically or as game) or in danger of extinction, whether the bison in North America or the giraffe in Africa. They were much less concerned with preserving the surrounding web of life. In most cases, indeed, early wildlife management policies had the opposite effect. Although not all of the species targeted for protection provided conventional hunting trophies – for example, by the end of the nineteenth century, many great ape populations received some form of protection – all were herbivores. Further, none offered significant resistance to human domination of their territory. (If they did, policies could be reversed. For example, hippopotami, which enjoyed protection in some parts of southern Africa, were slaughtered with official encouragement in Uganda, where their belligerent attitude toward river traffic interfered with trade.[7]) Predators inclined to kill the species designated for protection received no protection themselves, either physical or legal. On the contrary, in many settings people simply replaced large predators at the top of the food chain and showed no mercy to their supplanted rivals.

Deep ancient roots can be unearthed for holistic or ecological thinking. But although most of the British pioneers of game preservation had enjoyed the classical education prescribed for privileged Victorian boys, the works of Charles Darwin may have offered more readily accessible arguments for understanding biological assemblages as interconnected wholes. Darwin provided many illustrations of the subtle and complex relationships among the organisms that shared a given territory. For example, in *Origin*, he explained the frequency of several species of wild flowers in southern England as a function of the number of domestic cats kept in nearby villages. Like most cats, the ones he chronicled had no direct interest in the flowers, but more cats meant fewer field mice, which preyed on bee hives – therefore fewer mice meant more bees to fertilise the flowers.[8] According to the *Oxford English Dictionary* (OED), the first English use of the word 'ecology' was in 1876, in a translation of the work of Ernst Haeckel, who coined it. The first decades of the twentieth century saw the emergence of ecology as a subdiscipline of biology.

Nevertheless, it was not until the second half of the twentieth century that individual species were routinely considered as components of larger systems by wildlife managers, and that the standard unit of management became the ecosystem rather than the species. In consequence, large predators were redefined as essential components (even indicators) of a healthy environment rather than blots on the landscape. They often began to receive legal protection, however belated and ineffective. And there has been a movement to reintroduce them to areas that have been ostensibly preserved in their wild form or that are in process of restoration. Thus in recent decades wolves have reoccupied several of their former habitats in the western United States, both as a result of carefully coordinated reintroduction by humans, as in Yellowstone National Park, and as a result of independent (but unimpeded) migration from Canada.

You Know It When You See It

I have been using several terms as if their meanings were clear and definite, when in fact they are contested and ambiguous. As has often been repeated, the cultural critic Raymond Williams characterised 'nature' as 'perhaps the most complex word in the English language'.[9] The term 'wilderness' is similarly problematic. In the context of preservation or restoration, it often collocates with words like 'pristine' and 'untouched', and therefore connotes a condition at once primeval and static. This connotation suggests that the first task of landscape stewards is to identify this ur-condition, but even a moderately long chronological perspective suggests that any such effort is bound to be quixotic. The environment in which modern animals have evolved has never been stable. Less than twenty thousand years ago much of North America and Eurasia was covered by glaciers. After their gradual release from the burden of ice and water, most northern lands continued to experience significant shifts in topography and climate, and, therefore, in flora and fauna. These natural changes have been supplemented for thousands of years by the impact of human activities. The theoretical and political problems presented by 'wilderness' are knottier still. In a groundbreaking essay published more than a decade ago, William Cronon argued that wilderness and civilisation (or 'garden') were not mutually exclusive opposites, but rather that they formed part of a single continuum. Far from being absolute, 'the one place on earth that stands apart from humanity', wilderness, was itself 'a quite profoundly human creation'.[10] Cronon's formulation sparked (and continues to spark) agonised resistance on the part of those environmentalists who base their commitment on the notion of untouched nature.

If wildness in landscape has been effectively (if controversially) problematised, the same cannot be said for wildness in animals. The OED defines the adjective 'wild' unambiguously, and it emphasises its zoological application. The first sense refers to animals: 'Living in a state of nature; not tame, not domesticated: opp. to TAME'. In a standard lexicographical ploy, 'tame' is defined with equal confidence and complete circularity as (also the first sense) 'Reclaimed from the wild state; brought under the control and care of man; domestic; domesticated. (Opp. to wild.)'. According to the OED, both 'wild' and 'tame' have persisted for a millennium, remaining constant in form as well as core meaning, while the language around them has mutated beyond easy comprehension, if not beyond recognition. And not only were both words used, in their modern senses and as clear opposites to each other, from the beginning of written English, but they have much deeper philological roots. Cognates of 'wild' exist in many extant and extinct Germanic languages, which implies derivation from an ancient common root. Further, according to the speculations that the OED finds most plausible, this root may itself be cognate with the ancestor of the Latin 'ferus'. Similarly, 'tame' has a long history within English and other Germanic languages, and the OED positively derives it from the much older form that also gave rise to the Latin root of 'domesticated'.

But this robustness on the level of abstraction has cloaked imprecision and ambiguity on the level of application or reference. Outside the dictionary these terms are harder to pin down and their interrelationships are more complex. Like Cronon's wilderness and garden, the wild and the tamed or domesticated exist along a continuum. In a world where human environmental influence extends to the highest latitudes and the deepest seas, few animal lives remain untouched by it. At least in this sense, therefore, few can be said to be completely wild – for example, it would be difficult so to characterise the wolves that were captured, sedated, airlifted to Yellowstone, and then kept in 'acclimatisation pens' to help them adapt to their new companions and surroundings. To put it another way: what do the lives of a free-living seal, a bear who raids trashcans in national parks, and a raccoon who chews through suburban roofs to shelter from the rain have in common (see Figure 10.3)? And as the valence of the wild has increased and its definition has become more obviously a matter of assertion rather than description, the boundaries of domestication have also blurred.

Not that they were ever especially clear. As twenty-first century

Figure 10.3: Bears at Yellowstone National Park. U.S. National Park Service photograph no. 11162. Photograph by R. Robinson.

wolves belong to a long line of animals whose wildness has been compromised, tameness has conversely also existed on a sliding scale. Even now, it would be difficult to produce a simple understanding of 'domesticated' by conflating the experiences of a pocketbook dog, a free-range pig, a barn cat, and a laboratory mouse. And although medieval farmers and hunters may have had no trouble distinguishing livestock animals from game or vermin, it would have been difficult to extract any general definition from their practices. The impact of domestication varied from kind to kind, as well as from creature to creature. The innate aggression of the falcons and ferrets who assisted human hunters was merely channelled, not transformed; when they were not working, they were confined like wild animals in menageries. Then, as now, people exerted much greater sway over their dogs than over their cats, who were mostly allowed to follow their own instincts with regard to rodents and reproduction. Medieval cattle, the providers of labour as well as meat, milk, and hides, led more constrained lives than did contemporary sheep, and pigs were often left to forage in the woods like the wild boars that they closely resembled.

With hindsight, even these relatively tame cattle could appear undomesticated, especially as wildness gained in glamour. Thus changes in the animals' physical circumstances were complicated by changes in the way they were perceived. In the late eighteenth century, for example, a few small herds of unruly white cattle, who roamed like deer through the parks of their wealthy owners, were celebrated as aboriginal and wild. As the Earl of Tankerville, whose Chillingham herd was the most famous, put it, his 'wild cattle' were 'the ancient breed of the island, inclosed long since within the boundary of the park'.[11] The 'ancient breed' was sometimes alleged to be the mighty aurochs (the extinct wild ancestor of all domestic cattle, which had been eliminated in Britain by Bronze Age hunters; the last one died in Poland in the seventeenth century), which gave these herds an ancestry distinct from that of ordinary domestic cattle. To increase or underscore their distinctiveness, the white cattle were never milked, and if their meat was required for such ceremonial occasions as the coming-of-age of a human heir, they were hunted and shot, not ignominiously slaughtered (see Figure 10.4). Through the nineteenth century, their autochthonic nobility continued to inspire the effusions of such distinguished poets and painters as Sir Walter Scott and Sir

Figure 10.4: Head of the Chillingham Bull shot by H.R.H. Prince of Wales in 1872. Engraved, by permission, from a photograph taken by the London Stereoscopic Company. John Storer, *The Wild White Cattle of Great Britain* (London: Cassell, Petter, and Galpin, 1879), 169.

Edwin Landseer, as well as the expenditure of newly wealthy landowners eager to bask by association in the prestige of wild nobility and ancient descent.

But even at the height of their renown, it was clear that their claims to wildness included a large measure of wishful thinking. Skeptics persuasively wondered whether, even assuming that the nineteenth-century emparked herds lived in a state of nature, that state represented a historical constant or a relatively modern restoration.[12] Many who investigated the background of the herds concluded that they were feral at best (at wildest, in other words) – that they were the descendants of domesticated animals, whether originally owned by Roman settlers or by later farmers. Modern anatomical and genetic research has confirmed these doubts, definitively connecting the emparked herds with the ordinary domestic cattle of the medieval period.[13] But so great is the continuing appeal of wildness, and so limited the persuasive force of scientific evidence, that a recent president of the Chillingham Wild Cattle Association has nevertheless asserted that 'although there is still much that is not known about the origins

Figure 10.5: Chillingham Bull, photography reproduced courtesy of Stephen J.G. Hall.

of the Chillingham Wild Cattle, one fact that is certain is that they were never domesticated' (see Figure 10.5).[14]

Only a few people possessed the resources necessary to express their admiration for the wild, and their somewhat paradoxical desire to encompass it within the domestic sphere, on such a grand scale. But numerous alternative options emerged for those with more restricted acres and purses. An increasing variety of exotic animals stocked private menageries. The largest of these were sufficiently extensive to have also included a cattle herd, if their owners had been so inclined – for example, those of George III or the thirteenth Earl of Derby, which accommodated large animals like kangaroos, cheetahs, zebras, and antelopes. Smaller animals required more modest quarters, and parrots, monkeys, canaries, and even the celebrated but ill-fated wombats owned by the poet Dante Gabriel Rossetti could be treated as pets. Breeders attempted to enhance or invigorate their livestock with infusions of exotic blood. If they were disinclined or unable to maintain their own wild sire, they could, in the 1820s and 1830s, pay a stud fee to the newly established Zoological Society of London for the services of a zebu or a zebra.

More expansively, the mid-nineteenth century saw the estab-

Figure 10.6: Herdwick Rams. Frank W. Garnett, *Westmorland Agriculture, 1800-1900* (Kendal: Titus Wilson, 1912), 156.

lishment of acclimatisation societies in such far-flung locations as Australia, North America, and Algeria, as well as in various European nations. A few of these societies, especially those of Australia and New Zealand, had as their primary purpose the replacement of indigenous faunas considered to be inadequate, with superior assemblages from Europe. But most of them wanted merely to supplement existing fauna with exciting exotic species.[15] Sometimes their aims seemed very practical (if not necessarily practicable). Thus large meaty ungulates like the eland were frequent targets of acclimatisers, who hoped to increase the meat supply (or perhaps only dietary variety). Camels were transplanted to remote sandy locations in Texas, where they soon disappeared, and in Australia, where their numerous descendants persist as an unappreciated invasive species.[16] But there was no pragmatic explanation for the choice of some prospective immigrants; for example, the notorious efforts to introduce such English birds as the titmouse, chaffinch, skylark, pheasant, sparrow, and starling to the United States.[17] The inspiration for such efforts lay in the realm of aesthetics or sentiment.

At the same time that these diverse activities reflected enhanced appreciation of wildness in animals, they also signalled an alteration in the nature of the category itself. That is, one attribute that connected all the indications of increased enthusiasm was a diminution in the very qualities conventionally associated with wildness. Animals in cages led very different lives from their unconfined relatives, even if they seemed especially appealing when they were growling and slavering over their food. Metaphorical pets, like those physically curled up in front of the fire, engaged a special emotional range in the hearts of their owners. Stud animals contributed their own wild essence to a domesticated next generation. Exotic domestic dogs were exhibited in zoos, and wild cats were entered in cat shows. In general, as the idea of wildness became more attractive, its concrete manifestations became less easy to distinguish from their traditional opposites.

The popular appeal of wild animals has continued to increase as they have become more accessible, either in the flesh or in the media. So entangled have wildness and domesticity become that it is now necessary to warn visitors to North American parks that roadside bears may bite the hands that feed them, and it is now possible for domesticated animals to represent nature. This extended symbolic reach was demonstrated in 2001, when foot and mouth disease struck British livestock.[18] Because the disease spreads rapidly and easily, the government prescribed a cull not only of all infected herds and flocks, but of all apparently healthy livestock living in their vicinity. Although outbreaks were widespread, the greatest number of cases occurred in the Lake District, the starkly dramatic landscape that had been disparaged by Daniel Defoe and praised by Thomas de Quincey; it is now the site of England's largest national park. Video and print coverage of the cull, which took the spectacular form of soldiers shooting flocks of sheep and then immolating them in enormous pyres, thus featured some of the nation's most cherished countryside as background.

The ovine victims also had iconic status. Most of them belonged to the local Herdwick breed (see Figure 10.6), and at first the intensive cull seemed to threaten its very survival. What was at stake was not merely adaptation to a demanding environment, since several other British hill breeds look very like the Herdwicks and share their physical and emotional toughness. The Herdwicks' special claim to consideration was their connection to their native ground, itself a kind of national sacred space. Not only were the sheep acknowledged to possess detailed topographical information about the hills they inhabited, but their owners claimed that they transmitted it mystically down the generations, from ewe to lamb. So well recognised was their attachment to their home territories, that when a farm was sold, the resident Herdwicks were conventionally included in the bargain, on the theory that if they were taken away, they would soon manage to return. And despite strong historical indications that the ancestral Herdwicks had arrived in the vicinity of the Lake District by ship, and the further fact that all British sheep descend from wild mouflons originally domesticated in the eastern Mediterranean region, they were traditionally celebrated as indigenous, 'peculiar to that high, exposed, rocky, mountainous district'.[19] An article in the preeminent Victorian agricultural journal asserted that the Herdwicks possessed 'more of the characters of an original race than any other in the county' and that they showed 'no marks of kindred with any other race'.[20] Twenty-first-century journalists reporting on the threatened toll of foot and mouth disease adopted similar rhetoric.

As the sheep were nativised, they were also naturalised. A reporter for the *Independent* newspaper feared that if the Herdwicks disappeared, the whole ecology of the region might be changed 'beyond recognition'.[21] And, since the dramatic bare uplands of the Lake District have been maintained by nibbling flocks for at least a millennium, his concern was not completely unreasonable. Thus whether technically indigenous or not, and although they are incontestably domesticated, the Herdwicks have become compelling symbols of the apparently untamed landscape they inhabit – more compelling than the numerous wild birds and small mammals with which they share it. Like the landscape itself, they seem wilder than they are; that is, they appear to be independent and free ranging, but their lives (and, indeed, their very existence) are ultimately determined by human economic exigencies. They are both accessible (that is, there are a lot of them and they are everywhere, not only in the fields, but grazing and napping beside the roads and even on top of them) and also inaccessible (that is, they are skittish, and tend to retreat when approached). The armed assault on the Herdwick sheep was therefore perceived as an attack on both the domesticated countryside and the unspoiled natural landscape. In both the sheep and their environment the wild and the tame had inextricably merged.

Survival Stakes

If vernacular usage illustrates the increasing slippage between wildness and tameness in animals, scientific classification has made a similar point from the opposite direction. The species concept has a long and vexed history. The study of natural history (or botany and zoology) requires that individual kinds be labelled, but for many plants and animals (those that, unlike giraffes, for example, have very similar relatives) it has often been difficult for naturalists to tell where one kind ends and the next begins. Darwin's theory of evolution by natural selection provided a theoretical reason for this difficulty, and his shrewd observations that 'it is in the best-known countries that we find the greatest number of forms of doubtful value' and that 'if any animal or plant...be highly useful to man...varieties of it will almost universally be found recorded' offered a more pragmatic explanation.[22] The classification of domesticated animals has epitomised this problem. That is, none of them has become sufficiently different from its wild ancestor to preclude the production of fertile offspring (the conventional if perennially problematic definition of the line between species), and some mate happily (or at least effectively) with more distant relatives. Nineteenth-century zoo keepers enjoyed experimenting along these lines, and zoo goers admired the resulting hybrids between horses and zebras, domestic cattle and bison, and dogs and wolves.[23]

Despite these persuasive demonstrations of kinship, however, from the eighteenth-century emergence of modern taxonomy, classifiers have ordinarily allotted each type of domestic animal its own species name. While recognising the theoretical difficulties thus produced, most modern taxonomists have continued to follow conventional practice. Domestic sheep are still classified as *Ovis aries* while the mouflon is *Ovis orientalis*, and dogs as *Canis familiaris* while the wolf is *Canis lupus*. The archaeozoologist Juliet Clutton-Brock explains this practice as efficient (it would be unnecessarily confusing to alter widely accepted nomenclature) as well as scientifically grounded, at least to some extent (most domestic animal populations are reproductively isolated from wild ones by human strictures, if not by biological ones).[24] But it also constitutes a simultaneous acknowledgement of the artificiality of the distinction between wild animals and domesticated ones, and of its importance and power. Vernacular understandings can trump those based on anatomy and physiology.

The implications of making or not making such distinctions extend beyond the intellectual realm. They construct the physical world at the same time that they describe it. With the advent of DNA analysis in recent decades, the taxonomic stakes have risen, so that even animals that look and act wild may be found genetically unworthy. For example, the Scottish Wildcat Association was established in 2007 to protect the small remaining British sub-population of the very widely distributed species ancestral to domestic cats. The targeted animals strongly resemble domesticated tabbies, although they tend to be larger and more irascible. Perhaps for this reason, the distinction between pure wild animals and those contaminated by miscegenation featured prominently on its website: 'In 2004 a team of scientists…estimated that 400 wildcats remained, the other 5000 or so being feral domestic cats or hybrid mixes of domestic and wildcat'.[25] The Association advocated 'improving legal protection, launching a public awareness campaign, supporting the captive breeding program and creating special reserves for wildcats which would in turn benefit many other species'. In consequence, the Scottish wild cat was declared a 'priority species' (in Scotland). It therefore became eligible to benefit from the establishment of a stud book, a captive breeding programme, and other measures that blur the cultural boundary between wild and domesticated, even as they attempt to reinforce the genetic boundary that separates them.

And the reification of the wild can be a matter of life and death, as well as of definition and genealogy. Although the howls of the wolf may retain their primordial menace, the wolves who make them have long vanished from most of their vast original range, and they are threatened in much of their remaining territory. To persist or to return, they need human protection, not only physical, but legal and taxonomic. Thus efforts to preserve the red wolf, which originally ranged across the southeastern United States, have been complicated by suggestions that it is not a separate species, but a hybrid of the grey wolf and the coyote. No such aspersions have been cast upon the pedigree of the grey wolf, but nevertheless every attempted grey wolf restoration has triggered human resistance (by means of guns and traps, as well as press conferences and lawsuits). Local challenges to their endangered status inevitably follow even moderate success. If domestic dogs were returned to their ancestral taxon, wolves would belong to one of the commonest species in the lower forty-eight states, rather than one of the rarest. Their survival as wild animals depends on the dog's continuing definition as domesticated.

An earlier version of this essay appeared in *Daedalus*, 137 (2008): 22–30.

Notes

1 Charles Darwin, *The Variation of Animals and Plants under Domestication* (London: John Murray, 1868).
2 Darwin, *On the Origin of Species*, facsimile (1859; Cambridge, MA: Harvard University Press, 1964), 83.
3 Wolf Hollow website, 2010, http://www.wolfhollowipswich.org/ (accessed 23 July 2010).
4 Edward Lockwood, *Natural History, Sport, and Travel* (London: W.H. Allen, 1878), 237.
5 Daniel Defoe, *A Tour through the Whole Island of Great Britain* (1724–6; New Haven, CT: Yale University Press, 1991), 291.
6 Thomas De Quincey, *Literary Reminiscences from the Autobiography of an English Opium Eater*, in *The Works of Thomas De Quincey*, 3 vols (1821; Boston: Houghton Mifflin, 1851), iii. 310–11.
7 Harriet Ritvo, *The Animal Estate: The English and Other Creatures in the Victorian Age* (Cambridge, MA: Harvard University Press, 1987), 284–9.
8 Darwin, 1859, 73–4.
9 Raymond Williams, *Keywords* (New York: Oxford University Press, 1976), 184.
10 William Cronon, 'The Trouble with Wilderness; or Getting Back to the Wrong Nature', in: Cronon (ed.), *Uncommon Ground: Rethinking the Human Place in Nature* (New York: Oxford University Press, 1996), 69–90: 69.
11 C.A. Tankerville and L. Hindmarsh, 'On the Wild Cattle of Chillingham Park', *Athenaeum*, 565, 25 August 1838, 611–12: 611.
12 For an extended discussion of the history of this debate, see Harriet Ritvo 'Race, Breed, and Myths of Origin: Chillingham Cattle as Ancient Britons', *Representations*, 39 (1992), 1–22.
13 For summaries of recent research, see Stephen J.G. Hall,

'Running Wild', *Ark*, 16 (1989), 12–15, 46–9; Stephen J.G. Hall, 'The White Herd of Chillingham', *Journal of the Royal Agricultural Society of England*, 150 (1989), 112–19; and Stephen J.G. Hall and Juliet Clutton-Brock, *Two Hundred Years of British Farm Livestock* (London: British Museum, Natural History, 1989).

14 I. Bennet, 'Chillingham Cattle', *Ark*, 18 (1991), 22.

15 Ritvo, 1987, 232–42.

16 'Camel Fact Sheet', 2009, Department of the Environment, Water, Heritage, and the Arts, Government of Australia, http://www.environment.gov.au/biodiversity/invasive/publications/camel-factsheet.html (accessed 26 July 2010).

17 'American Acclimatization Society', *New York Times*, 15 November 1877, 2.

18 For an extended discussion of this episode see Harriet Ritvo, 'Counting Sheep in the English Lake District: Rare Breeds, Local Knowledge, and Environmental History', in Dorothee Brantz, ed., *Beastly Natures: Animals, Humans, and the Study of History* (Charlottesville, VA: University of Virginia Press, 2010), 264-279.

19 J. Bailey and G. Culley, 'General View of the Agriculture of Northumberland, Cumberland, and Westmorland', facsimile (1805; Newcastle: Frank Graham, 1972), 245.

20 William Dickinson, 'On the Farming of Cumberland', *Journal of the Royal Agricultural Society of England*, 13 (1852), 264.

21 Ian Herbert, 'Foot and Mouth Crisis: Cumbria', *Independent*, 27 March 2001, 5.

22 Darwin, 1859, 50.

23 Harriet Ritvo, *The Platypus and the Mermaid, and Other Figments of the Classifying Imagination* (Cambridge, MA: Harvard University Press, 1997), 92–5.

24 Juliet Clutton-Brock, *A Natural History of Domesticated Mammals* (London: British Museum, Natural History, 1987), 194–7.

25 Scottish Wildcat Association, 2009, http://www.scottishwildcats.co.uk/wildcat.html (accessed 26 July 2010).

11

The Development of Emotional Life

Michael Lewis

For this chapter I have chosen to speak of Darwin's work on emotional development. In *The Expression of the Emotions in Man and Animals* (1872), and in his baby diaries, Darwin foresaw many of the issues which were to occupy the subsequent decades in research on emotional development.[1] His idea of emotions as adaptive action patterns – that is, innate ways of behaving measured in the face and in physiological responses to specific contexts – has been well received and, although controversial, still receives considerable support.[2] His interest in the emotions which emerged later in infancy, which he called self-conscious emotions, was not as well developed. He thought that the blushing response indicated the emergence of self-conscious emotions when a child was around three years of age. Although this was long after the early action patterns such as fear, joy, and sadness were seen, he understood that a person's thoughts about what others think about them was the elicitor of such emotions as embarrassment and blushing. Even so, he did not develop a theory of the role of cognition or consciousness in the emergence of these complex emotions. Nor, in fact, was he able to distinguish between such different emotions as guilt, shame and embarrassment. This was in part due to his desire to show the evolution of emotional life. His belief was that human emotional life evolved from the emotional expressions in animals. Given that these deep cognitive capacities are unlikely to be available for nonhuman animals, Darwin did not attempt to discriminate between the various kinds of self-conscious emotions.

In this chapter I will discuss the new research evidence on these emotions, which has now become available through the use of more sophisticated measurement systems. For example, while Darwin used blushing as his major measure of self-conscious emotions, I will use other cognitive and bodily expressions which allow me to consider these complex and varied self-conscious emotions and how they develop over the course of the early life of the child. First, I will briefly outline Darwin's views on the development of emotional life. Then I shall focus on his consideration of what he called the self-conscious emotions. Following this, I will articulate a theory of emotional life I first discussed in a 1979 book called *Social Cognition and the Acquisition of Self,* later developed in a book on *Children's Emotions and Moods* (1983), and finally explored in some detail in *Shame, the Exposed Self* (1992).[3]

Darwin's View of Emotional Development

Darwin saw the early emergence of some emotional expressions through the observations of his first-born son. Darwin wrote, for example, '…of the various expressions which he exhibited, for I felt convinced, even at this early period, that the most complex and fine shades of expression must all have had a gradual and natural origin'.[4] Darwin was referring to the early emotions that he saw which included anger, happiness, interest, surprise, sadness, fear, and disgust. He said of them that they 'had a gradual and natural origin', which meant that they had evolved. To argue this point, he showed that in many non-human animals there appeared facial, bodily, and vocal expressions that matched these emotions. Anger and sadness were two of his examples. He thought that 'Emotions and sensations have often been classed as exciting or depressing';[5] that is, Darwin saw not only emotions such as anger which prompted a 'moving towards' or 'approach', but withdrawal emotions, those which lead to 'moving away from' or 'not moving', such as fear and sadness. He wrote of them:

> When all the organs of the body and mind,– those of voluntary and involuntary movement, of perception, sensation, thought, &c.,– perform their functions more energetically and rapidly than usual, a man or animal may be said to be excited [approach], and, under an opposite state, to be depressed [withdrawal].[6]

In classifying specific emotions in this way, he saw joy and anger as approach emotions:

> Anger and joy are from the first exciting emotions, and they naturally lead, more especially the former, to energetic movements, which react on the heart and this again on the brain. A physician once remarked to me as a proof of the exciting nature of anger, that a man when excessively jaded will sometimes invent imaginary offences and put himself into a passion, unconsciously for the sake of reinvigorating himself; and since hearing this remark, I have occasionally recognized its full truth'.[7]

Further on he suggests that anger habitually leads to action. However, he also contrasted anger with sadness and saw sadness as a withdrawal emotion. For example, he wrote,

> As soon as the sufferer is fully conscious that nothing can be done, despair or deep sorrow takes the place of frantic grief. The sufferer sits motionless, or gently rocks to and fro; the circulation becomes languid; respiration is almost forgotten, and deep sighs are drawn... As associated habit no longer prompts the sufferer to action, he is urged by his friends to voluntary exertion, and not to give way to silent, motionless grief.[8]

Clearly here Darwin is talking about early emotions. It is important for us to keep in mind that, over 125 years ago, Darwin saw these emotions as action patterns that involve not only the face, but the body and physiological responses, so that when he refers to the organs of the body he is certainly leading us to believe that these early action patterns have evolved gradually. He is also emphasising, most importantly, that these action patterns were evolved, adaptive responses to particular events or context. When we think about this work we can see that these two ideas proposed that these evolved action patterns served as motives, connecting action to particular contexts in the world. Moreover, these action patterns were open to the pressures of adaptation. Thus, particular action patterns were either innate or needed little learning to be connected to particular contexts. It is this singular idea of Darwin's which has motivated much work. Darwin's idea of approach and withdrawal in particular has been followed by Walter B. Cannon, John B. Watson, and others who were intrigued by the concept of bipolar mechanisms of emotions as motives for both approach and withdrawal.[9] T.C. Schneirla, in a work still often quoted, wrote 'the aspect of towardness or awayness is common in animal behavior'. He believed that approach and withdrawal behaviours are 'applicable to all motivated behavior in animals'.[10]

This idea that specific action patterns connect to specific contexts as evolved adaptive responses, or as motives that lead the infant into commerce with its world, has received considerable attention. There are some who have argued that there has been little support in data for the idea that specific action patterns are related to specific contexts, and indeed, research in physiology has found little support for unique physiological responses to unique situations.[11] Moreover, little difference by situation has been shown to exist in early infancy, leading some to believe that particular action patterns are learned responses to particular situations and are governed by rules of socialisation rather than evolved evolutionary adaptive responses. However, there are some of us who believe that there are indeed adaptive specific action patterns to specific contexts, but that issues of measurement, physiology, and individual differences may make this difficult to observe. One particularly fruitful area has been the observation of facial expressions.

Utilising Darwin's observations, first Silvan S. Tomkins, then Carroll E. Izard and then Paul Ekman, Wallace V. Friesen and Phoebe Ellsworth, have argued that there are unique facial action patterns which are elicited by unique situations and are found universally across diverse cultures.[12] Moreover, following Darwin, they have argued that they have evolved as adaptive responses to action in the world. The issue of action patterns unique to specific situations is a complex issue and clearly beyond the scope of this particular chapter. The reader is referred to several articles of interest on this topic.[13] However, part of the problem of linking unique facial and bodily action patterns to unique contexts in early infancy is establishing what constitutes a unique situation. Working with my colleagues on a series of papers, we have examined a particular situation which, from a Darwinian point of view, should give specific responses. In particular, we have taken infants as young as two months of age and attached a string to their wrists which they can readily learn to pull to obtain access to interesting pictures and sounds. Upon learning to pull the string to obtain these rewards, we prevent the reward from occurring. In other words, we have blocked the learned response of pulling the string leading to a reward. Since the infant has learned in the past that it does lead to a reward, this constitutes a blocked goal. According to Darwin, anger was a response to a blocked goal and he argued further that the motive of anger was to allow the child or adult to push through the barrier to achieve the blocked goal. He saw active responses that included facial expressions of anger and bodily movement that would facilitate obtaining the blocked goal, as adaptive responses. In our work we have found that this is exactly the case. When infants are blocked from reaching a desired goal, they predominately show anger expressions, they pull vigorously to obtain the interesting pictures and sounds, and they show specific physiological responses. Of particular interest is the fact that there are individual differences in young infants' responses to this research paradigm. While the vast majority of infants show anger and make active attempts to overcome the blocked goal, there are a proportion of infants who, having learned to pull the string to get to the goal, when the goal is withdrawn or blocked, show sadness and stop pulling. From a Darwinian point of view, this is a nonadaptive response, so we see earlier individual differences in the strength of the action pattern to achieve the blocked goal.

It is particularly interesting to look at the stress hormones of these two groups of children. The children who vigorously attempt to overcome the blockage show no increases in stress hormones, whereas those children who show sadness and give up, show significant increases in stress hormones. These data have some important features usually not found in studies trying to examine unique facial expressions in unique contexts. In particular we have been very careful to try to describe unique situations, and we have used Darwin's idea of a blocked goal, and anger as a response, as a way of demonstrating that unique action patterns in unique situations can be found at quite early ages. Nevertheless, the literature in general fails to show such concordances. However, it is my strong belief that this is due to ill-defined contexts, rather than the lack of unique action patterns in unique contexts.

Darwin made many observations of his son and they are of considerable interest in the study of social development. He was sensitive to the child learning to lie, which he saw happened early in life. If we think about the issue of lying and deception, we can see that although there may be action patterns that have evolved and are adaptive, the fact that the human child does have the capacity to deceive or to lie, in effect means that at some point in development we should be able to gain control over these action patterns and, therefore, suppress their expression. In addition, given that we have some control over them, these action patterns should be open to socialisation factors such that different norms should lead to the different types of expression. So, for example, smiling and laughter, which involve the lower part of the face, as in a broad smile, in some cultures is deemed part of the action pattern that should not be displayed. Thus, in Japanese culture today, children will often hide the lower facial expression of laughter so as not to show their mouth or teeth, whereas American children do not show such behaviour.

The work on deception and lying leads almost immediately into a consideration of the idea of a self, and a self able to reflect upon itself and thus to control its own behaviour, in particular, its emotional expression, if not the entire action pattern itself. Darwin was interested in the development of the idea of 'me' which he called self-consciousness, and it is to this that we turn our attention next.

Self-Conscious Emotions

Darwin was particularly interested in what came to be called self-conscious emotions. In chapter 13 of *Expression*, he considered such topics as self-attention, shame, shyness, modesty, blushing. It was the topic of blushing that most occupied his attention. This is a particularly human or species-specific behaviour, since it does not appear that any other animal besides humans shows a blushing response. Darwin's interest in blushing also involved his interest in its development, and, as we shall see, Darwin suggested that blushing emerged in the child at about three years of age, observations he made by observing his son. In my own work on the emergence of self-conscious emotions, three years of age is a good approximation for what has subsequently been shown to be the case.[14] Darwin's interest in the emergence of self related to when he thought children came to recognise themselves in mirrors. Here, Darwin was somewhat misled by his observations, for while even infants show interest in looking at images in the mirrors, the interest in mirrors prior to the middle of the second year of life is mostly due to interest in faces features in general. Infants, like children and adults, are captivated by and show great interest in images of infants and very young children. This interest in babyishness can be seen across species and does not reflect self-recognition but reflects interest in looking at images of babies. Thus there is no reason to believe that infants and toddlers before fifteen months of age show interest in themselves. In addition to an interest in facial features, infants also have a sensitivity to contingency, that is, for two events which occur close in time, or The Law of Effect. Thus when there is movement and a corresponding movement reflected in the mirror, it is this contingency which attracts their attention, not a recognition of the self.

In our work on self-recognition in mirrors, we have been able to develop a system which gets around the two problems of babyishness and contingency. In our procedure, called the Mirror Rouge procedure, we surreptitiously paint the child's nose with odourless red rouge, and place the child in front of a mirror. Prior to fifteen months, no infant looks at the image in the mirror and touches their nose. After fifteen months, infants, in increasing numbers, start to touch their noses when they see the image in the mirror with rouge. Earlier than fifteen months, some infants touch the rouged nose in the mirror and some even go behind the mirror to see if there is another child. However, starting at fifteen months, and by two years, all children developing normally recognise themselves in the mirror by touching their nose when they see their image in the mirror. There are other behaviours which seem to confirm this emergence of self-recognition or self-consciousness, that is, the mental representation of the idea of 'me'. For example, personal pronoun use such as 'me' and 'mine' emerges at the same time as does mirror recognition. And, finally, pretend play which involves others also emerges. The idea of pretend suggests that children know that they are not what they are pretending to be. This is a kind of double negative which only conscious minds can do. If they actually thought it was real it would not be pretend. These three measures of the emergence of the mental representation of 'me' is associated with many other behaviours and the reader is referred to my work for a full account of how the emergence of this representation, or what we have called consciousness, is related to social and emotional development.[15] Nevertheless, Darwin's interest in the emergence of self and, more particularly, self-attention, can be seen in his work on blushing.

The emergence of self, the mental representation of the idea of 'me', serves a particularly important function in children's emotional life. From a developmental perspective, one can argue that the infant knows: that is, it has certain intuitive knowledge. Infants have some idea of quantity; for example, that something is one of many, and these ideas are part of their evolutionary past. In fact, even animals have this knowledge. However, the emergence of the mental representation or the idea of 'me', which we have shown is related to maturation of particular brain regions,[16] leads me to make the epistemological distinction between this early knowledge as 'I know' and the latter knowledge 'I know that I know'. The change in this knowledge in the human child's life occurs between fifteen to twenty-four months of age in normally developing children. This capacity brings with it the emergence of self-conscious emotions, of which I have considered two classes: the early self-conscious emotions which emerge at the same time as self-representation emerges, and an evaluative self-conscious class of emotions which emerge approximately one to two years later. By three years of age, the child is able not only to represent itself but has developed cognitive capacity to evaluate its emotions, thoughts, and actions. While Darwin did not develop this idea of the development of mental capacities, he was attuned to the idea that these self-conscious emotions develop and that they do not exist earlier. When considering the blushing response, he considered self-attention or, from our point of view, self-consciousness as well as shame, shyness, and modesty. Before outlining the research related to the development of these capacities we will first describe Darwin's thoughts on these issues.

In trying to understand the blushing response Darwin wrote, 'The Nature of the Mental States which induce Blushing. – These consist of shyness, shame, and modesty; the essential element in all being self-attention'.[17] You can see that a representation of myself, self-attention, is necessary for all of these. He goes on to say,

> Many reasons can be assigned for believing that originally self-attention directed to personal appearance, in relation to the opinion of others,

> was the exciting cause; the same effect being subsequently produced, through the force of association, by self-attention in relation to moral conduct. It is not the simple act of reflecting on our own appearance, but the thinking what others think of us, which excites a blush.[18]

With these words Darwin is presenting a fairly elaborate theory of self-development which needs further comment. He states two things, first that shyness, shame, and modesty are mental states which induce blushing, making no distinction between them, and arguing that the essential element of blushing is self-attention. The second point he makes is that there is some development which occurs, although he does not specify when. While the initial cause of blushing is self-appearance in relation to the opinion of others, the second reason, occurring later in development, has to do with the self's action, namely its moral behaviour, or the child's violating particular moral values. This is particularly interesting for me since this is what I have proposed as part of a developmental theory about the self-conscious emotions. My argument, considered below, distinguishes between exposure embarrassment and evaluative embarrassment, the first having to do with being the object of another's attention and the latter related to transgressions of moral standards and rules.

Darwin was particularly sensitive to the issue of blushing and not only as a measure of self-attention. He utilised the relation between praise and blushing, and quite clearly captured an important aspect of self-attention. He writes, 'But undoubtedly praise and admiration are highly efficient [that is, in causing a blush]: a pretty girl blushes when a man gazes intently at her, though she may know perfectly well that he is not depreciating her'.[19] Thus praise causes self-attention rather than fault:

> Many children, as well as old and sensitive persons blush, when they are much praised. Hereafter the question will be discussed, how it has arisen that the consciousness that others are attending to our personal appearance should have led to the capillaries, especially those of the face, instantly becoming filled with blood.[20]

Notice, however, that he is struck by personal appearance rather than the more general idea of attending to self as a general reaction. He goes on to say that:

> My reasons for believing that attention directed to personal appearance, and not to moral conduct, has been the fundamental element in the acquirement of the habit of blushing, will now be given. They are separately light, but combined possess, as it appears to me, considerable weight. It is notorious that nothing makes a shy person blush so much as any remark, however slight, on his personal appearance. One cannot notice even the dress of a woman much given to blushing, without causing her face to crimson. It is sufficient to stare hard at some persons to make them, as Coleridge remarks, blush.[21]

Again, Darwin makes quite clear that blushing has to do with how others respond to us. Moreover, he has considered embarrassment as measured by blushing to have two features: appearance and moral conduct.

In 'Self-development and Self-conscious Emotions' (1989) and in a series of experiments, my colleagues and I present data to show these two kinds of embarrassment, one which we call self-exposure embarrassment and the other which we call self-evaluative embarrassment. The former emerges at the same time as does self-recognition and other measures of self-representation, somewhere between fifteen and twenty-four months, while the latter emerges almost a year later, towards the end of the second year of life. In self-exposure embarrassment, the embarrassment is caused simply by being the object of others' attention, and we have found that complimenting, pointing to, and any behaviour by others which causes the child to pay attention to itself, induces this response.[22] While Darwin used blushing as a measure of embarrassment, it has its limitations, since not all children blush. We have developed more sophisticated measures of embarrassment including smiling behaviour, head gaze (both looking towards and looking away), and actions such as touching one's body. This self-exposure embarrassment arises early, indeed arises simultaneously with the emergence of a mental state/representation of self. The second embarrassment which Darwin talks about in terms of moral behaviour is what we have called self-evaluation embarrassment; that is, it is embarrassment caused by others evaluating oneself for failing some kind of standard, goal, or rule (SRG). Embarrassment is related, not only to moral failure, as in right and wrong, but also occurs, for example, in achievement tasks where the child fails a goal related to solving a problem. Some have considered this form of embarrassment similar to shame, but it is certainly less intense and its measurements of both body behaviour and facial expression are quite different from that when measuring shame. Self-evaluative embarrassment or moral failure embarrassment emerges later than embarrassment of exposure. The distinction between the two is most evident in the embarrassment related to praise which occurs earlier than the other type of embarrassment. The later embarrassment is added to the earlier one, so that by three years of age both are present.

By using the measure of blushing, Darwin limited himself in his consideration of the various types of self-conscious emotions. Moreover, he also limited himself in who he might have seen embarrassment in – although he did not collect much data but relied on examples. Blushing behaviour, that is, the blood flow into the capillaries of the face, shows large individual differences, with fair-skinned and fair-haired people blushing much more easily than darker-skinned and darker-haired people, although there is no difference in actual embarrassment that they show.by other means.[23] Looking at self-conscious emotions by looking at blushing has given way to looking at self-conscious emotions by looking at bodily as well as facial behaviour, and this has led to a considerable advancement in our understanding of the self-conscious emotions; something which Darwin did not do. Because he did not focus on mental operations in relation to the self-conscious emotions, Darwin was restricted in his approach. Nevertheless, he was interested in some individual differences.

Let us return to the individual differences Darwin saw in the self-conscious emotions. In particular, he spoke to the topic of individual differences in shyness:

> As shyness apparently depends on self-attention, we can perceive how right are those who maintain that reprehending children for shyness, instead of doing them any good, does much harm, as it calls their attention still more closely to themselves. It has been well urged that 'nothing hurts young people more than to be watched continu-

> ally about their feelings, to have their countenances scrutinized, and the degrees of their sensibility measured by the surveying eye of the unmerciful spectator'. Under the constraint of such examinations they can think of nothing but that they are looked at, and feel nothing but shame or apprehension.[24]

These individual differences appear to be related to such factors as being closely scrutinised and also therefore to individual differences in socialisation. Not much attention was given to such factors which today we recognise as temperament differences.

Up to a certain point in the development of emotional life, Darwin seemed to be right in terms of his observations. Where his work was limited was in respect of the theoretical perspective under which he worked. For Darwin, emotional development in humans had the same evolutionary history as other characteristics found in the biological world. So, since Darwin's theory required a gradual change brought on by the adaptive advantage of some differences over others, he had to find a connection between animal and human behaviour. Darwin's problem when he thought about the self-conscious emotions had to do with the very term self-conscious. The idea of self-conscious emotions implies a self, a certain level of cognition about the self, others, and the relation between the self and others. Moreover, it involves an understanding of standards, rules, and goals. While Darwin appears to recognise this need, he was confronted with the difficulty of explaining how these human skills emerged and differed from those of animals or even young children. Observation of great apes provided some information about an intermediate stage. For example, chimpanzees, gorillas, and orangutans show self-recognition. Such leaps into moral behaviour around standards unique to humans would have been difficult for Darwin to make, and remain so, under his idea of gradualism. Without such an explanation Darwin's own analysis of emotional life and development was limited. It must be recalled that Darwin's whole discussion of the self-conscious emotions has to do with his interest in the phenomenon of blushing, which we have said does not occur in any other species but humans. If there is this discontinuity in intellectual life between humans, perhaps some great apes, and the rest of the biological world, then emotions which are dependent upon this intellectual capacity should be possible only in humans, and possibly great apes. This uniqueness must have been difficult for Darwin to accommodate. Perhaps focusing on the behaviour of blushing rather than on the mental operations themselves suited his theoretical needs.

Post-Darwinian Times: A Theory of Emotional Life

As I have suggested throughout this chapter it is in our understanding of the self-conscious emotions that I can add to Darwin's observations. We have seen some of the limitations of the Darwinian view of emotional development in the superior range of the intellectual abilities of humans. This limitation is likely due to a complex set of factors, not the least of which was Darwin's struggle to eliminate the idea that humans held some unique and therefore godlike position in the biological world. Here I refer particularly to issues of measurement and theory based upon the idea that thoughts, rather than actions in the world, can be the stimuli for complex emotions, especially those we talk about as self-conscious emotions.

The measurement issues are quite simple. With increases in technology, facial expressions can be scored with ease. In addition, detailed measurement of physiological change including brain activity and behavioural movements is now possible. Thus, for example, the measurement of anger, fear, or happiness now involves a whole new set of measures not restricted to behavioural activity. Moreover, elaborate theories and data in regard to young children's intellectual development have accrued in the last fifty years. The growth of new measures has made it possible to think about and measure different self-conscious emotions and to relate them to intellectual development as well as to such diverse physiological measures as stress hormones and lateralisation of brain activity. A description of intellectual development would require many pages in itself, and so the reader is referred to another work to gain some insight into what has happened in our understanding of infants' development. It is called *The Scientist in the Crib* (1999).[25] The implications of our new understanding are broad, showing that very young infants are capable of extraordinary thinking-like abilities, much more so than we thought. In fact such measurement has led to the replacement of the Piagetian idea of development (starting with the newborn having little capacity and developing complex abilities over time), by one in which modules of very complex abilities exist at birth or very soon after. Development then becomes an elaboration of them. For our discussion, this means that we have come to be able to measure self-representation directly and to show in a much more precise way that the emergence of this capacity is related to the emergence of these later self-conscious emotions. The growth of technology has allowed us to measure different self-conscious emotions in unique ways. Thus while Darwin was limited to blushing as his way of understanding these emotions, we have elaborate measurement systems and can distinguish embarrassment from shame, and shame from guilt. We can distinguish differences in early embarrassment as measures of shyness and study individual differences in them. Details of these new measurement procedures are given in my book *Shame, The Exposed Self* (1992). However, I think it is of value here to discuss my theory of thinking about the self and how these types of thought become the stimuli for the elicitation of specific self-conscious emotions.

My colleagues and I have found and reported elsewhere that the emergence through brain maturation of the left temporal region of the human brain facilitates, or is at least associated with, the emergence of the mental representation of the idea of 'me'.[26] In other words brain maturation is related to the idea that there is a 'me' in time and space. The emergence of this capacity allows for the emergence of a whole new class of emotions, those I have called the exposed self-conscious emotions that include embarrassment, empathy, and envy. In each of these cases a rudimentary idea of 'me' is necessary in order for these emotions to emerge. What I am suggesting is that the stimulus for these new emotions has to do with how I think about myself. In empathy, I must place myself in the role of the other to know how the other might be thinking, feeling, or going to act. In the case of embarrassment, I become embarrassed when I become an object of the attention of another. It has no evaluative component, it is simply that I am the object of the other's attention. Finally, in envy, I desire that which you have. The emergence of these self-conscious emotions can and have been measured, and have been found to emerge at approximately the same time in development as when a representation of the self, the mental state necessary to produce this set of emotions, emerges.

When we come to the self-conscious evaluative emotions, the

cognitive elaboration becomes even more pronounced. Since I have described this in some detail in *Shame, The Exposed Self*, I will discuss them only briefly here. While Darwin did not develop an elaborate discussion of the various self-conscious evaluative emotions, often called moral or secondary emotions, considerable progress has nevertheless been made since. Psychoanalysts began to consider the emotion of shame as opposed to guilt in the early 1970s. The topic of shame did not become popular until Helen Lewis's *Shame and Guilt in Neurosis* appeared in 1971.[27] Since then, the impact of emotions such as embarrassment, shame, guilt, hubris and pride have received much interest. Under the influence of the attributional school of thought in which ideas about oneself led to specific emotional states, the idea of attributions about oneself as the elicitors of particular self-conscious evaluative emotions started to appear.

From a developmental point of view, I concluded as far back as 1979 that the way we think about ourselves might influence the emotions that we feel.[28] In my theory, I viewed shame as an evaluation by a person, whether it be a preschooler or an adult, of their own success or failure in regard to some thought, feeling, or action. Such an evaluation is obviously itself a cognitive act and the child has to be old enough to engage in it. The ensuing emotions that occur from failure are dependent upon a particular attribution – whether it is about 'me' as the core 'me', or whether it is about my action. In the attribution literature this evaluation has been considered by a variety of names, including 'global' versus 'specific' self-attribution or 'performance' versus 'task' attribution. If it is the case that one evaluates one's action, thought, or feeling and finds one has failed one's own standard, negative emotions will be the consequence. In the case of shame, when the attribution is about the entire self as no good, the evaluation is considered to be a global or performance attribution. In simple terms, the self is broken. The feelings associated with this attribution are feelings of wanting to die, hide, and disappear. The behaviours seen are of body collapse and blank facial expression. In older children and adults, it is often accompanied by the inability to speak or by speaking in a certain stutter-like behaviour using many self references such as 'I, I, I, I don't feel well'. Conversely, if the adult or child makes an attribution that is specific or task related, then the self is not broken and the adult or child is likely to feel guilt. While guilt and regret are also the consequence of failure, they are not the consequence of a global attribution. Specific attributions allow the emotion felt to act as a motivation to try and correct the failure. Thus, while the case of a global attribution leads to shame and bodily collapse, specific attributions lead to action in an attempt to correct the problem. Let me give you an example of our work. Children, ages three years and older, are given a task to perform and given a time limit to perform it. A large clock is shown to them. We tell them that if the bell sounds before they finish the task they have failed the task, while if they finish before the bell sounds they have succeeded. Unknown to them we adjust the time so that sometimes a child succeeds and sometimes they fail. During the task we videotape the children's facial and bodily behaviour and record what they say.

When they succeeded, we saw the children smile, puff up their chests, raise their hands, and often said such things as 'hurrah' or 'I did it'. These are measures of pride. When the children failed, two distinct responses were shown. For shame, children stopped moving, their chests collapsed, and they stopped talking and smiling. For guilt there was another pattern. When the bell had sounded, the children did not stop but tried to continue in order to finish. This response of repair is consistent with my theory that guilt is associated with the attribution of 'my action, not me, is broken'. When the action rather than the self is broken, repair is possible, unlike when the self is broken where repair is often not possible or easily repaired.

To make this clearer I offer the following as examples of how our thoughts about ourselves are the stimuli for the self-conscious evaluative emotions. Shame is the product of a complex set of cognitive activities: individuals' evaluation of their actions in regard to their standards, rules, and goals and their global evaluation of the self. The phenomenological experience of the person having shame is that of a wish to hide, disappear, or die.[29] It is a highly negative and painful state that also results in the disruption of ongoing behaviour, confusion in thought, and an inability to speak. The physical action accompanying shame is a shrinking of the body, as though to disappear from the eye of the self or the other. Because of the intensity of this emotional state, the global attack on the self-system, all that individuals can do when presented with such a state is to attempt to rid themselves of it. However, since it is a global attack on the self, people have great difficulty in dissipating this emotion. There are specific actions individuals employ when shamed in efforts to undo the shame state, such as reinterpretation, self-splitting (multiple personalities), or forgetting (repression).[30]

Shame is not produced by any specific situation, but rather by the individual's interpretation of the event. Even more important is that shame is not necessarily related to the event's being public or private. Although many hold that shame is a public failure, this need not be so. Failure, attributed to the whole self, can be either public or private. Shame may be public, but it is as likely to be private. Each of us can think of private events when we say to ourselves, 'I'm ashamed of having done that'. Shame can centre on moral action as well. Thus, when persons violate some moral standard, rule or goal, they are ashamed.

The emotional state of guilt or regret is produced when individuals evaluate their behaviour as failure but focus on the specific features or actions of the self that led to the failure. Unlike the focus on the global self in shame, the focus in guilt is on the self's actions and behaviours that are likely to repair the failure. From a phenomenological point of view, individuals are pained by their failure, but this pained feeling is directed to the cause of the failure or the object of harm. Because the cognitive-attributional process focuses on the action of the self rather than on the totality of self, the feeling that is produced – guilt – is not as intensely negative as shame and does not lead to confusion and to the loss of action. In fact, the emotion of guilt always has associated with it a corrective action that the individual can take (but does not necessarily take) to repair the failure. Rectification of the failure, and preventing it from occurring again, are the two possible corrective paths. Whereas in shame we see the body hunched over itself in an attempt to hide and disappear, in guilt we see individuals moving in space as if trying to repair their action.[31] The marked postural differences that accompany guilt and shame are helpful both in distinguishing these emotions and in measuring individual differences. We might point to blushing as a measure also distinguishing guilt from shame; however, because of the variability in the likelihood of individuals to blush, the use of blushing is not an accurate index.

Because in guilt the focus is on the specific, individuals are capable of ridding themselves of this emotional state through action. The corrective action can be directed towards the self as well as towards the other; thus, unlike shame, which is a melding of the self as subject and object, in guilt the self is differentiated from the object. As such, the emotion is less intense and more capable of dissipation. Moreover, it should be unrelated to maladaptive behaviour. The problem is that guilt may merge into shame, thus two types of guilt might be considered, guilt and maladaptive guilt. The expectation would be that guilt would not be correlated with shame, while maladaptive guilt would be.[32] Thus, should the corrective action not be forthcoming, either in thought, feeling, or deed, it is possible that a guilt experience can convert into one of shame. Here, then, is another difference between shame and guilt. We can be ashamed of our guilty action, but we cannot be guilty about being ashamed, suggesting a difference in level and a different in direction in the experiencing of these emotions. The emotion of guilt lacks the negative intensity of shame. It is not self-destroying, and as such can be viewed as a more useful emotion in motivating specific and corrective action. However, because it is less intense, it may not convey the motivation necessary for change or correction.

Hubris is defined as exaggerated pride or self-confidence often resulting in retribution. It is an example of pridefulness, something dislikeable and to be avoided. Hubris is a consequence of an evaluation of success in regard to one's standards, rules, and goals (SRGs) where the focus is on the global self. In this emotion, the individual focuses on the total self as successful. Hubris is associated with such descriptions as 'puffed up'. In extreme cases, it is associated with grandiosity or with narcissism.[33] Claudia M. Mueller and Carol S. Dweck have shown that too much praise of children may result in negative achievement performance; the assumed mechanism may be that such high levels of praise may elicit a global self focus rather than a specific focus, or to use Dweck's terms, lead to a self/performance focus rather than a task focus.[34] The self focus can lead in time to shame over failure. Given the powerful negative state, the desire to avoid it can lead to not taking responsibility for the failure on one hand, and on the other, the assumption of success where others see failure. These two features have been associated with narcissistic disorders.[35] Hubris occurs because of the global or performance focus. Because of the global nature of the attributions of this emotion, it is likely to be transient, and in order to be able to maintain this state, the individual must either alter standards or reevaluate what constitutes success. Unlike shame, it is highly positive and emotionally rewarding; that is, one feels good about oneself.

Hubris is, however, an emotion difficult to sustain, since there is no specific action that precipitates the feeling. Because such feelings are addictive, people prone to hubris derive little satisfaction from the emotion. Consequently, they seek out and invent situations likely to repeat this emotional state. They can do this either by altering their SRGs or by reevaluating what constitutes success in their actions, thoughts, or feelings. From the outside, other people observe the individual having hubris with some disdain. Hubristic people have difficulty in their interpersonal relations, since their own hubris is likely to interfere with the wishes, needs, and desires of others, in which case there is likely to be an interpersonal conflict. Moreover, given the contemptuousness associated with hubris, the other persons are likely to be shamed by the nature of the actions of the persons having this emotion. The three problems associated with hubris, therefore, are: one, it is a transient but addictive emotion; two, it is not related to a specific action, and thus requires altering patterns of goal setting or evaluation of what constitutes success; and three, it interferes with interpersonal relationships because of its contemptuous and insolent nature. In its extreme form it is quite similar to narcissism.

Pride, however, is the consequence of a successful evaluation of a specific action. The phenomenological experience is joy over an action, thought, or feeling well done. Here, the focus of pleasure is specific and related to a particular behaviour. In pride, the self and object are separated, as in guilt.[36] Unlike shame and hubris, where subject and object are fused, pride occurs when people focus on their action; the person is engrossed in the specific action that gives it pride. Deborah J. Stipek, Susan Recchia and Susan McClintic have likened this state to achievement motivation. Because this positive state is associated with a particular action, individuals have available to themselves the means by which they can reproduce the state.[37] Notice that, unlike hubris, pride's specific focus allows for action. Because of the general use of the term 'pride' to refer to 'hubris', 'efficacy', and 'satisfaction', the study of pride as hubris has until recently received little attention.

However, more work now is beginning to appear. Jessica L. Tracy, Richard W. Robins and Kristen H. Lagattuta have shown that this recognition of pride occurs relatively early, by four years, and has a set of specific behavioural markers.[38]

From a behavioural, as well as a phenomenological, point of view, shame and guilt can be differentiated; however, shyness and embarrassment appear to be bound up with these as well. Shyness is sheepishness, bashfulness, a feeling of uneasiness or psychological discomfort in social situations, and oscillation between fear and interest, or between avoidance and approach. In this description, shyness is related to fear and is a nonevaluative emotion centered around the individual's discomfort response to others. Such a description fits the notion of shyness as an emotional response which is elicited by experiences of novelty or conspicuousness. For some, shyness and fear are closely related and represent a fearfulness towards others. A way of distinguishing shyness from shame is that it appears much earlier than either shame or guilt. Such an approach to shyness seems reasonable because it fits with other notions relating the self to others, or what we might call the 'social self'. Shyness may be a dispositional factor not related to self. Rather, it may simply be the discomfort of being in the company of other social objects – in other words, the opposite of sociability.

For some, embarrassment is closely linked to shame. The most notable difference between embarrassment and shame is the intensity level. While shame appears to be an intense and disruptive emotion, embarrassment is clearly less intense and does not involve the disruption of thought and language that shame does. Second, in terms of body posture, people who are embarrassed do not assume the posture of one wishing to hide, disappear, or die. In fact, their bodies reflect an ambivalent approach and avoidance posture. Repeated looking and then looking away, accompanied by smiling behaviour, seem to index embarrassment. Rarely in a shame situation do we see gaze aversion accompanied by smiling behaviour. Thus, from a behavioural point of view, these two emotions appear to be different. Finally, while shame can occur in the physical absence of another, embarrassment almost always occurs in the company of others. Phenomenologically, embarrassment is less clearly differentiated from shame than from guilt. People often report that embarrassment is 'a less intense experience of shame'. Similar situations that invoke shame are found to invoke embarrassment, although its intensity, duration, and disruptive quality are not the same. However, in order to understand embarrassment, I have suggested earlier that there are two types of embarrassed behaviour: one, embarrassment as exposure, and two, embarrassment as less intense shame.

In this discussion we have not had a chance to mention that deciding on whether you have succeeded or failed in a particular action, thought, or goal to begin with requires that you have been socialised to certain standards, values, and goals which then guide your evaluation. Individuals, based on their socialisation (culture and historical time), will make evaluations and have standards that are quite different. What is constant is not the actual context of socialisation but rather that whatever the standards, rules, and goals may be, thinking about them and about oneself are the stimuli for these complex emotions. The content does vary so that, for example, one child can get an A- on an exam and the parents are very unhappy that she did not get an A, while another child who gets a C has parents who are happy that the child passed. In the latter case the child has succeeded *vis-à-vis* his/her parents' standards, while in the former case the child has failed even though her grade was higher. Thus, along with the improved measurement has come the understanding that there is a deep and probably permeable connection between thinking and feeling. Both influence each other so that their connection is like a fugue with the convergent parts and pieces entwining and moving in a pattern which has no beginning and no end. It is in this fugue that the emotional life of humankind is embedded and it is this fugue which emerged in the evolution of our emotional life.

Notes

1 Charles Darwin, *The Expression of the Emotions in Man and Animals* (1872; Chicago: University of Chicago Press, 1965).

2 See Carroll E. Izard, *Human Emotions* (New York: Plenum, 1977); and Michael Lewis, *The Development of Emotional Life* (New York: Guilford Press, 2013).

3 Lewis and Jeanne Brooks-Gunn, *Social Cognition and the Acquisition of Self* (New York: Plenum, 1979); Lewis and Linda Michalson, *Children's*

Emotions and Moods: Developmental Theory and Measurement (New York: Plenum, 1983); Lewis, *Shame, The Exposed Self* (1992; New York: The Free Press, 1995).

4 *The Autobiography of Charles Darwin: 1809–1882*, ed. by Nora Barlow (1959; New York: W.W. Norton & Co., 1993), 131–2.

5 Darwin, 1872, 79.

6 *Ibid.*, 79.

7 *Ibid.*, 79–80.

8 *Ibid.*, 80–1.

9 See Walter B. Cannon, 'The James-Lange Theory of Emotions: A Critical Examination and an Alternative Theory', *American Journal of Psychology*, 39 (1927), 106–24; and John B. Watson, *Behavior: An Introduction to Comparative Psychology* (New York: Holt, 1914).

10 T.C. Schneirla, 'An Evolutionary and Developmental Theory of Biphasic Processes Underlying Approach and Withdrawal', in Marshall R. Jones (ed.), *Nebraska Symposium on Motivation* (Lincoln, NE: University of Nebraska Press, 1959), Vol. 7, 1–42: 42.

11 Lewis, 2013.

12 See Silvan S. Tomkins, *Affect, Imagery, Consciousness, Vol. 1: The Positive Affects* (New York, Springer, 1962); Izard, 1977; and Paul Ekman, Wallace V. Friesen and Phoebe C. Ellsworth, *Emotion in the Human Face* (Cambridge, England: Cambridge University Press, 1982).

13 See Lewis, 'Inside and Outside: The Relation between Emotional States and Expressions', in Lewis (ed.), Special Issue on Infant Emotional Development, *Emotion Review*, 3 (2011), 131–7; Steven M. Alessandri, Margaret W. Sullivan and Lewis, 'Violation of Expectancy and Frustration in Early Infancy', *Developmental Psychology*, 26 (1990), 738–44; Lewis, Steven M. Alessandri and Margaret W. Sullivan, 'Violation of Expectancy, Loss of Control, and Anger in Young Infants', *Developmental Psychology*, 26 (1990), 745–51; Lewis, Margaret W. Sullivan, Douglas S. Ramsay and Steven M. Alessandri, 'Individual Differences in Anger and Sad Expressions During Extinction: Antecedents and Consequences', *Infant Behavior and Development*, 15 (1992), 443–52; Lewis, Douglas S. Ramsay and Margaret W. Sullivan, 'The Relation of ANS and HPA Activation to Infant Anger and Sadness Response to Goal Blockage', *Developmental Psychobiology*, 48 (2006), 397–405.

14 See Lewis, 1992.

15 Lewis, 'The Emergence of Consciousness and its Role in Human

Development', in Joseph LeDoux, Jacek Debiec, and Henry Moss (eds.), *The Self: From Soul to Brain* (New York: Annals of the New York Academy of Sciences, 2003), Vol. 1001, 1–29.

16 See Lewis and Dennis P. Carmody, 'Self-representation and Brain Development', *Developmental Psychology,* 44 (2008), 1329–34.

17 Darwin, 1872, 325.

18 *Ibid.*, 325.

19 *Ibid.*, 325.

20 *Ibid.*, 325–6.

21 *Ibid.*, 326.

22 See Lewis, Sullivan, Catherine Stanger and MayaWeiss, 'Self Development and Self-Conscious Emotions', *Child Development,* 60 (1989), 146–56.

23 See Lewis, Kiyoko Takai-Kawakami, Kiyobumi Kawakami and Margaret Wolan Sullivan, 'Cultural Differences in Emotional Responses to Success and Failure', *International Journal of Behavioral Development*, 34 (2010), 53–61.

24 Darwin, 1872, 331.

25 See Alison Gopnik, Andrew N. Meltzoff and Patricia K. Kuhl, *The Scientist in the Crib: Minds, Brains, and How Children Learn* (New York: William Morrow & Co., 1999).

26 See Lewis and Carmody, 2008; Carmody and Lewis, 'Regional White Matter Development in Children with Autism Spectrum Disorders', *Developmental Psychobiology*, 52 (2010), 755–63.

27 Helen B. Lewis, *Shame and Guilt in Neurosis* (New York: International Universities Press, 1971).

28 See Lewis and Brooks-Gunn, 1979.

29 See Lewis, 1992.

30 See Lewis, 1992, for a fuller discussion.

31 K.C. Barrett and C. Zahn-Waxler, 'Do Toddlers Express Guilt?', poster presented at the meeting of the *Society for Research in Child Development*, Toronto, April 1987.

32 June Price Tangney and Ronda L. Dearing, *Shame and Guilt* (New York: Guilford Press, 2002).

33 Andrew P. Morrison, *Shame: The Underside of Narcissism* (Hillsdale, NJ: Analytic Press, 1989).

34 Claudia M. Mueller and Carol S. Dweck, 'Praise for Intelligence can Undermine Children's Motivation and Performance', *Journal of Personality & Social Psychology*, 75 (1998), 33–52.

35 Melvin R. Lansky and Andrew P. Morrison, *The Widening Scope of Shame*

(Hillsdale, NJ: Analytic Press, 1997).

36 Jessica L. Tracy and Richard W. Robins, 'Show your Pride: Evidence for a Discrete Emotion Expression', *Psychological Science,* 15 (2004), 194–7.

37 Deborah J. Stipek, Susan Recchia and Susan McClintic, 'Self-Evaluation in Young Children', *Monographs of the Society for Research in Child Development*, 57 (1992), 85–95.

38 Jessica L. Tracy, Richard W. Robins and Kristin H. Lagattuta, 'Can Children Recognize Pride?', *Emotion,* 5 (2005), 251–7.

Afterword

The Emotional and Moral Lives of Animals: What Darwin Would Have Said

Marc Bekoff

What Did Charles Darwin Say about the Emotional and Moral Lives of Animals?

> Nevertheless the difference in mind between man and the higher animals, great as it is, certainly is one of degree and not of kind.[1]

> We must also admit that there is a much wider interval in mental power between one of the lowest fishes, as a lamprey or lancelet, and one of the higher apes, than between an ape and man; yet this interval is filled up by numberless gradations.[2]

> To return to our immediate subject: the lower animals, like man, manifestly feel pleasure and pain, happiness and misery. Happiness is never better exhibited than by young animals, such as puppies, kittens, lambs, &c., when playing together, like our own children.[3]

> Any animal whatever, endowed with well-marked social instincts, the parental and filial affections being here included, would inevitably acquire a moral sense of conscience, as soon as its intellectual powers had become as well-developed, or nearly as well-developed, as in man.[4]

> ...every one has heard of the dog suffering under vivisection, who licked the hand of the operator; this man, unless the operation was fully justified by an increase of our knowledge, or unless he had a heart of stone, must have felt remorse to the last hour of his life.[5]

Clearly, Darwin had much to say about the emotional and moral lives of animals and from time to time some words about how animal sentience and their moral sentiments should inform the ways in which we choose to interact with our animal kin. His quotations frame much of what follows.[6]

Bill, a Troubled Scientist, and Reno, his Companion: Isn't a Dog at Home the Same as a Dog at Work?

A few years ago I visited a major American university where I was asked to give a talk on the nature of animal emotions. My colleague, Bill, came up before the lecture to say hello and I asked him how his dog Reno was doing. For more than five minutes Bill extolled Reno's deep emotional life: he loves to play with his friends, misses Bill when he's gone, has separation anxiety and tears up garbage and books when he's alone, and a few days ago Reno got jealous when Bill gave his daughter attention. I was thrilled to hear these and other stories, as Bill always had been one of those researchers who claimed that we really didn't know what animals were feeling and in some ways we couldn't really be sure that they were feeling anything. While animals do act 'as if' they experience an array of various emotions it's too early to make any grandiose pronouncements about their feelings.

'Well, You Know What I Meant'

My excitement was short-lived. Following my talk Bill accused me of being too anthropomorphic and too sure of myself. He did it in a sort of light-hearted academic way. When he was done, I simply asked him to recall for the audience the conversation that we had before my lecture about Reno's emotional life. Bill turned slightly red, and said:

> Well, you know what I meant when we talking before, I was just letting my hair down and telling stories about Reno. I really don't know that he enjoys playing with his friends or that he really is depressed when I leave him alone. And, I feel rather certain that he wasn't really jealous of the attention I was giving to my daughter, he just behaved as if he were.

To be quite frank I don't really know what Bill meant. Isn't a dog at home the same as a dog at work? Don't they have similar emotional lives? He was quite comfortable telling me about Reno's feelings in one context, but not the other, a distancing mechanism that's used by many people who live with dogs, cats, birds, rats, hamsters and other animals at home but who then confront them or similar animals in their laboratories or other research projects.

In a project concerned with the reintroduction of endangered black-footed ferrets in the western United States, thousands of black-tailed prairie dogs and golden hamsters are bred solely to be fed alive to these predators by people who are ethical vegetarians and vegans, so that the ferrets can practice their predatory skills before they are released.[7] Bill and others don't seem to realise that there is a disconnect in how they understand animal emotions in their private and professional lives. They ignore the fact that humans and many other animals share ancestral parts of the brain that are important in positive and negative emotions, so it's not at all surprising that Reno and the hamsters would indeed experience and express an array of different emotions. The story of Bill and Reno highlights just how complicated our relationships with other animals can be. It shows how the behaviourist party line infiltrates and influences discussions about animal emotions even when it makes no sense to argue this sanitised and reductionistic view of beastly passions.[8]

Grieving Magpies

A few years ago, my friend Rod and I were riding our bicycles around Boulder, Colorado, when we witnessed a fascinating encounter among five magpies. Magpies are Corvids, a very intelligent family of birds. One magpie had obviously been hit by a car and was lying dead on the side of the road. The four other magpies were standing around him. One approached the corpse, gently pecked at it, just as an elephant would nose the carcass of another elephant, and stepped back. Another magpie did the same thing. Next, one of the magpies flew off, brought back some grass, and laid it by the corpse. Another magpie did the same. Then, all four magpies stood vigil for a few seconds and one by one flew off.

Were these birds thinking about what they were doing and what were they feeling? Were they showing magpie respect for their friend?

Were they merely acting *as if* they cared?[9] Were they just animal automatons? I feel comfortable answering these questions, in order: yes, yes, no, no. I'd never seen anything like this before and hadn't read any accounts of grieving magpies. When I published an essay about this incident[10] I received numerous emails from people who had seen the same ritual in magpies, ravens, and crows. We can't know what they were actually thinking or feeling, but reading their action there's no reason not to believe these birds were saying a magpie farewell to their friend.

Nasty Nick, Rueben, and Ruth: Baboons of a Different Kind

Emotional baboons make a case for strong intersubjectivity, that is, the sharing of subject states. Nick, an olive baboon living in the forest troop in the southeast corner of Masai Mara National Reserve in Kenya, joined the group when he was an adolescent. According to Robert Sapolsky you could almost see contempt on his face. Being unashamedly anthropomorphic, Sapolsky noted that Nick dominated his age group and that '…he was confident, unflinching, and played dirty'.[11] During a fight, another adolescent male, Rueben 'stuck his ass up in the air', a sign of submission, but Nick went over as if to examine his bottom and then slashed Reuben's behind with his canines. To quote Sapolsky,

> The guy simply wasn't nice…He harassed the females, swatted at kids, bullied ancient Gums and Limp. On one memorable day, he took exception to something that poor nervous Ruth had done and chased her up a tree. Typically, at this point, the female takes advantage of one of those rare instances when it pays to be smaller than the males – she goes to the farthest end of a flimsy branch and hangs on for dear life, depending on the fact that the heavier male can't crawl out to where she is and bite her. And, typically, the male, thwarted, positions himself to at least trap the female, keeping her screaming on the precarious branch until he gets bored. So Ruth gallops up the tree, Nick after her, and Ruth leaps out to a safe edge. Nick promptly climbs onto a stronger, thicker branch directly above her. And then urinates on her head.[12]

Empathic Elephants: Babyl and her Friends

While I was watching a group of wild elephants living in the Samburu Reserve in Northern Kenya, I noted that one of them, Babyl, walked very slowly. I learned that she was crippled and that she couldn't travel as fast as the rest of the herd. However, the elephants in Babyl's group didn't leave her behind when they travelled. When I asked the elephant expert Iain Douglas-Hamilton about this he said that these elephants always waited for Babyl, and they'd been doing so for years. They'd walk for a while, then stop and look around to see where Babyl was. Depending on how she was doing, they'd either wait or proceed. Iain said the matriarch even fed her on occasion. Why did the other elephants in the herd act this way? Babyl could do little for them, so there seemed no reason or practical gain for helping her. Reciprocity wasn't in the picture at all. The only obvious conclusion we could see is that the other elephants cared for Babyl, and so they adjusted their behaviour to allow her to remain with the group. Friendship and empathy go a long way.

Angry Elephants

Joyce Poole and Peter Gramli, who have studied wild African elephants for decades in the Amboseli National Park in Kenya, note that if elephants are moved from one place to another the distressed elephants may take revenge if family groups are split. In 'Elephant Breakdown', a landmark paper in *Nature*, Gay Bradshaw and her colleagues, including Poole, showed that elephants are very sensitive beings and that social trauma can affect their physiology, behaviour, and culture over generations.[13] Indeed, some elephants might suffer from post-traumatic stress disorder (PTSD) that leads to chronic stress and undermines socialisation. And these severely traumatised mammoths might also get annoyed, hold grudges, and get even with humans who abuse them.

An article in the *New Scientist* by Caroline Williams followed up on what might be happening to elephants when their families are broken up, noting that unprovoked attacks are getting more common.[14] In Bunyaruguguru in western Uganda, people are now afraid to ride their bicycles because elephants are getting even. Poole notes:

> They are certainly intelligent enough and have good enough memories to take revenge...Wildlife managers may feel that it is easier to just shoot so-called 'problem' elephants than face people's wrath. So an elephant is shot without realising the possible consequences on the remaining family members, and the very real possibility of stimulating a cycle of violence.[15]

Play and Joy: Happy Hounds and Laughing Rats

Social play is an excellent example of a cooperative behaviour in which many animals partake, and one that they enjoy immensely. Play is characterised by deep intersubjectivity, cooperation, and empathy. The cooperation that's needed for animals to engage in social play, and the emotions experienced while playing, might also be important in the evolution of social morality and fairness or 'wild justice'.[16] The four basic rules of social play across diverse species are: ask first, be honest, follow the rules, and admit when you're wrong.

Animals relentlessly seek play out and when a potential partner doesn't respond to a play invitation they often turn to another individual or to inanimate objects or to themselves and engage in self-play.[17] Specific play signals also are used to initiate and to maintain play.[18] The play mood is highly contagious; just seeing animals playing can stimulate play in others. Animals seek out play because it is fun. Consider my field notes on two dogs playing:

> Jethro runs towards Zeke, stops immediately in front of him, crouches or bows on his forelimbs, wags his tail, barks, and immediately lunges at him, bites his scruff and shakes his head rapidly from side-to-side, works his way around to his backside and mounts him, jumps off, does a rapid bow, lunges at his side and slams him with his hips, leaps up and bites his neck, and runs away. Zeke takes wild pursuit of Jethro and leaps on his back and bites his muzzle and then his scruff, and shakes his head rapidly from side-to-side. They then wrestle with one another and part, only for a few minutes. Jethro walks slowly over to Zeke, extends his paw toward Zeke's head, and nips at his ears. Zeke gets up and jumps on Jethro's back, bites him, and grasps him around his waist. They then fall to the ground and engage in 'mouth wrestling'. Then they chase one another and roll over and play once again.

There can be little doubt that Jethro and Zeke were enjoying themselves. Studies of the chemistry of play support the idea that play is fun. Stephen Siviy has shown that dopamine (and perhaps serotonin and norepinephrine) are important in the regulation of play, and that large regions of the brain are active during play.[19] Rats show an increase in dopamine activity when anticipating the opportunity to play. Jaak Panksepp and Jeff Burgdoff discovered a close association between opiates and play and also claim that rats enjoy being playfully tickled.[20] *Science* magazine has even published his results on animal laughter and joy in which Panksepp claims, 'Young rats, in particular, have a marvelous sense of fun'.[21]

Improbable Partners and Emotional Intelligence

While it is clear that individuals of many species form close social bonds with other members of their species, there are examples of improbable between–species relationships that yield to explanations encompassing intersubjectivity.

A gracious behemoth: a winking whale

In December 2005, a fifty-foot, fifty-ton female humpback whale got tangled up in crab lines, the weight of which was making it difficult for her to keep her blow hole above the water. A courageous team of divers freed her and their wonderful and empathic stories about this encounter are well worth recalling for they bring up many issues with which I'm concerned. After being freed the whale nuzzled each of her saviours in turn and flapped around in what whale experts said was 'a rare and remarkable encounter'.[22] James Moskito, one of the rescuers, recalled 'It felt to me like it was thanking us, knowing it was free and that we had helped it'. He went on 'It stopped about a foot away from me, pushed me around a little bit and had some fun'. But this wasn't all he had to say. 'When I was cutting the line going through the mouth, its eye was there winking at me, watching me…it was an epic moment of my life'.[23] One of the other rescuers, Mike Menigoz, also deeply touched by the encounter, said 'You hate to anthropomorphize too much, but the whale was doing little dives and the guys were rubbing shoulders with it…I don't know for sure what it was thinking, but it's something I will always remember. It was just too cool'.[24] In this encounter we see sympathy, empathy and the acknowledgement that the female whale was suffering. We also see that it's very natural to attribute 'human' feelings to her. We also see gratitude on the part of the rescued whale.

Writing about the importance of eyes makes a compelling case for how some of our intuitions are ultimately borne out by hard science. It turns out that the eyes are of paramount importance in knowing that another human is feeling fear, and people tend to look at the eyes more when a face is fearful.[25] A study by Patrik Vuilleumier, of a woman with a specific deficit in recognising fearful facial expressions due to damage to her the part of her brain called amygdala, showed that that she couldn't perceive fear because she didn't look spontaneously towards the eyes. Rather, she judged the face as having a neutral expression. It's also likely that the eyes are not only important in perceiving fear but also other emotions. The results of this study made me think that perhaps one reason that so many people can't look into the eyes of an animal who is afraid or otherwise suffering is because they know just what the animal is feeling and it's easier to deny this if one doesn't look at their eyes and feel the fear emanating from the poor beast.

A hippo and a tortoise

A one-year-old hippopotamus, named Owen by his caretakers, formed a close bond with a century-old male tortoise, named Mzee (Swahili for old man), after floods in Kenya left Owen dehydrated and alone. Owen had survived the waves caused by the tsunami that struck the coast of Kenya on 26 December 2004. Owen was found by wildlife rangers near the Indian Ocean and brought to a wildlife sanctuary in Mombasa. Owen sleeps and eats with his tortoise friend and they have become inseparable. More than a year later they are still close buddies.[26]

A hamster and a snake

At Tokyo's Mutsugoro Okoku Zoo a dwarf hamster named Gohan (a tasty rice dish in Japanese) was offered as a meal to a rat snake named Aochan. Aochan had refused to eat frozen mice and zookeepers figured that a Gohan would be more appetising. However, rather than eating Gohan, Aochan made friends with her. They share a cage and have remained friends ever since. Gohan even naps on Aochan's back. Even after Aochan began eating frozen rodents he shows no interest in eating his friend. Kazuya Yamamoto, a keeper at the zoo, said, 'Aochan seems to enjoy Gohan's company very much'.[27]

Dogs are Happy, Not 'Happy'

Animals, in some ways, are a neutral palette on which we paint our needs, feelings and view of the world.[28]

> Although there are many species differences in the emotional systems that we share as ancestral gifts with other animals...the field of neuroscience will be more productive if it remains open to the similar nature of human and animal affective experiences.[29]

What's exciting about studying the emotional lives of animals, their pains, passions and virtues, what they like and don't like, is that all sorts of 'big questions' are raised and sometimes we don't even know this is happening. By simply asking what a dog feels as she plays with her buddies, or what an elephant feels when he's lost his friends or caged alone in a zoo, we confront 'big' questions about empathy, theory of mind, self-cognisance, anthropomorphism, and the role that scientific data, evolutionary biology, and common sense and intuition play in our coming to a better understanding and appreciation of the subjective, intersubjective, and passionate lives of animals. Studying animal emotions is exciting, challenging, and frustrating. Numerous lines of independent evidence from many different disciplines show that many animals have rich and deep emotional lives.[30] They're not neutral palettes. Scientific data,[31] evolutionary biology, and common sense show clearly that animals have a point of view and preferences about what they do and don't like.

When taken as a whole, the data leads to no other reasonable conclusion. We know that dogs, cats, and numerous other animals like to play, feel joy as they run here and there, and get lonely and sad. We all know that elephants don't like having their families broken up, pigs don't like having their tails cut off (docked), their teeth ground down on grinding stones, or their testicles cut off, all without anesthesia. And coyotes don't like being trapped, wolves don't like being chased and shot, moon bears in China don't like being kept immobilised in crush cages for upwards of twenty years so their bile can be extracted, nor do chimpanzees or other animals like spending their lives in impoverished cages where they can hardly move.

Evolutionary Continuity: Denying Emotions and Moral Intelligence to Animals is Bad Biology

Denying emotions and moral intelligence to animals is bad biology. Darwin's well-accepted ideas about evolutionary continuity, that differences among species are differences in degree rather than kind, argue strongly for the presence of animal emotions, empathy, and moral intelligence, as they do in investigations of comparative anatomy and physiology. Basically, *if we have something, 'they' have it too.* It's no longer a question of *if* certain animals (let's focus on mammals for the moment) have rich and deep emotional lives, it's a matter of *why* they have evolved and what they are good for. We no longer have to put quotation marks that reflect our uncertainty around words such as 'happy' or 'sad' when we say that Fido is happy or sad. He's just happy or sad in his own ways, just as we're happy or sad in our own ways. Emotions have evolved because they function as 'social glue', playing major roles in the formation and maintenance of social relationships among animals and people, among subjects, not objects, with intersubjective lives. Emotions also serve as 'social catalysts', causing individuals to follow different courses of action depending on the situation in which they find themselves.[32]

What's especially exciting and important is that non-invasive neural imaging techniques such as functional Magnetic Resonance Imaging (fMRI) and Positron Emission Tomography (PET) scans are allowing researchers to localise brain activity when animals are in specific social situations.[33] Just as we've learned that the function of mirror neurons can help explain empathy and the sharing of social intentions and desires,[34] so too do fMRIs and PET scans provide solid neurobiological support for much of what we 'knew' all along based on ethological observations.

In practice, continuity allows us to connect the 'evolutionary dots' among different species to highlight similarities in evolved traits including individual feelings and passions. For example, all mammals (including humans) share neuroanatomical structures (for example, the amygdala hippocampus) and neurochemical pathways in the limbic that are important for feelings. We know that elephants suffer from PTSD. During recent years there have also been many other surprises. We know that mice[35] and rats[36] are empathic rodents – they feel the pain of other mice – and that chickens also display empathy,[37] and whales possess spindle cells, which are important in processing emotions. Before this discovery it was thought that only humans and other great apes possessed spindle cells.

Emotions and empathy are keys to survival. Without them, animals – both human and nonhuman – would perish. That's how important they are. It's bad biology to argue against the existence of animal emotions. It's not important if dog joy or chimpanzee grief isn't the same as human joy or human grief. What is important is to allow that different species may experience their emotions in different ways and that humans shouldn't be the template against which we compare other animals. There's dog joy and dog grief, and chimpanzee joy and chimpanzee grief, and there are even differences among members of the same species.

Wild Justice: The Moral Lives of Animals

Do animals have a sense of morality? Do they know right from wrong? In our book, *Wild Justice: The Moral Lives of Animals*, the philosopher Jessica Pierce and I argue that the answer to both of these questions is a resounding 'yes'. 'Ought' and 'should' regarding what's right and what's wrong play important roles in the social interactions of animals, just as they do in ours. Darwin would not have been the least surprised if he knew what we have discovered in the past decade or so about moral intelligence in animals. Darwin believed that animals, like humans, could be moral beings. He suggested that human morality is continuous with similar social behaviour in other animals. Darwin paid special attention to the capacity for sympathy, which he believed was evidenced in a large numbers of animals.

More recently, Jane Goodall noted in her book *The Chimpanzees of Gombe*,

> it is easy to get the impression that chimpanzees are more aggressive than they really are. In actuality, peaceful interactions are far more frequent than aggressive ones; mild threatening gestures are more common than vigorous ones; threats per se occur much more often than fights; and serious, wounding fights are very rare compared to brief, relatively mild ones.[38]

After carefully analysing the social interactions of various primate species, primatologists Robert Sussman and Paul Garber came to the conclusion that the vast majority of social interactions are affiliative (positively social or prosocial) rather than agonistic or divisive (see also Sussman and C.R. Cloninger for more extensive discussions of the prevalence of affiliative and cooperative behaviour across diverse species).[39] Grooming and bouts of play predominate the social scene, with only an occasional fight or threat of aggression. In prosimians, the most ancestral of existing primates, an average of 93.2 percent of social interactions are affiliative. In New World monkeys who live in the tropical forests of southern Mexico and Central and South America, 86.1 percent of interactions are affiliative, and likewise, for Old World monkeys who live in South and East Asia, the Middle East, Africa, and Gibraltar, 84.8 percent are affiliative. Unpublished data for gorillas obtained by Sussman and Garber show that 95.7 percent of their social interactions are affiliative.

Consider the following scenarios.[40] A teenage female elephant nursing an injured leg is knocked over by a rambunctious hormone-laden teenage male. An older female sees this happen, chases the male away, and goes back to the younger female and touches her sore leg with her trunk. Eleven elephants rescue a group of captive antelope in KwaZula-Natal; the matriarch elephant undoes all of the latches on the gates of the enclosure with her trunk and lets the gate swing open so the antelope can escape. A rat in a cage refuses to push a lever for food when it sees that another rat receives an electric shock as a result. A male Diana monkey who learned to insert a token into a slot to obtain food helps a female who can't get the hang of the trick, inserting the token for her and allowing her to eat the food reward. A female fruit-eating bat helps an unrelated female give birth by showing her how to hang in the proper way. A cat named Libby leads her elderly deaf and blind dog friend, Cashew, away from obstacles and to food. In a group of chimpanzees at the Arnhem Zoo in The Netherlands individuals punish other chimpanzees who are late for dinner because no one eats until they're all present. A large male dog wants to play with a younger and more submissive male. The big male invites his younger partner to play and when they play, the big dog restrains himself and bites his younger companion gently and allows him to bite gently in return. Do these examples show that animals display moral behaviour, that they can be compassionate, empathic, altruistic, and fair? Yes they do. Animals not only have a sense of justice, but also a sense of empathy, forgiveness, trust, reciprocity, and much more as well.[41]

Fairness in Animals

The social lives of many animals are strongly shaped by affiliative and cooperative behaviour. Fairness is also an important part of the social life of animals. Researchers Sarah Brosnan, Frans de Waal and Hillary Schiff discovered what they call 'inequity aversion' in Capuchin monkeys, a highly social and cooperative species in which food sharing is common.[42] These monkeys, especially females, carefully monitor equity and fair treatment among peers. Individuals who are short-changed during a bartering transaction, by being offered a less preferred treat, refuse to cooperate with researchers. The Capuchins expect to be treated fairly. Research by Friedericke Range and her colleagues in Austria also shows that dogs expect to be treated fairly. Dogs won't work for food if they see other dogs getting more than they do for performing the same task.[43]

The study of moral behaviour in animals also has implications for our understanding of human morality. Animal morality may open the door to a whole new area of philosophical investigation and suggests new ways of thinking about old problems. For example, our work speaks to the question of moral agency in nonhuman animals and the related question of whether agency should continue to be a central category in philosophical ethics. Research from psychology and neuroscience suggests that most moral behaviour occurs below the radar of consciousness; so 'agency' cannot be said to be the primary defining characteristic of human moral behaviour.

Animals are incredibly adept social actors: they form intricate networks of relationships and live by rules of conduct that maintain social balance, or what we call social homeostasis. Humans should be proud of their citizenship in the animal kingdom. We're not the sole occupants of the moral arena. But at the same time, the notion of *wild justice* embraces human uniqueness. Saying that animals have moral behaviour doesn't degrade human morality, as some people fear. In fact, it helps us see the myriad ways in which human morality is special. Humans are exquisitely social and have incredibly complex and nuanced moral behaviours. Studying moral behaviour in nonhuman animals can help us understand ourselves better and can highlight the ways in which we are unique.

The Public Lives of Animals: Intuitions about Emotions

> Much of what the average man 'knows' about his own dog, and about dogs in general is, of course, quite unknown to the animal psychologist.[44]

> Because uncertainty never disappears, decisions about the future, big and small, must always be made in the absence of certainty. Waiting until uncertainty is eliminated is an implicit endorsement of the status quo, and often an excuse for maintaining it…Uncertainty, far from being a barrier to progress, is actually a strong stimulus for, and an important ingredient of, creativity.[45]

The lives of animals are rather public. The problem of 'other minds' is often blown out of proportion. While there might not always be full disclosure and there will always be mysteries, there is much transparency. If we carefully study animal behaviour we can get a strong grasp on what animals are experiencing and feeling. This might sound unbelievably naïve and simplistic to some of my colleagues, but I think that simplicity is good because it will generate future studies that don't obscure the real questions at hand. The notion of uncertainty also creeps in, but the study of animal emotions is no different from other scientific endeavours. All branches of science are burdened by uncertainty, but this doesn't mean that we don't have enough information to offer solid explanations with strong predictive value. We need to be able to live with uncertainty and give up control. This is not to be anti-science.

There's also concern among some people that science moves too fast. Certain kinds of science can progress rapidly with new results appearing almost daily. However, there is no substitute for taking the time to watch animals as they go about their daily activities and this takes time and lots of patience. Sitting on a hillside watching coyotes move about when it's 15 degrees below zero isn't all that appealing. Daily, seasonal, and yearly cycles can't be rushed. Neither can the ways in which individuals negotiate their social relationships. Researchers, frequently lab-bound, who impose a time limit on the gathering of knowledge about intersubjectivity among animals, need to be more patient.

We also need time to assess just how accurate our intuitions are. Indeed, there are studies that show that there is a lot of agreement among people concerning what animals are feeling and these intuitions are quite good at predicting what an individual is likely to do in the future. For example, Françoise Wemelsfelder and Alistair Lawrence discovered that even people who have little experience observing animals usually agree with one another on what an animal is most likely feeling.[46] Their intuitions are borne out because their characterisations of animal emotional states predict future behaviour quite accurately. Highly accurate prediction offers one of the strongest arguments for the existence of emotions and feelings in nonhuman species.

In another study, Wemelsfelder and her colleagues asked eighteen naïve observers to independently observe and describe in their own words the behavioural expression of twenty pigs. They discovered that the observers showed 'significant agreement in their spontaneous assessment of pig expressions, which suggests that these assessments were based on commonly perceived and systematically applied criteria'.[47] And, in another very important study, Wemelsfelder and M. Farish noted:

> Qualitative judgements of welfare tend to be habitually regarded as 'subjective', ie as an anthropomorphic and unreliable; however from this review it appears that such judgements reflect the overall demeanour of an animal's behaviour, and as such may provide a useful empirical basis for the investigation of emotional expressive repertoire of sheep.[48]

Wemelsfelder and her colleagues' extensive research shows that much of an animal's subjective experience of suffering or well-being is public, not hidden or private. What animals do expresses their personal and subjective experiences. People are also very good at figuring out what wolves are doing and feeling, especially when they could see the wolves' faces. R.E. Anderson, J. Ryon, and J.C. Fentress discovered that naïve observers were very good at distinguishing between friendly, playful interactions and fighting in wolves.[49]

When all the jargon is boiled down, it means that what people sense is likely what animals are feeling. Scientific data supports this conclusion yet some people still won't accept it. Surely, we can be wrong, but just because we're wrong some of the time doesn't mean we're wrong all of the time. No one has ever compared how accurate science sense is when compared to common sense and intuition in the study of animal emotions.

Anthropomorphic Double-talk: The Abundant Evidence of our Own Senses

There are many different ways of describing and explaining what animals do. How researchers choose to summarise what they see, hear, or smell when studying animals depends on the questions in which they're interested. There isn't only one correct way to describe or to explain what animals do or feel. Using words that describe such human characteristics as thinking, joy, grief, embarrassment, and jealousy is to be anthropomorphic, a practice that irks many of my colleagues although they freely engage in it. Anthropomorphism is the attribution of human characteristics to nonhuman creatures and inanimate objects.

As humans who study other animals, we can only describe and explain their behaviour using words with which we're familiar from a human point of view. So, when I tried to figure out what was happening in my late dog Jethro's head I had to be anthropomorphic but I tried to do it from a dog-centered – 'dogocentric' – point of view. Just because I tell you Jethro was happy or jealous, this doesn't mean he was happy or jealous as are humans or, for that matter, as are other dogs. Being anthropomorphic is a tool to make the thoughts and feelings of other animals accessible to humans. While we surely make errors from time to time, we're pretty good about making accurate predictions in the realm of the mental.[50]

Sapolsky is clearly anthropomorphic in his writing about baboons and places much of the debate about anthropomorphism in a very useful perspective. He wrote:

> Do I get grief for the fact that in communicating, say, about the baboons I'm doing so much anthropomorphizing? One hopes that the parts that are blatantly ridiculous will be perceived as such. I've nonetheless been stunned by some of my more humorless colleagues – to see that they were not capable of recognizing that. The broader answer, though, is I'm not anthropomorphizing. Part of the challenge in understanding the behavior of a species is that they look like us for a reason. That's not projecting human values. That's primatizing the generalities that we share with them.[51]

There's no need to think of anthropomorphism as something for which its practitioners should be embarrassed or punished. We're just doing what comes naturally, we don't lose important information that the animals are sharing, and it's very useful for learning more about beastly passions. There's no reason to discount or dismiss the abundant evidence of our own senses.

If we decide against anthropomorphism we have no alternatives other than to talk about animals as a bunch of hormones, neurons, or muscles absent any context for what they're doing and why. Gordon M. Burghardt notes that denying our own intuitions about an animal's experience is 'sterile and dull'.[52] If we don't anthropomorphise we lose important information. The renowned psychologist, Donald Hebb, also made some important observations about anthropomorphism. Hebb noted that keepers' anthropomorphic accounts proved 'an intuitive and practical guide to behavior', enabling them to best interact with the captive chimpanzees for whom they cared.[53]

Not everyone agrees. Behaviourist Clive Wynne believes that anthropomorphic explanations are extremely imprecise and he privileges reductionistic stimulus-response explanations over explanations that appeal to such notions as consciousness, intentions, and beliefs.[54] However, Wynne doesn't scientifically support his position.[55] Animal welfarist Marian Dawkins is also critical of anthropomorphism and suggests that being carefully 'biocentrically' or 'critically' anthropomorphic is anti-science.[56] This overlooks my argument that:

> The way human beings describe and explain the behavior of other animals is limited by the language they use to talk about things in general. By engaging in anthropomorphism – using human terms to explain animals' emotions or feelings – humans make other animals' worlds accessible to themselves...But this is not to say that other animals are happy or sad in the same ways in which humans (or even other conspecifics) are happy or sad.[57]

Of course, I cannot be absolutely certain that Jethro, my companion dog, is happy, sad, angry, upset, or in love, but these words serve to explain what he might be feeling. However, merely referring acontextually to the firing of different neurons or to the activity of different muscles in the absence of behavioural information and context is insufficiently informative. Using anthropomorphic language does not have to discount the animal's point of view. Anthropomorphism allows other animals' behaviour and emotions to be accessible to us. Thus, I maintain that we can be biocentrically anthropomorphic and do rigorous science. Dawkins also claims I argue 'that there are no limits to how we interpret animal behaviour'.[58] Nothing could be further from the truth. We can use human terms to describe, interpret, and explain the behaviour of other animals and learn a lot doing it with care.

Surely, given the complexity and flexibility of behaviour, no explanatory scheme will be correct *all* of the time. And, for example, more importantly, Wynne ignores the fact that the utility and accuracy of various sorts of explanations haven't been assessed empirically. So, we really don't know if his flavour of explanations is better for understanding and predicting behaviour than those he eschews. Until the data is in we all must be careful in claiming that one sort of explanation is always better than others. It's poor scholarship and sets a poor example for students to take a narrow-minded approach in the absence of supportive data. Suffice it to say, many who favour mechanistic explanations often pay little attention to the detailed arguments they are attempting to refute. Many also haven't spent much time watching free-ranging animals or have little familiarity with the subtleties of animal behaviour.[59] I agree with Henry Pollack when he says that 'unfamiliarity breeds uncertainty'.[60]

It's Okay to Say an Elephant is 'Doing Well' but it's not Okay to Say She 'Isn't Doing Well'

Our strong and irresistible anthropomorphic tendencies also reflect strong intersubjectivity. An illustrative example of what I call 'anthropomorphic double-talk' concerns the fate of Ruby, an elephant who, on occasion, was shipped from zoo to zoo in the United States as if she were a piece of furniture. Some people argue against the use of anthropomorphism without seeming to know they're using it themselves – or hope that others don't see that they are. For example, the former Director of Conservation and Science for the American Zoo and Aquarium Association (AZA), Michael Hutchins, recently claimed that we mustn't be anthropomorphic and that it's bad science to attribute human-like feelings to animals. Hutchins said: 'Animals can't talk to us so they can't tell us how they feel'.[61] He was critical of people who claimed that Ruby, a forty-three-year-old female African elephant at the Los Angeles Zoo, wasn't 'doing well' in captivity, that she was unhappy. How could Hutchins be critical of those who claim Ruby is unhappy unless he believes that she is happy, and isn't he then being anthropomorphic? What Hutchins means is that he and others in their camp can be anthropomorphic but their foes can't. They can say that an animal is 'doing well' but others can't say the elephant is *not* doing well. Indeed, the director of the Los Angeles Zoo, John Lewis said about Ruby, 'She seems to be fine'.[62] The bottom line is that one can't talk about how animals are doing – well or not so well – without being anthropomorphic; critics need to acknowledge this.

Most people who live with and work closely with animals take it as a matter of fact that many animals have emotions and they know and admit they're being anthropomorphic. For example, mahouts, people who work with elephants, know that one ignores an elephant's 'mood' at one's own peril. Mary Midgley says it well:

> Obviously the mahouts may have many beliefs about the elephants which are false because they are 'anthropomorphic' – that is, they misinterpret some outlying aspects of elephant behaviour by relying

> on a human pattern which is inappropriate. But if they were not doing this about the basic everyday feelings – about whether their elephant is pleased, annoyed, frightened, excited, tired, sore, suspicious or angry – they would not only be out of business, they would often simply be dead.[63]

Midgley is right. If we ignore what other animals are feeling we lose not only much knowledge about who these amazing beings are, but perhaps also about our own lives.

Animal Emotions, Morality, and Why They Matter

Our relationship with other animals are complex, ambiguous, challenging, and frustrating,[64] and we must continually reassess how we should interact with our nonhuman kin. Part of this reassessment involves asking difficult questions. Thus, I often ask researchers who conduct invasive work with animals, or people who work on factory farms, 'Would you do that to your dog?'. Some are startled to hear this question, but it's a very important one to ask because if someone won't do something to their dog that they do daily to other dogs or to mice, rats, cats, monkeys, pigs, cows, elephants, or chimpanzees we need to know why. We know a lot about animal emotions, sentience, and moral behaviour. While we obviously have much more to learn, what we *already know* should be enough to inspire changes in the way we treat other animals. We need to turn our knowledge into action. We must not simply continue with the status quo because that is what we've always done. What we know has changed, and so should our relationships with animals. Quite often what we accept as 'good welfare' isn't 'good enough'.

When people tell me that they love animals because they're feeling beings and then go on to abuse them I tell them that I'm glad they don't love me. Recognising that animals have emotions is important because animal feelings matter. Animals are sentient beings who experience the ups and downs of daily life, and we must respect this when we interact with them. The word 'animals' doesn't refer only to the companions with whom we live, care for, and love: billions of other domesticated animals live on farms and in slaughterhouses and provide us with food and clothing, and wild animals are continually faced with trying to share our evermore crowded world.

It is important to note that while Darwin wrote about 'higher' and 'lower' species, it was part of the *Zeitgeist* when he lived. Speaking about 'higher' and 'lower' species is no longer well-accepted because these terms lead to speciesistic thinking, such as 'higher' meaning 'more valuable or smarter' and 'lower' meaning 'not as valuable or dumber'.[65] Animals do what they need to do to be card-carrying members of their species and cross-species comparisons in levels of intelligence or feelings are of limited use.[66]

Humans have enormous power to affect the world any way we choose. Daily, we silence sentience in innumerable animals in a wide variety of venues. However, we also know that we're not the only sentient creatures with feelings and, with the knowledge that what hurts us hurts them, comes the enormous responsibility and obligation to treat other beings with respect, appreciation, compassion, and love. There's no doubt whatsoever that, when it comes to what we can and cannot do to other animals, it's their emotions that should inform our discussions and our actions on their behalf. Some say life is too short to ponder if animals have feelings, and I agree. However, I also feel that life is too long to ponder whether animals have emotions. The suffering that numerous animals endure while some researchers try to figure out if they're feeling anything at all is unacceptable, cruel, and needless given what we know about animal emotions and sentience.

Where to from Here? Darwin Knew Much of What We've Denied and Now Know

That many animals have subjective and rich and deep emotional and moral lives, that other animals are in their thoughts and feelings, that they have a personal point of view on the world they share with other individuals, is beyond question. While it's true that we'll likely never know *all* there is to know about animal passions, their emotional lives are public and transparent if we pay attention.[67] What we know about animal emotions has changed a great deal in the last five years because of rapidly accumulating data from research in cognitive ethology and social neuroscience.

When I first began my studies three decades ago, focusing on the question 'What does it feel like to be a dog or a wolf?', many researchers were skeptics who spent their time wondering *if* dogs, cats, chimpanzees, and other animals felt anything. I'm glad I wasn't their dog! But now there are fewer and fewer skeptics, and the question of real importance is not *whether* animals have emotions but *why* animal emotions have evolved the way they have. There is a growing interest in animal emotions from researchers in diverse disciplines ranging from biology, psychology, anthropology, to law, philosophy and theology.[68]

Prestigious scientific journals now publish essays on joy in rats, grief in elephants, and empathy in mice, and no one blinks. In fact, the paradigm has shifted to such an extent that the burden of 'proof' now falls to those who still argue that animals don't experience emotions. A majority of scientists agree with what seems like common sense to most everyone else – many animals have rich and deep emotional lives. Recently, a group of renowned scientists meeting at the University of Cambridge seemingly reinvented the wheel by declaring that animals are conscious. Of course we've known this for a long time but perhaps their endorsement of the obvious will help the plight of other animals who are abused by the billions in a wide variety of venues.[69] Let's hope that the recognition that other animals are conscious and feeling beings isn't just grand-standing. What they want, need, and feel is deeply important to them as it should be to us and we should use what we know about them to give them the very best lives we can.

It is important to study animals where they are able to express a full range of behaviour and have normal social interactions, and not rely heavily on controlled studies in impoverished environments, the nature of which often preclude getting reliable answers to the questions of interest. Since the main question is *why* emotions have evolved and what they're good for, an agenda is set for an exciting journey into the hearts and minds of animals. We know that many animals have deep and rich emotional lives and we can feel what they're feeling if we open the door to these sorts of experiences. Research on mirror neurons leads in this direction.

It's all too easy to deny the subjective lives of animals and the passions they experience because so they're supposedly so very private and inaccessible, and go on to treat animals as if they're objects at our disposal for whatever we deem necessary for our own well-being. Admitting we know a lot about the subjective and intersubjective lives of animals doesn't make them more special or unique and us less special or unique. Rather, this appreciation helps us to learn more about who they are and who we are in the grand scheme of things. Knowledge about animal cognition and sentience clearly is important in making decisions about how we treat nonhuman beings and is being factored into an ever-growing discipline called compassionate conservation[70] in which those interested in the well-being of individual animals are talking more and more with those who are more interested in saving species, populations, and ecosystems. Darwin would be happy to know this. No animal is an island. But we can make animals an island if we so choose. Often we do this by insulating ourselves from the other beings with whom we are supposed to coexist. We would do well to continue to follow Darwin's lead.

I thank Angelique Richardson for inviting me to the wonderful meeting in Exeter at which a version of this paper was presented. This essay is a revised and updated version of a paper first published in the *Journal of Consciousness Studies,* 13 (2006): 115–131, copyright: Imprint Academic, Exeter, UK.

Notes

1 Charles Darwin, *The Descent of Man and Selection in Relation to Sex* (1871; New York: Random House, 1936), 494.

2 *Ibid.*, 445.

3 *Ibid.*, 448.

4 Darwin, *The Descent of Man and Selection in Relation to Sex* (1871; New York: Penguin, 2004), 71–2.

5 Darwin, 1871 (1936), 449.

6 See also Darwin, *The Expression of the Emotions in Man and Animals,* 3rd edn, with an Introduction, Afterword, and Commentaries by Paul Ekman (1872; New York: Oxford University Press, 1998). The complete works of Darwin can be found at http://darwin-online.org.uk/.

7 See http://www.blackfootedferret.org/ (accessed 4 July 2012).

8 See: Marc Bekoff (ed.), *The Smile of a Dolphin: Remarkable Accounts of Animal Emotions* (Washington, D.C.: Random House/Discovery Books, 2000); Bekoff, *Animal Passions and Beastly Virtues: Reflections on Redecorating Nature* (Philadelphia, PA: Temple University Press, 2006);

Bekoff, 'Animal Emotions and Animal Sentience and Why They Matter: Blending "Science Sense" with Common Sense, Compassion and Heart', in Jacky Turner and Joyce D'silva (eds), *Animals, Ethics, and Trade* (London: Earthscan 2006), 27–40; Bekoff, *The Emotional Lives Of Animals* (Novato, California: New world Library, 2007); Bekoff, *Animals Matter* (Boston, MA: Shambhala, 2007); Bekoff, *The Animal Manifesto: Six Reasons For Expanding Our Compassion Footprint* (Novato, California: New world Library, 2010).

9 See Bekoff, *Emotional Lives*, 2007, for a dismissal of the 'as if' disclaimer.

10 Bekoff, 'Animal Emotions, Wild Justice and Why They Matter: Grieving Magpies, a Pissy Baboon, and Empathic Elephants', *Emotion, Space and Society* (2009), doi:10.1016/j.emospa.2009.08.001

11 Robert M. Sapolsky, *A Primate's Memoir* (New York: Touchstone, 2002), 234.

12 Sapolsky, 2002, 234.

13 G.A. Bradshaw, Allan N. Schore, Janine L. Brown, Joyce H. Poole, and Cynthia J. Moss, 'Elephant Breakdown', *Nature*, 433 (2005), 807.

14 Caroline Williams, 'Elephants on the Edge Fight Back', *New Scientist,* 18 February 2006, 39.

15 Bradshaw et al, 2005, 39.

16 See Bekoff, 'Wild Justice and Fair Play: Cooperation, Forgiveness, and Morality in Animals', *Biology & Philosophy*, 19 (2004), 489–520; Bekoff and Jessica Pierce, *Wild Justice: The Moral Lives of Animals* (Chicago: University of Chicago Press, 2009); Jessica Pierce and Bekoff, 'Moral in Tooth and Claw', *The Chronicle of Higher Education*, 18 October 2009, http://chronicle.com/article/Moral-in-ToothClaw/48800/ (accessed 4 July 2012); and Pierce and Bekoff, 'Wild Justice Redux: What We Know About Social Justice in Animals and Why it Matters', *Social Justice Research*, 25 (2012), 122–39.

17 See Robert Fagen, *Animal Play Behavior* (New York: Oxford University Press, 1981); Bekoff and John A. Byers (eds), *Animal Play: Evolutionary, Comparative, and Ecological Approaches* (New York, Cambridge University Press, 1998); and Gordon M. Burghardt, *The Genesis of Animal Play: Testing the Limits* (Cambridge, MA: Bradford, 2005).

18 See Bekoff, 'The Communication of Play Intention: Are Play Signals Functional?', *Semiotica*, 15 (1975), 231–9; Bekoff, 'Social Communication in Canids: Evidence for the Evolution of a Stereotyped Mammalian Display', *Science*, 197 (1977), 1097–99; Bekoff, 'Play Signals

as Punctuation: The Structure of Social Play in Canids', *Behaviour*, 132 (1995), 419–29; Bekoff, 'Wild Justice and Fair Play', 2004; Bekoff, *Animals at Play: Rules of the Game* (Philadelphia, PA: Temple University Press, 2008); and Bekoff and Pierce, *Wild Justice*, 2009.

19 Stephen M. Siviy, 'Neurobiological Substrates of Play Behavior: Glimpses into the Structure and Function of Mammalian Playfulness', in Bekoff and Byers (eds), *Animal Play*, 221–42.

20 Jaak Panksepp and Jeff Burgdoff, '"Laughing" Rats and the Evolutionary Antecedents of Human Joy?', *Physiology and Behavior*, 79 (2003), 533–47.

21 Jaak Panksepp and Jeff Burgdoff, 'Beyond a Joke: From Animal Laugher to Human Joy', *Science*, 308 (2005), 62–3.

22 Quoted in Peter Fimrite, 'Daring Rescue of Whale off Farallones', *San Francisco Chronicle*, 14 December 2005, http://www.sfgate.com/bayarea/article/Daring-rescue-of-whale-off-Farallones-Hump-back-2557146.php (accessed 4 July 2012).

23 Quoted in Fimrite, 2005.

24 Quoted in Fimrite, 2005.

25 Patrik Vuilleumier, 'Cognitive Science: Staring Fear in the Face', *Nature*, 433 (2005), 22–3.

26 Jennifer Ludden, 'A Hippo and Tortoise Tale', *NPR*, 17 July 2005, http://www.npr.org/templates/story/story.php?storyId=4754996 (accessed 4 July 2012).

27 'Hamster, Snake Best Friends at Tokyo Zoo', *NBC News*, 24 January 2006, http://www.msnbc.msn.com/id/10903211/ (accessed 4 July 2012).

28 John Capitanio, quoted in Patricia Ward Biederman, 'Soft Heart Under Her Thick Skin?', *Los Angeles Times*, 16 November 2004, http://articles.latimes.com/2004/nov/16/local/me-ruby16 (accessed 4 July 2012).

29 Jaak Panksepp, 'Feeling the Pain of Social Loss', *Science*, 302 (2003), 239.

30 Jonathan Balcombe, *Pleasurable Kingdom: Animals and the Nature of Feeling Good* (London: Macmillan, 2006); Bekoff, *Minding Animals: Awareness, Emotions, and Heart* (New York: Oxford University Press, 2002); Bekoff, *Animal Passions*, 2006; Bekoff, 'Animal Emotions and Animal Sentience', 2006; Bekoff, *Emotional Lives*, 2007; Bekoff, *Animals Matter*, 2007; Bekoff, 2010; Bekoff and Pierce, *Wild Justice*, 2009; and Franklin D. McMillan, with Kathryn Lance, *Unlocking the Animal Mind: How*

Your Pet's Feelings Hold the Key to His Health and Happiness (Emmaus, PA: Rodale, 2004).

31 Or, 'science sense': see Bekoff, *Animal Passions and Beastly Virtues*, 2006; and Bekoff, 'Animal Emotions and Animal Sentience', 2006.

32 Bekoff, *Minding Animals*, 2002; Bekoff, *Animal Passions*, 2006; Bekoff, 'Animal Emotions'; and Bekoff, *Emotional Lives*, 2007.

33 See, for example, James K. Rilling, James T. Winslow, and Clinton D. Kilts, 'The Neural Correlates of Mate Competition in Dominant Male Rhesus Macaques', *Biological Psychiatry*, 56 (2004), 363–75; and Jean A. King, Timothy S. Garelick, Mathew E. Brevard, Wei Chen, Tara L. Messenger, Timothy Q. Duong and Craig F. Ferris, 'Procedure for Minimizing Stress for fMRI Studies in Conscious Rats', *Journal of Neuroscience Methods*, 148 (2005), 154–60.

34 For reviews, see: Vittorio Gallese and Alvin Goldman, 'Mirror Neurons and the Simulation Theory of Mind-Reading', *Trends in Cognitive Science*, 2 (1998), 493–501; Sandra Blakeslee, 'Cells that read minds', *The New York Times*, 10 January 2006, http://www.nytimes.com/2006/01/10/science/10mirr.html?_r=1&ex=1144814400&en=5a616f2a0b0e9493&ei=5070 (accessed 4 July 2012); and David Dobbs, 'A Revealing Reflection', *Scientific American Mind*, April/May 2006, 22–7.

35 Dale J. Langford, Sara E. Crager, Zarrar Shehzad, Shad B. Smith, Susana G. Sotocinal, Jeremy S. Levenstadt, Mona Lisa Chanda, Daniel J. Levitin and Jeffrey S. Mogil, 'Social Modulation of Pain is Evidence of Empathy in Mice', *Science*, 312 (2006), 1967–70.

36 Inbal Ben-Ami Bartal, Jean Decety and Peggy Mason, 'Empathy and Pro-social Behavior in Rats', *Science*, 334 (2011), 1427–30.

37 J.L. Edgar, J.C. Lowe, E.S. Paul, and C.J. Nicol, 'Avian Maternal Response to Chick Distress', *Proceedings of the Royal Society, B*, 278 (2011), 3129–34.

38 Jane Goodall, *The Chimpanzees of Gombe* (Cambridge, MA: Harvard University Press, 1986), 357.

39 Robert W. Sussman, Paul A. Garber and Jim M. Cheverud, 'Importance of Cooperation and Affiliation in the Evolution of Primate Sociality', *American Journal of Physical Anthropology*, 128 (2005), 84–97; and Robert W. Sussman and C. Robert Cloninger (eds), *Origins of Altruism and Cooperation* (Dordrecht: Springer, 2011).

40 Cited in Bekoff and Pierce, *Wild Justice*, 2009.

41 Bekoff and Pierce, *Wild Justice*, 2009; Pierce and Bekoff, 2012.

42 Sarah F. Brosnan, Hillary C. Schiff and Frans B.M. de Waal, 'Tolerance for Inequity May Increase with Social Closeness in Chimpanzees', *Proceedings of the Royal Society, B*, 272 (2005), 253–58.

43 Freiderike Range, Lisa Horn, Zsófia Viranyi and Ludwig Huber, 'The Absence of Reward Induces Inequity Aversion in Dogs', *Proceedings of the National Academy of Science*, 106 (2009), 340–5.

44 C.J. Warden and L.H. Warner, 'The Sensory Capacities and Intelligence of Dogs, with a Report on the Ability of the Noted Dog "Fellow" to Respond to Verbal Stimuli', *Quarterly Review of Biology*, 3 (1928), 1–28.

45 Henry N. Pollack, *Uncertain Science...Uncertain World* (New York: Cambridge University Press, 2003).

46 Françoise Wemelsfelder and Alistair.B. Lawrence, 'Qualitative Assessment of Animal Behaviour as an On-farm Welfare-Monitoring Tool', *Acta Agriculturae Scandinavica*, 30 (2001), 21–5 (supplement).

47 Françoise Wemelsfelder, Tony E.A. Hunter, Michael T. Mendl and Alistair B. Lawrence, 'Assessing the "Whole Animal": A Free-Choice-Profiling Approach', *Animal Behaviour*, 62 (2001), 209–20.

48 Françoise Wemelsfeder and M. Farish, 'Qualitative Categories for the Interpretation of Sheep Welfare: A Review', *Animal Welfare*, 13 (2004), 261–68.

49 R. E. Anderson, J. Ryon and J. C. Fentress, 'Human Perception of Friendly and Agonistic Wolf Interactions', *Aggressive Behavior*, 17 (1991), 58.

50 Bekoff, *Animal Passions*, 2006; Bekoff, 'Animal Emotions and Animal Sentience', 2006.

51 Quoted in Douglas Cruikshank, 'Robert Sapolsky', *Salon*, 14 May 2001, http://www.salon.com/2001/05/14/sapolsky/ (accessed 4 July 2012).

52 Gordon M. Burghardt, 'Animal Awareness: Current Perceptions and Historical Perspective', *American Psychologist*, 40 (1985), 905–19.

53 D.O. Hebb, 'Emotion in Man and Animal: An Analysis of the Intuitive Process of Recognition', *Psychological Review*, 53 (1946), 88–106.

54 Clive D.L. Wynne, *Do Animals Think?* (Princeton, NJ: Princeton University Press, 2004).

55 Bekoff, 'The Great Divide' [review of Wynne, 2004], *American Scientist*, September–October 2004, http://www.americanscientist.org/bookshelf/pub/the-great-divide (accessed 4 July 2012).

56 Marian Dawkins, *Why Animals Matter* (New York, Oxford University Press, 2012); Bekoff, 'Animal Emotions: Exploring Passionate Natures', *BioScience*, 50 (2000), 861–70; Gordon M. Burghardt, 'Amending Tinbergen: A fifth

Aim for Ethology', in Robert W. Mitchell, Nicholas Thompson and H. Lyn Miles (eds), *Anthropomorphism, Anecdote, and Animals: The Emperor's New Clothes?* (Albany, NY: SUNY Press, 1997), 254–76.

57 Bekoff, 'Animal Emotions', 2000.

58 Dawkins, 2012, 22.

59 Bekoff, *Animal Passions*, 2006.

60 Pollack, 2003.

61 Quoted in Biederman, 2004.

62 Quoted in Biederman, 2004.

63 Mary Midgley quoted in Bekoff, *Emotional Lives*, 2007, 127–8.

64 Bekoff, 2010.

65 *Ibid.*

66 Bekoff, *Animals Matter*, 2007; Bekoff, 2010.

67 *Ibid.*

68 Bekoff, 'Animal Passions and Beastly Virtues: Cognitive Ethology as the Unifying Science for Understanding the Subjective, Emotional, Empathic, and Moral Lives of Animals', *Zygon* 41 (2006), 71–104.

69 See http://fcmconference.org/img/CambridgeDeclarationOnConsciousness.pdf (accessed 2 August 2012); see also http://www.psychologytoday.com/blog/animal-emotions/201208/scientists-finally-conclude-nonhuman-animals-are-conscious-beings (accessed 3 August 2012).

70 Bekoff (ed.), *Ignoring Nature No More: The Case for Compassionate Conservation* (Chicago: University of Chicago, 2013); and Bekoff, *Rewilding our Hearts* (Novato, California: New World Library, 2013).

Index

Notes: unless otherwise indicated 'Darwin' refers to Charles Darwin; references to illustrations are in ***bold****.*

www.ingramcontent.com/pod-product-compliance
Lightning Source LLC
LaVergne TN
LVHW020632300926
849161LV00004B/677

* 9 7 8 9 0 4 2 0 3 7 4 7 2 *